God's World Science Series

God's World Science Series

God's
Orderly World

Grade 8

Rod and Staff Publishers, Inc.
P.O. Box 3, Hwy. 172
Crockett, Kentucky 41413
Telephone: (606) 522-4348

1   2   3   4   5   —   20   19   18   17   16   15   14   13   12   11

In Appreciation

For the greatness of God and His orderly world, we give thanks and praise. From the biggest galaxy to the smallest atom, we see the power and wisdom of God. To inspire our children with these wonders, that they might love and fear God, is a sacred privilege and responsibility. We are thankful for the freedom, ability, and resources to have Christian schools. We are thankful for the vision to publish textbooks that exalt God and are based on the truth of His Word. We are thankful for the church to whom God has given gifts to produce this science textbook, *God's Orderly World, Science 8.*

We are grateful to God for enabling Brother Lester Showalter and Sister Lucy Martin to do the original writing. Many were involved in reviewing, classroom testing, and revising the material. Brother Seth Rudolph and Brother Marvin Eicher were the editors. As each did his part in response to the Lord of the church, there was a blending of efforts, for which we are thankful.

The publishing of this text does not yet meet its objective. These pages have not served their purpose until they become a tool for increasing knowledge about God's created world and, with that knowledge, inspiring the rising generation to worship and serve their Creator. As God has blessed the efforts to produce this textbook, may He further bless the teachers and students who use it.

—The Publishers

Cover photo: The beautiful Whirlpool galaxy (M51, NGC 5194) is one of the most magnificent spiral galaxies in the sky. This galaxy of type Sc is 25 million light-years from the earth and can be found in Canes Venatici, the Hunting Dogs constellation. Very hot stars and glowing hydrogen gas are the major contributors to the pretty coloration. The pinwheel design demonstrates order in God's world. Numerous other examples of His orderly handiwork appear throughout this book.

Contents

Introduction

"Who is like unto thee, O LORD, among the gods? who is like thee, glorious in holiness, fearful in praises, doing wonders?" (Exodus 15:11). Whether we look up into the heavens or investigate the gravity that holds us to the earth, we see the ability of God to create wonders that we cannot explain. Even if we consider our own body, we must confess that we are "fearfully and wonderfully made: marvellous are [God's] works" (Psalm 139:14). A study of the wonders of the natural world should lead us to praise God.

God's Orderly World is designed to guide you in a systematic study of God's world. A diligent use of this book should help you in the following ways:

1. To grow in praise and reverence for God, the Creator of all things.
2. To develop an appreciation for God's natural gifts and to use them wisely as a good steward.
3. To learn practical truths that will be helpful in your use of God's natural world.
4. To see the relationship and agreement between God's world and His Word, the Bible.
5. To understand the error of false science and superstition, and to guard against having your faith in God and the Bible spoiled by the false theories of men.

To gain the most from your study, you should read this textbook carefully with understanding. This is not a storybook that you can read rapidly and still follow what it says. Train your mind to be alert and to concentrate on the meaning of each sentence. Avoid distracting thoughts. If you come to words that you do not understand, use a dictionary or the glossary in the back of the book.

Three sections at the beginning of each unit are intended to spark your interest and get you started in thinking about the subject of that unit. These sections are as follows:

> *Searching for Truth* gives a Bible perspective on the subject of the unit.
> *Searching for Understanding* is a set of questions that will reveal what you already know and what you do not know about the material before you. You do not need to write answers to these questions.
> *Searching for Meaning* lists the vocabulary words of the unit. These words are highlighted in the text where they are introduced. Give special attention to these words as you read.

The exercises in each unit are as follows:

> *Study Exercises* appear at intervals through the text. Answer these according to the explanations given and not from what you think the answers may be. Putting the answers in your own words is more valuable than merely copying statements from the text.
> *Review of Vocabulary* reviews the vocabulary words listed at the beginning of the unit.
> *Multiple Choice* reviews important ideas in the unit. Do not just guess the answers. Think through each question so that you can feel confident of what the best answer is. Sometimes one of the choices is worded to check whether you are thinking carefully. Beware that you do not get caught by such choices.

Approach your study of God's orderly world with the desire to find reasons to praise God for His handiwork and to use the things you learn in a way that pleases Him.

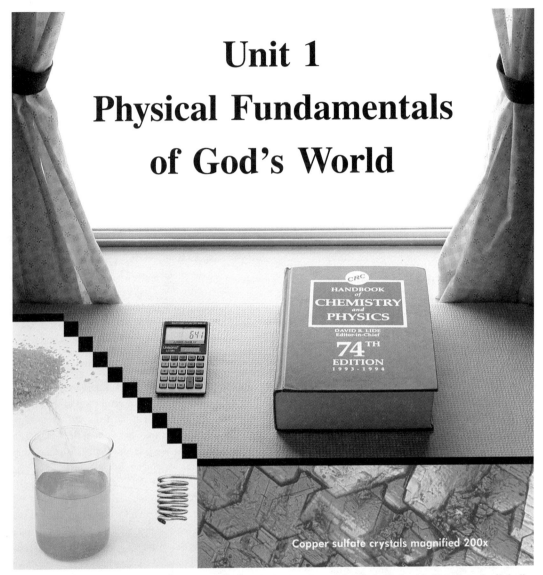

Unit 1
Physical Fundamentals
of God's World

Copper sulfate crystals magnified 200x

Matter and energy are physical fundamentals of God's created world. When God said, "Let there be light," a basic form of energy came into being. Light from the sun provides the energy for heat, weather, food, and electricity to run calculators. God made it possible to transform the energy available into the form of energy we need.

The matter God created is represented in the photo by the element copper, the compound copper sulfate, and the mixture copper sulfate solution. Thousands of different materials provide a vast array of properties that are useful for many purposes. Copper makes excellent electrical wire. Copper sulfate solution is used to treat hoof diseases in cattle. The copper sulfate crystals are beautiful specimens to show the order in matter.

Many facts have been discovered about both energy and matter. The accumulation of these facts is represented in the photo with the *Handbook of Chemistry and Physics*, a standard reference book for scientists. That handbook was consulted in the preparation of the *God's Orderly World* textbooks. The grandeur and wisdom evident in God's created world is truly awesome. We must not fail to recognize the Creator of all this wonder and give Him His rightful thanks and glory.

Searching for Truth

"Lo, these are parts of his ways: but how little a portion is heard of him? but the thunder of his power who can understand?" (Job 26:14).

These questions follow several verses about God's ability to "hang the earth upon nothing," to "divide the sea with his power," and to "garnish the heavens." As we study God's mighty works, we must remember that God is greater than anything we can observe in nature. The thunder, stars, earthquake, and wind are "parts of his ways," but how much greater is God who created them! As we seek to understand the matter and the order in the natural world, let us never forget the Creator. His power is beyond our comprehension.

Searching for Understanding

1. What power produces motion in the natural world?
2. On what basic principles does riding a bicycle depend?
3. Why does air rushing out the back of a balloon make the balloon move forward?
4. Into what groups could you put all the different materials?
5. Where would you begin if you wanted to study God's entire world?

Searching for Meaning

acceleration	emulsion	molecule
alloy	energy	momentum
atom	entropy	physical change
centrifugal force	fluid	physics
centripetal force	gas	potential energy
chemical change	inertia	solid
chemical formula	jet propulsion	solute
chemical symbol	kinetic energy	solution
chemistry	liquid	solvent
compound	mass	surfactant
element	matter	suspension

A chemical change occurs when oxygen combines with iron to produce layers of iron oxide (rust).

A physical change occurs when a drill bit peels off metal shavings.

The Beginning of Science

Science can be defined as the observation, investigation, and classification of things in God's creation. True science begins with the Bible, which describes how the world came into being. "For in six days the Lord made heaven and earth, the sea, and all that in them is" (Exodus 20:11).

Genesis, the book of beginnings, tells us that God created matter: "In the beginning God created the heaven and the earth." Then God created energy: "And God said, Let there be light: and there was light." Matter and energy are fundamentals of the natural world. In this unit you will learn some of the basic principles of matter and energy through a study of chemistry and physics.

The study of the properties and changes of matter is called *chemistry.* For example, iron is a hard metal that is attracted by a magnet. Hardness and magnetic attraction are properties of iron. When iron becomes wet in the presence of oxygen, it will rust. Rusting is a chemical change that takes place in iron. The properties and changes of materials such as iron, water, and carbon are studied in the branch of science called chemistry. A person who makes a special study of chemistry is called a chemist.

The study of energy and motion is called *physics.* For example, when you throw a ball, you are using energy to make the ball move from one place to another. As the ball travels through the air, the air must be pushed aside. This slows the motion of the ball. Gravity pulls down on the ball and makes it come to the ground. Energy, motion, gravity, heat, and electricity belong to the branch of science called physics. A person who makes a special study of physics is called a physicist.

Scientists have discovered many facts of chemistry and physics. Entire science encyclopedias are devoted to information about the natural world. Yet even if you knew everything in all the encyclopedias combined, you would not be able to explain how everything works. In your study of science this year, you will learn only a limited number of the main facts. This first unit introduces some of the basic principles of physics and chemistry.

Since the natural world is the handiwork of God, science will open up wonders that should lead you to praise God for His power and wisdom. You will also learn many facts that are useful in everyday life, whether you become a farmer, mechanic, carpenter, or housekeeper.

Energy Produces Changes

Energy is the power to produce changes in God's world. It can cause two kinds of changes: physical changes and chemical changes.

A *physical change* occurs in a material when its shape, form, or position is changed but no new material is produced. Moving an object from one place to another is a physical change. Breaking, melting, dissolving, mixing, and bending are also physical changes, since they involve only a change of shape or form but do not result in new materials.

A *chemical change* does not merely change the shape or position of materials; it results in new materials. For example, when wood burns, it combines with oxygen and produces water, carbon dioxide, and smoke. These products are no longer wood. Rusting, fermenting, and digestion are other examples of chemical changes. Photosynthesis is an important chemical change in which green plants change water and carbon dioxide into sugar and oxygen.

You can cut a match into very small pieces, crush the match, or soak the match in water, but these actions cause only physical changes. No new materials are produced. Even when you strike the match and generate heat, you still cause only physical changes. But as soon as the match becomes hot enough to burst into flame, a chemical change begins. The match becomes black and gives off smoke and gases. New materials are produced.

Energy is the power that causes physical and chemical changes to happen. Without energy, there could be no changes. You could not move yourself, and nothing else would move. There could be no fire and no digestion, and not even any physical life. But God created a moving, changing world. The sun heats the earth, water splashes over falls, people walk about, and plants grow, all because God created energy to produce physical and chemical changes.

———— Study Exercises: Group A ————

1. Both the Bible and true science begin with the ——— of the world, which is described in the Book of ———.

2. What branch of science deals with the properties and changes of matter?

3. A person who designs machines would need to know much about the branch of science called ———.

4. A person who makes a special study of the different chemicals and the way they react to form new chemicals is called a ———.

5. A person who makes a special study of energy and moving things is called a ———.

6. If there is a change in the natural world, what causes the change?

7. In what way is a chemical change different from a physical change?

8. Tell whether each of these illustrates a physical change (*P*) or a chemical change (*C*).
   a. melting ice
   b. stretching wire
   c. rusting iron
   d. igniting gunpowder
   e. evaporating water
   f. heating copper
   g. burning coal
   h. rotting of wood
   i. digesting starch
   j. photosynthesis
   k. grinding wheat
   l. dissolving salt in water

Two states of energy. When energy is in action and causing either physical or chemical changes, it is called ***kinetic energy.*** But if the energy is at rest and is not producing a change, it is called ***potential energy.*** Kinetic energy is sometimes called energy in motion, while potential energy is called stored energy.

Potential energy. A kilogram (2.2 pounds) of coal contains enough potential energy to change 13 liters (over 14 quarts) of boiling water into steam. But that will not happen until the coal is burned. The coal can stay in a bin for months or years without causing any changes. But as soon as the coal is burned, the energy becomes kinetic and causes changes.

A wound spring is another example of potential energy. You put energy into a mouse-trap when you set it at 8:00 P.M. The spring will store that energy until *snap!*—a mouse trips the trap, the potential energy changes to kinetic energy, and the mouse is caught.

Kinetic energy. When a child is swinging, kinetic energy repeatedly changes to poten-tial energy and back again. When the child is as high as he will go, he stops and has much potential energy because of his height above the ground. As gravity pulls him down, he speeds up so that he is going his fastest when he is at the lowest point. The potential energy he had at the top is now all the motion of kinetic energy.

All energy is either kinetic energy in action or potential energy in storage. These are the two main *states* of energy.

Many forms of energy. The energy of

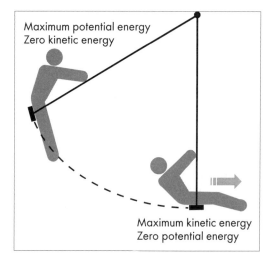

Maximum potential energy
Zero kinetic energy

Maximum kinetic energy
Zero potential energy

coal is *chemical* energy in the potential state. The chemical change of burning releases energy from the carbon in the coal. Boiling water has *heat,* which is the kinetic energy of the molecules of matter in motion. *Sound* is the kinetic energy of vibrating molecules as a wave motion travels through air or other matter.

Some Forms of Energy		
	Potential State	**Kinetic State**
Chemical	fuels, explosives, food	burning, explosions, digestion
Heat	(none)	warmth of matter
Sound	(none)	noise and music
Light	(none)	sunlight, glow of light bulbs
Electricity	static electricity	current electricity
Atomic	elements such as uranium	fission, fusion
Mechanical	water behind dam, raised weight, compressed spring	moving water, falling weight, moving car, running machinery

Light is a form of kinetic energy that can travel through empty space. *Electricity* is the form of energy that moves tiny electrons through wires. *Atomic* energy is released when atoms split apart, as in nuclear power plants or in atomic bombs.

The turning of a wheel and the flowing of water are examples of *mechanical* energy. Forces such as gravity, magnetism, and spring tension are also considered mechanical since they cause bodies of matter to move.

Energy is an essential part of the world. Energy makes things happen and accomplishes work. Whether you see an awesome display of electrical energy during a thunderstorm or hear a gentle rustling of leaves, you are observing the operation of energy that was originally created by God.

────────── Study Exercises: Group B ──────────

1. Give a simple definition of potential energy.

2. Give a simple definition of kinetic energy.

3. Tell whether each of these is an example of kinetic energy (*K*) or potential energy (*P*).
   a. waterfall
   b. light
   c. wound watch spring
   d. ton of coal
   e. electric spark
   f. weight hanging on a rope
   g. fire
   h. wind
   i. water behind a dam
   j. flashlight battery

4. Write *physical change, chemical change, potential energy,* or *kinetic energy* for each blank.

   Gasoline contains more *(a)* —— per pound than kerosene. As gasoline burns, a *(b)* —— takes place in the cylinder of an engine, and the piston is driven down. This *(c)* —— is then transmitted by gears to where it is needed. Thus the energy of gasoline can be harnessed to make a *(d)* ——, such as grinding feed or crushing rocks.

5. Which form of energy provides the power for almost all our transportation?

6. Which form of energy supplies light, heat, and motion in our homes?

7. Why can light and sound exist only as kinetic energy?

8. Which form of energy do we measure with a thermometer?

9. Engines, water wheels, windmills, and cranks are all made to produce the —— form of energy.

10. A form of energy released by extremely powerful bombs is —— energy.

11. The sun is continuously pouring out so much energy that even the small percentage received by the earth is enough to keep us warm, to feed living things through photosynthesis, and to cause weather changes. In Psalm 19:5, what two figurative comparisons are used to describe the "enthusiasm" of the sun?

Transformation of Energy

One form of energy can change into another form of energy. For example, the light energy that falls on a green leaf is changed into the chemical energy of sugar. When light energy strikes a dark object, it is absorbed and changed into heat. Photovoltaic cells in some calculators generate electricity when light shines on them.

Light can even be changed into mechanical energy, as you can see by using a radiometer (rā′·dē·om′·i·tər). This is a simple device with four vanes enclosed in a bulb from which most of the air has been removed. Each vane is white on one side and black on the other side. Most of the light striking the white sides of the vanes is reflected. But as light strikes the black sides, it changes into heat energy and warms the air next to the vane. The fast-moving particles of warm air push against the black sides of the vanes and cause the rotor to turn.

This example illustrates a chain of energy transformations. Light energy is transformed into heat energy, and the heat energy is transformed into mechanical energy. Many modern inventions are simply devices for transforming energy. A light bulb changes electrical energy into light energy. A gasoline engine changes fuel or chemical energy into mechanical energy. An electric motor changes electrical energy into mechanical energy. An electric bell changes electrical energy into sound energy. At an atomic power plant, atomic energy changes into heat energy and then into mechanical energy and then into electrical energy.

In studying energy transformations, scientists have discovered that every transformation involves heat. From their observations, they have formulated principles that are now called "laws of thermodynamics." *Thermo* means "heat," and *dynamics* means "energy or motion." The laws of thermodynamics describe how heat energy moves.

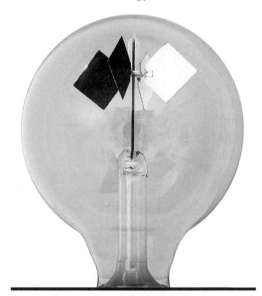

A radiometer

Two Important Laws of Energy

God created the universe with a certain amount of energy in it. Since the Creation, energy has been continually changing from one form to another. But the total amount of energy has been conserved; none has been lost, and no more has been created (except through certain miracles). This principle is stated in the first law of thermodynamics:

Energy cannot be created or destroyed.

Consider an electric generator driven by a gasoline engine. It transforms the chemical energy in fuel into heat energy, which is transformed into mechanical energy, which is transformed into electrical energy. The electrical energy is the useful output energy. But not nearly all the chemical energy in the fuel becomes electrical energy. Most of it is dissipated in the air as waste heat.

The waste heat is not destroyed energy; it is simply lost energy. The total heat energy plus the total electrical energy is equal to the chemical energy in the burned fuel. This fact can be stated as Total Energy Input = Energy Output + Energy Lost. The total energy is not increased or decreased; only God can create and destroy energy.

The second law of thermodynamics says:

> Energy always moves from a higher
> to a lower state of concentration.

For example, if you place a very hot piece of iron on a very cold piece of iron, some of the heat in the hot iron will move into the cold iron until they both become warm. This is because heat always flows from hot to cold and never in the other direction. And again, some energy is always lost in the transformation. No electric motor is 100% efficient in transforming all the incoming electrical

energy into useful mechanical energy.

An ice cube has some heat in it because heat could be removed to make the ice cube colder. Now if you put an ice cube into a cup of hot tea, would it be possible for some of the heat in the ice to move into the tea and the ice cube become colder and the tea hotter? Of course not! Neither will the light energy given off by the sun ever return to the sun. The energy in a gallon of gasoline, after it is burned, will never again be contained in a gallon of gasoline. That energy is spread out over miles of air that has been pushed aside, slightly heated, and filled with the sound of the passing vehicle.

Energy constantly decreases in concentration but increases in *entropy* (disorder and randomness). Heat spreads out; sound spreads and becomes fainter; the energy of burning fuel dissipates as heat, light, and smoke. Total entropy would exist after everything in the universe became the same temperature and no energy was available to do work.

The concept of entropy leads to two interesting conclusions. One is that a perpetual motion machine is impossible. Many men have tried to make a machine that never stops once it is started. But because of the second law

Entropy is much like the spreading out of one liquid in another liquid. The red food coloring spreads, never again to become concentrated on the toothpick.

of thermodynamics, the energy that runs the machine always becomes less and less until the machine stops.

Another important conclusion resulting from entropy is that orderly systems and complex forms could not have developed by chance. Since the natural tendency is always for disorder and randomness to increase, there must have been a Creator to make the concentrated sources of energy (such as the sun) and the complex forms (especially living things) that are in the world today.

────── Study Exercises: Group C ──────

1. Most of the light that strikes a black roof is changed into ——— energy. This example illustrates the ——— of energy.

2. Give the chain of energy transformations in a radiometer.

3. What energy transformation takes place in an electric stove?

4. Put the following words in the correct order to trace the energy from *sun* to *running*.

   sun, sugar, leg muscles, eating, light, food, green leaf, digestion, running

5. The total energy in the universe today is (greater than, equal to, less than) the energy that was there a year ago.

6. a. State the law that explains the reason for your answer to number 5.
   b. What is the name of that law?

7. If a hot object is placed against a cold object, in which direction will the heat energy move?

8. State the law that explains the reason for your answer to number 7.

9. The energy of the sun spreads out through space. This is an example of the increase of ———.

10. Since energy always spreads, a ——— ——— machine is impossible.

11. Tell whether each of the following statements relates to the *first* or the *second* law of thermodynamics.
    a. The energy of the sun is highly concentrated.
    b. As the sun gives off its energy, entropy increases.
    c. If there was ever a time when the sun did not exist, no sun could have come into being on its own.
    d. The energy given off by the sun will never return.
    e. New energy will never be added to the sun.

Group C Activity

Portraying transformation of energy. Demonstrate or make a chart of various devices that transform energy. Perhaps each student could explain a different one. Examples besides those given in the text are as follows: A microphone transforms sound energy into electrical energy. A loudspeaker transforms electrical energy into sound energy. A steam engine transforms heat energy into mechanical energy. The sun transforms atomic energy into light and heat energy. If you make a poster, you could use pictures of these items.

Three Important Laws of Motion

Motion is a very common change produced by energy. It takes energy to move an object from one place to another. God established three main laws that govern the action of moving objects: the law of inertia, the law of acceleration, and the law of action and reaction. The English scientist Isaac Newton (1642–1727) discovered and stated these laws; hence they are called Newton's three laws of motion.

Inertia

Newton's first law of motion, the law of *inertia,* can be stated like this:

> An object in motion tends
> to continue in straight-line motion,
> and an object at rest tends to remain at rest.

Here are some illustrations of this law.

Inertia of motion. When you throw a ball, it continues flying through the air, even though you do not continue pushing it. After you stop pedaling a bicycle, it continues moving forward. In both cases, the inertia of motion is in operation.

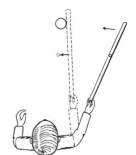

Lay a coin on a yardstick, and hold one end of it as shown in the diagram. Move the yardstick sideways in such a way that it stops suddenly by striking a post or other object. Repeat this experiment several times. On which side of the yardstick does the coin always fall? Why?

According to the law of inertia, an object in motion tends to remain in motion. The ball, bicycle, and coin tended to keep moving once they were in motion. A force must be applied to overcome the inertia of motion, such as the friction of brakes on a bicycle.

Inertia of motion tends to keep an object moving in a straight line. A marble rolling across a table tends to go in a straight line. You may grip a softball and swing your hand in a curve, but the moment that you release the ball, it travels straight from your hand. Gravity slows the ball and works against its straight-line motion.

A multiple exposure of a thrown ball shows gravity overcoming inertia and bringing the ball down.

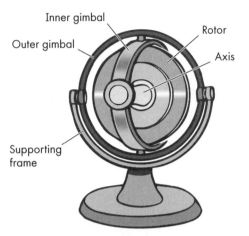

Inner gimbal

Rotor

Outer gimbal

Axis

Supporting frame

The rotor of this gyroscope is spun by hand. The combination of hinges allows the axis to turn in any direction.

Inertia also makes a rotating object maintain its plane of rotation. A spinning top stands on its point because the inertia of its rotation is greater than the force of gravity that would make it fall over. You can balance yourself much better on a moving bicycle than on one that is standing still, because of its spinning wheels.

The spinning disk of a gyroscope illustrates this principle. Gyroscopes can be used to guide ships and airplanes through dense fog. For example, an airplane would have two gyroscopes, one spinning vertically and one spinning horizontally. Pressure sensors on the gyroscopes detect changes in the tilt and direction of the plane and tell the pilot whether or not he is flying on a straight course. Not all modern autopilot systems use actual gyroscopes today, but they still use the same basic principles.

Inertia of rest. You can demonstrate the inertia of rest by again using the coin and yardstick. This time give the yardstick a sudden jerk to the right. On which side does the coin fall? Why?

You can use the inertia of rest to get a paper out from under a book. If you pull suddenly on the paper, inertia will let you remove it without touching the book.

Place a card on top of an empty water glass. Place a large coin on top of the card, directly over the center of the glass. Now give the card a sharp snap on the edge. With a little practice, you will be able to make the card fly away while the coin drops into the glass, due to the inertia of rest.

Tie a strong cord tightly around a stone the size of your fist. Use a light string tied to the strong cord to hang the stone from a substantial support. Now fasten another piece of light string to the cord on the underside of the stone. If you jerk sharply on the string below the stone, which piece of light string will tear—the upper or lower one? Do not jump to conclusions about God's world. Try it and see. Inertia of rest is the reason for the result.

The law of inertia has two parts: An object in motion tends to remain in motion, and an object at rest tends to remain at rest. These two parts can be stated as one: An object resists changing its state of rest or motion. The object will continue to do what it is doing unless a force is applied to it. The following paragraphs discuss what happens when a moving object is forced to travel in a curve.

Centrifugal force. An interesting result of inertia is called ***centrifugal force.*** This is the outward force produced when an object is made to travel in a curved path. Moving in a curve is contrary to the inertia of the object, which would keep it moving in a straight path.

The illustration shows that three different forces are at work when an object follows a curved path. *Inertia* keeps the ball moving, though inertia alone would make the ball travel in a straight path. ***Centripetal force*** is the inward pull that makes the object follow a curved path. *Centrifugal force* is the outward force that results from the interaction of inertia and centripetal force.

Curves in roads are often slanted to provide centripetal force, which helps to control the inertia of moving vehicles. You need to lean a bicycle slightly when making a turn so that the bicycle will not be pulled over by the outward force. Automatic washers spin clothes to remove the water by centrifugal force. The faster the curved motion, the greater the outward force. This fact must be considered in the construction of high-speed wheels and turbines. As you can see, centrifugal force is an interesting scientific principle that has many important practical applications.

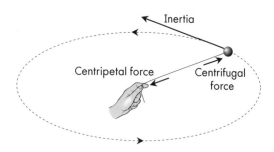

———— Study Exercises: Group D ————

1. When an object is moving, it has a tendency to (move faster, stop, move at the same speed).

2. When an object is at rest, it has a tendency to (resist being moved, begin moving in a straight line, begin moving in a curve).

3. Cars need (wheels, brakes, windshields) because of the inertia of motion.

4. Which of these activities does *not* depend on the inertia of motion?
   a. playing softball
   b. hammering a nail
   c. writing with a pen
   d. painting with a paint sprayer

5. If a box is placed in the center of a truck bed and the truck suddenly moves forward, the box is likely to slide backward. Why?

6. Write the complete statement of the law of inertia.

7. Answer with the names of the three forces at work when an object moves in a curved path.
   a. Inward pull that makes the object move in a curved path.
   b. Force that tends to make the object move in a straight line.
   c. Outward force produced because the object moves in a curved path.

8. Which one of these does *not* involve centrifugal force?
   a. automatic washer
   b. David's sling
   c. brakes on a bicycle
   d. leaning when running around a curve
   e. spinner on a seed spreader
   f. planets in orbit

Group D Activities

1. *Demonstrating inertia.* Make a stack of ten dominoes or similar flat, narrow pieces. Practice striking the bottom domino with a knife blade to remove only the bottom one from the stack. Can you remove them all, one at a time, from the bottom up? What law allows you to do this?

2. *Using inertia in a practical way.* Inertia is an advantage when fastening the head of an ax or a sledgehammer to its handle. The simplest method is to bump the end of the handle against something solid, such as a large rock or a concrete floor. The bump stops the handle suddenly; but due to the inertia of motion, the heavy head keeps moving and is driven onto the handle.

3. *Investigating gyroscopes.* A large gyroscope can be made by fastening a handle about 6 inches long (15 cm) to the end of a bicycle wheel axle. (An eye bolt with its nut can be threaded onto the axle.) The demonstrations you can do with this gyroscope are more impressive than with the toy variety. While holding the handle level in one hand, use the other hand to make the wheel spin rapidly. Then open your hand, and let the handle rest on your palm. Why does the wheel not fall down? Why is it hard to turn the axle in different directions? Instead of resting the handle on your palm, hold it with a string. How does the gyroscope tend to move? Do research on the reason for the behavior of the gyroscope.

4. *Demonstrating centrifugal force.* Fill a small pail about one-fourth full of water. Tie a length of strong cord to the handle. Swing the pail over your head in a large horizontal circle. The water will not fall out if you swing it fast enough, because centrifugal force holds it against the bottom of the pail.

   For another demonstration of centrifugal force, get a piece of ½-inch (1-cm) plastic pipe about 6 inches long (15 cm). Use a round file to smooth the sharp edge of the inner rim at one end, which will be the upper end. Fasten a soft weight (such as a rubber ball) to one end of a heavy string about 15 inches long (40 cm). Run the string down through the pipe, and tie a heavy weight (such as a large nut) to the other end of the string. Holding the pipe upright, swing the ball in a small circle and watch it lift the weight. Can you cause the ball to circle at just the right speed so that the weight is lifted without touching the pipe?

5. *Using toys that depend on inertia.* The yo-yo is one such toy. Another is the button motor, which you can make by passing the end of a yard-long string through one hole in a large button and bringing it through the opposite hole. Tie the ends of the string together to make a loop. Put your fingers through each end of the loop, with the button in the center.

To start your motor, swing the button in a circle a number of times to wind the string. Then begin alternately pulling and relaxing the string to make the button whirl first one way and

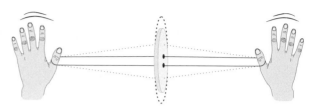

then the other. As it is whirling, move your hands so that the string is in various diagonal positions. Why does the button twist at a strange angle?

Acceleration

Newton's second law of motion describes how the application of force changes the speed of moving objects. Any change of speed is called *acceleration.* We normally use *acceleration* for speeding up and *deceleration* for slowing down. In this chapter, however, deceleration is considered as negative acceleration.

Suppose a wagon is standing still on level concrete, and it takes 4 pounds of force to make it move. If you push it with 5 pounds of force, the wagon will not simply move but will accelerate (speed up) as long as the force is applied. If you then start pulling back with 5 pounds of force, the wagon will accelerate in the opposite way (slow down).

Of course, friction slows down a moving object and hinders acceleration. But here we are studying the motion of objects without considering the effect of friction.

The second law of motion states that two things affect the amount of acceleration.

The greater the amount of force,
the greater the acceleration;
and the greater the amount of mass,
the smaller the acceleration.

Pushing a wagon with 5 pounds of force gives it a small acceleration. Pushing with 30 pounds of force accelerates the wagon much faster because the greater the force, the greater the acceleration.

If the amount of force does not change, greater mass will cause smaller acceleration. *Mass* is the amount of matter (measured in pounds). If the wagon is empty, 30 pounds of force will make it accelerate fast. But if a 20-pound child sits in the wagon, 30 pounds

Which wagon receiving 5 pounds of push will accelerate faster?

Momentum and Kinetic Energy
(Adding momentum multiplies the kinetic energy.)

Vehicle	at 30 mph	at 60 mph
½-ton trailer	15 (0.5 × 30)	60 (2^2 × 15)
3-ton van	90 (3 × 30)	360 (2^2 × 90)
40-ton truck	1,200 (40 × 30)	4,800 (2^2 × 1,200)

The ratings above are not in standard units; they are simply figures for comparison. Note that when the speed is doubled, the kinetic energy is multiplied by 2^2, or 4. A vehicle moving at 60 mph has 4 times as much kinetic energy as it has at 30 mph.

of force will result in less acceleration—because the greater the mass, the smaller the acceleration.

The principles that apply to making objects speed up also apply to making them slow down. Greater force means greater deceleration (negative acceleration), and greater mass means smaller deceleration (negative acceleration).

Momentum. It is easy to stop a Ping-Pong ball that someone throws to you. But a large softball can hurt your hands when you catch it. Both balls have the inertia of motion, but the softball has more momentum because its mass is greater. It is also easy to catch a softball that someone tosses gently to you. But if the

ball has been hit with a bat and is coming at high speed, you want to have a glove on your hand to catch it. The fast-moving ball has more momentum than the slow-moving ball.

Momentum (mō·men′·təm) is the strength of motion in a moving object. It is the product of the mass and speed of the object. Some moving objects have very little momentum. A marble rolling slowly on a table does not have much momentum; a very small force will stop it. But if the same marble rolls over the edge of the table and drops to the floor, it gains speed. This increases its momentum and makes it hit the floor with a sharp knock.

Momentum can also be defined as the measure of how hard it is to stop or to change the speed of a moving object. A large moving ship has great momentum because it has so much mass. That is why ships must come up to a dock very slowly, and why docks have large pilings for ships to push against when they come to a stop. If a large ship should come to the dock too fast, the result would be serious damage to both the dock and the ship because of its momentum.

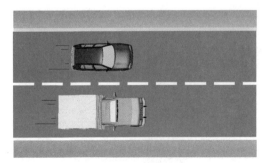

An object with greater mass has more momentum than an object with less mass.

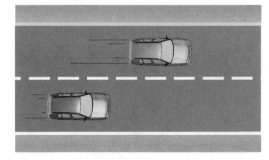

A fast-moving object has more momentum than a slow-moving object.

Speed (mph)	Driver Reaction Distance (feet)	Braking Distance (feet)	Total Stopping Distance (feet)
20	22	18–22	40–44
25	28	25–31	53–59
30	33	36–45	69–78
35	39	47–58	86–97
40	44	64–80	108–124
45	50	82–103	132–153
50	55	105–131	160–186
55	61	132–165	193–226
60	66	162–202	228–268
65	72	196–245	268–317
70	77	237–295	314–372
75	83	283–353	366–436
80	88	334–418	422–506

Speed limits are directly related to momentum. A car in motion has much momentum, or kinetic energy. The faster it is going, the more energy it possesses. Even with excellent brakes and tires, a car going 50 miles per hour (80 km/h) requires over 100 feet (30 m) to come to a complete stop. An understanding of these facts should help you to appreciate speed laws and other traffic regulations.

Momentum directly affects braking distances. When the speed is multiplied by 2, the braking distance increases about 4 times (2^2). When the speed is multiplied by 3, the braking distance increases about 9 times (3^2).

Study Exercises: Group E

1. Write the correct word for each description. Some words come from earlier parts of the unit.
   a. Amount of matter.
   b. Resistance to change in the state of rest or motion.
   c. Change in the speed of an object.
   d. Force of a moving object.
   e. Energy in action.
   f. Force produced when an object is made to move in a curved path.

2. If an object is not affected by friction and you push that object with a constant force of one pound, it will
   a. keep going faster and faster.
   b. keep going at the same speed.
   c. go faster and then stay at that speed.
   d. begin moving and then come to a stop.

3. Which two of the following statements are true?
   a. With a given force, the greater the mass of an object, the faster the acceleration.
   b. The same force will cause a light object to accelerate faster than a heavy object.
   c. The amount of force and acceleration increase together.
   d. The more force is applied to an object, the less the object will accelerate.

4. What two things determine the momentum of a moving object?

5. Which object in each pair would have the greater momentum?
   a. a car going 45 miles per hour or a car going 65 miles per hour
   b. an empty truck or a loaded truck (both going at the same speed)
   c. a truck going 50 miles per hour or a person on a bicycle going 15 mph
   d. a bullet going 1,000 miles per hour or an airplane going 500 miles per hour

 e. a fast-pitched softball or a hard-hit softball

 f. a fast-pitched softball or a hard-hit table tennis ball

6. Why does a car suffer greater damage in a collision at 65 miles per hour than at 45 miles per hour?

7. Explain how momentum is related to the importance of using seat belts when traveling in an automobile.

Group E Activities

1. *Demonstrating the second law of motion.* Find a small toy wagon with wheels that have very little friction, and put it on a smooth, level tabletop. Set up the wagon and a small cup as shown in the photo. Put a little weight in the cup; then release the wagon and observe how it gains speed slowly. Add weight to the cup, and notice how the wagon accelerates rapidly. Now add weight to the wagon with the same cup weight as before. What effect does that have on the acceleration? Experiment with various weights in the cup and the wagon.

2. *Using a momentum chair.* For this demonstration, you will need a chair with a seat that can turn in a full circle with little friction. While you sit in the chair with your arms and legs extended, have someone begin turning you slowly. Then bring your arms and legs in close to the center of the chair. You will begin spinning very rapidly.

   This demonstrates that momentum is equal to mass times speed and that the momentum stays the same, even though the speed appears to change. With arms and legs extended, your body weight is distributed over a wider area and therefore travels in a larger circumference with each rotation. Your momentum is your weight times the feet per second of travel in this circumference.

   Pulling in your arms and legs does not reduce your mass, but it does reduce the circumference within which the mass is distributed. You rotate faster to maintain the same number of feet per second of travel—and thus the same momentum—that you had with your arms extended. **Caution:** Be sure to begin the turning slowly; or when you pull in your arms, you will rotate so fast that you might become very dizzy.

3. *Using hammers to demonstrate momentum.* Experiment with the effect of weight on momentum by driving nails with hammers of different sizes. Try using a tack hammer to drive a large nail into a board. Now drive the same nail with a 16-ounce hammer. What makes the difference?

   Drive a nail while holding the hammer close to the head and then while holding the hammer near the end of the handle. What makes the difference?

   What two things about a sledgehammer give it enough momentum to break rocks?

Action and Reaction

Newton's third law of motion is usually stated as follows:

> **For every action, there is an equal and opposite reaction.**

This means that for every force exerted in one direction, an equal force is exerted in the opposite direction. For example, if a 3-pound book is lying on a table, it is pushing down on the table with a force of 3 pounds while the table pushes up with a force of 3 pounds. When you fly a kite, the kite is pulling on the string in one direction while you pull on the string with an equal force in the opposite direction.

When the object exerting a force is free to move, that object will move in the opposite direction. In a rotary lawn sprinkler, for example, the water rushing in one direction from the nozzles causes the rotor to turn in the opposite direction. As a large shotgun sends the shot in one direction, the shooter feels the kick of the gun in the opposite direction. If you jump off a small boat, you move in one

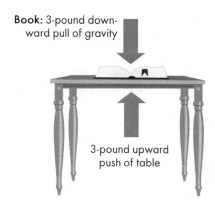

Book: 3-pound downward pull of gravity

3-pound upward push of table

direction while the boat moves in the opposite direction.

You take the law of action and reaction into account when you play ball. When you want to throw the ball as hard as you can, you brace one foot against the ground so that the force of throwing does not push you over backward. When you stand ready to bat, you spread your feet apart to resist the backward reaction of the bat as you swing it forward.

A force has the same effect on the object exerting it as on the object receiving it. If you

If a single ball comes rolling from the left, it hits the first ball of the group and stops. At the same instant, the ball at the opposite end leaves the group at the same speed as the ball that hit the group.

If two balls come rolling from the left and hit the group, two balls leave from the right.

What happens when six moving balls hit five stationary ones?

jump forward off a wagon that weighs the same as you do, the wagon will move backward as fast as you move forward. If the wagon is much lighter than you, it will move swiftly backward while you move only a small bit forward. (The result can be a bad fall.) But if you jump off the back of a loaded truck, you will move forward while the truck moves little if any. The reason is that the action of your small mass causes little reaction in the much larger mass.

Jet propulsion. A very important application of the third law of motion is jet propulsion. Blow up a balloon and release it. As it forces air out in one direction, the balloon moves in the opposite direction. *Jet propulsion* is the forward thrust caused by a backward-escaping fluid. Rocket and jet engines use the principle of jet propulsion. They send out a powerful exhaust in one direction, which produces an equal force on the engine (and everything attached to it) in the opposite direction. A visit to a large airport will give you an opportunity to see jet propulsion in practical use.

The simple fact of gases rushing out the back of a jet engine does not provide much forward push. The diagram shows two other means by which jet propulsion gives forward thrust. One is the high pressure inside the engine. The other is the rapidly expanding exhaust in the nozzle.

The high pressure in the combustion chamber is nearly equal on all sides. The pressure pushing right is counteracted by the pressure pushing left. But the forward pressure has very little pressure to counteract it on the opposite side.

The exhaust rushing from the combustion chamber expands rapidly in the nozzle area. This expanding gas adds to the forward thrust inside the engine. The design of the nozzle can greatly increase or decrease the efficiency of the jet propulsion.

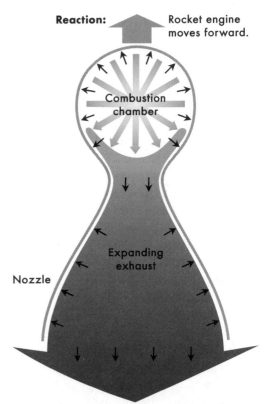

Action: Gases escape out the nozzle.

Firemen need to know about jet propulsion when they use a fire hose. The rapid escape of a great quantity of water from the nozzle produces a strong push backward. The firemen must hold on to the nozzle very securely. Two or three men may be needed to withstand the jet propulsion, or the nozzle may be mounted on a hydraulically controlled boom.

God's world is a very active world. The many forms of energy and their transformation, action, and interaction make the world a dynamic cosmos. But energy does not operate without restraint; God ordained laws that govern the way these forces work. The energy shows that God is powerful; the laws show that He is wise. To God be the glory for His orderly world.

─────── Study Exercises: Group F ───────

1. The force of a reaction is ──── and ──── to the force of the original action.

2. Suppose a certain action is a force of five pounds pushing northward. What is the reaction?

3. Why is it difficult to jump ashore from a small rowboat?

4. When you swing a ball in a circle on the end of a string, your pull on the string is called centripetal force. What is the name of the force that is opposite the centripetal force?

5. A lawn sprinkler rotates rapidly because of (jet propulsion, momentum, centrifugal force).

6. Which of these does *not* provide part of the forward thrust of a jet engine?
    a. gases rushing from the back of the engine
    b. expanding exhaust gases in the nozzle
    c. air coming into the front of the engine
    d. high pressure inside the engine

7. a. What is the action of a rocket engine?
    b. What is the reaction of a rocket engine?

8. What would happen if a fireman was holding a hose nozzle loosely when the water was turned on?

9. With all the energy in the world, what keeps it from destroying the world?

Group F Activity

Demonstrating action and reaction. If you have a momentum chair, as described in the Group E activities, you can use it to experiment with action and reaction. While sitting on the chair, hold a weight, such as a hammer, with your arm extended in front of you. Move the weight in a clockwise direction. In what direction does the chair move? With your arm extended to the side, throw a softball to someone. Can you explain the results?

You can also demonstrate action and reaction with a collision ball apparatus (sometimes called a Newton's cradle), as shown in the picture. Pull one ball aside and release it. Pull two balls aside and release them. Pull one ball from either side, and release them at the same time. Pull one ball from one side and two balls from the other side. Can you predict what will happen in each case before you do it?

God Created Matter

"Through faith we understand that the worlds were framed by the word of God, so that things which are seen were not made of things which do appear" (Hebrews 11:3). The Old Testament clearly states and the New Testament affirms that God created the world. Any other explanation about the origin of the world is false, and those who believe it soon find themselves struggling in a sea of unanswerable questions.

According to the natural laws God ordained, matter can never be created. That means there is no natural way that matter can exist or could ever have been made. But God is supernatural, above nature, and He can do things that science cannot explain.

Science can explore and explain nature and natural phenomena. But the Creation was a miracle. Science cannot investigate miracles; therefore, science will never be able to explain the origin of matter. We accept by faith that "the worlds were framed by the word of God."

Characteristics of matter. Matter is commonly defined as anything that has weight and takes up space. This includes your body, your books, the water you drink, and the air you breathe. You speak of the weight of your body in pounds or kilograms. A book could weigh 1 pound, or you could ask for 1 pound of water. Did you know there is such a thing as 1 pound of air? At sea level, a pound of air would be about 12 cubic feet (340 cm³).

If you have a sensitive scale, you can use it to demonstrate that air has weight. Deflate a basketball or football, and weigh it. Then fill up the ball with compressed air, and weigh it again. You will find that the ball is slightly heavier.

Books, water, and air all take up space. To demonstrate that air takes up space, invert a

Air takes up space. Some water had been allowed into the glass to help show the air/water division.

clear glass and push it downward into some water. You will see that the glass does not fill with water (though the air inside is slightly compressed). This is because the air in the glass is taking up space. The space that an object takes up is called its volume. So we can say that matter is anything that has weight and volume.

Here is the most accurate definition: *Matter* is anything that has mass and volume. Weight is the measure of the force of gravity on a piece of matter. The greater the distance from the earth, the smaller the weight; but the mass (amount of matter) does not change. Though weight is directly related to mass, weight changes with the force of gravity whereas the mass of an object is not affected by gravity.

Matter does not include forms of energy, such as light or sound. Energy has no weight or volume. We do not speak of a pound of light or sound, or a cubic foot of light or sound. When you turn on the light in a room, no air must leave to make added space for the light.

States of matter. Matter exists in three states called solid, liquid, and gas. Matter that has a definite shape and a definite volume is called a *solid.* A book is an example of a solid. It has a certain shape and size that does not change.

Matter that has a definite volume and takes the shape of its container is called a *liquid.* Water is an example. Although a pint of water does not change in volume, it can take very different shapes depending on the shape of container it is in.

Matter that takes both the shape and the volume of its container is called a *gas.* Air is an example of a gas. Since gases can expand or be compressed, they will completely fill the volume of a container. You could force 5 cubic inches of air into 1 cubic inch. If you enlarged the space to 25 cubic inches, it would still be full of air but the pressure would be lower.

Liquids and gases are called *fluids* because they flow. For example, both air and water can pass through a hose. Water flows across the countryside as streams and rivers. Air flows over the countryside as wind.

The Three States of Matter		
Solid at room temperature	**Liquid** at room temperature	**Gas** at room temperature
iron, sugar, salt, quartz, sand, gold	water, alcohol, mercury, gasoline, benzene, ether	oxygen, hydrogen, nitrogen, helium, carbon dioxide
Ice: 32°F (0°C) or below	Water: 32°F (0°C) to 212°F (100°C)	Steam: 212°F (100°C) or above
Definite volume Low compressibility Definite shape	Definite volume Low compressibility Indefinite shape (takes shape of container)	Indefinite volume High compressibility Indefinite shape (takes shape and volume of container)

——————— Study Exercises: Group G ———————

1. How do we know that God created matter out of nothing?

2. What law of matter gives evidence that God created matter?

3. Write a definition of matter.

4. What two facts about air prove that it is matter?

5. Write the correct word to complete each statement.
    a. Since God did not use natural laws to create matter, we know that He is ———.
    b. The amount of space that a piece of matter takes up is its ———.
    c. The measure of the pull of gravity on a piece of matter is its ———.
    d. The amount of matter in an object is its ———.

6. Solid, liquid, and gas are called the three ——— of matter.

7. The following table summarizes the characteristics of the different states of matter. Complete it by writing *definite* or *indefinite* for each blank space.

	Volume		Shape
Solid	a.		d.
Liquid	b.		e.
Gas	c.		f.

8. Tell whether each substance is a *solid, liquid,* or *gas* at room temperature. Be sure you know what the substance is.
    a. leather
    b. oil
    c. steam
    d. butter
    e. oxygen
    f. molasses
    g. mercury
    h. neon
    i. rubber

9. What property do both liquids and gases have that makes them fluids?

Group G Activities

1. *Studying solids.* There are two kinds of solids: true and amorphous (ə·môr'·fəs). True solids have a crystalline structure and a definite melting point. Amorphous solids have no definite melting point but become softer as they are heated until they begin to flow. Using a propane torch, test the following materials to discover whether they are true or amorphous solids. (The stars indicate materials that will catch fire if placed directly in the flame. Put a small amount in an old spoon, and heat slowly.)

    sulfur*   paraffin wax*   aluminum (use a piece of wire)
    butter*   rubber*   glass (use glass tubing)
    solder   lead   acrylic plastic*

2. *Investigating plasma.* There is a fourth state of matter called plasma. Do some research to find the properties of this state of matter and where it exists in the natural world.

3. *Experimenting with compression.* For all practical purposes, liquids are not compressible. Fill one syringe with water and another with air. Experiment with putting pressure on each while holding the end shut. Make each only half full of the fluid. Try enlarging the space of each. Why is it important to completely bleed a hydraulic brake line of all air?

Twenty Common Elements and Their Chemical Symbols

aluminum...Al	gold..............Au	lead.................Pb	silicon..........Si
calcium.......Ca	helium..........He	mercury..........Hg	silver..........Ag
carbon.........C	hydrogen........H	nitrogen.............N	sodium......Na
chlorine......Cl	iodine..............I	oxygen...............O	sulfur............S
copper.......Cu	iron...............Fe	phosphorus.......P	zinc............Zn

Elements. If someone should set out to group and display samples of all the different materials in the universe, he would face a gigantic task. This task was really the problem that faced scientists for many years. They are still working to discover and develop new and useful kinds of matter. Nylon, Teflon, stainless steel, and penicillin are only a few of the many materials that scientists have developed for everyday use within the last century.

For years scientists were seeking to answer this question: "Of what materials are all other materials made?" Such basic materials are called *elements.* For many years, people thought that water was an element. But electricity will divide water into oxygen and hydrogen, so water is not an element. Oxygen and hydrogen cannot be divided into other materials, so they are elements.

Scientists have discovered 93 elements on the earth. They include such gases as hydrogen, oxygen, chlorine, and helium. Most of the elements are solids, such as carbon, iron, sulfur, copper, and gold. Only two of the elements are liquid at room temperature: bromine and mercury. All elements are made of only one kind of atom. An *atom* is the smallest particle of an element. The element oxygen is made of only oxygen atoms. The element iron is made of only iron atoms.

The elements have symbols that provide a quick way to write them. Sometimes a *chemical symbol* is a single capital letter. If two letters are used, the first is capitalized and the second is not. The table above shows the symbols for some common elements.

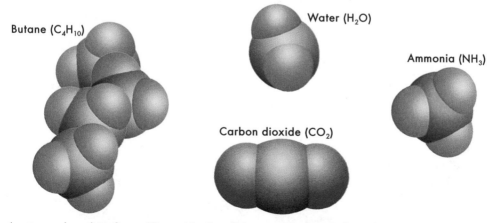

Butane (C_4H_{10})

Water (H_2O)

Ammonia (NH_3)

Carbon dioxide (CO_2)

A compound consists of a specific combination of elements. The formula for each compound gives the chemical symbol of each element and how many atoms of that element are in a molecule of the compound.

Compounds. Many materials are formed by the combining of two or more elements. Such materials are called **compounds.** Water is a compound of hydrogen and oxygen. Table salt, rust, and carbon dioxide are other examples of compounds. A quick way to name a compound is by using a **chemical formula** that shows the elements in the compound. The chemical formula for table salt is NaCl. This formula tells us that table salt is made of sodium and chlorine.

The smallest particle of a compound is a **molecule.** The formula tells how many atoms of each element are in the molecule. The formula for water is H_2O (read "H two O"). This tells us that each molecule of water has two atoms of hydrogen and one atom of oxygen.

In writing formulas, if there is only one atom of an element, no number is used. For more than one atom, a subscript number is written after the element symbol (such as the small *2* in *H_2O*). Some compounds have complicated formulas. Your mother probably has baking soda in her cupboard. The formula for baking soda is $NaHCO_3$ (1 sodium atom, 1 hydrogen atom, 1 carbon atom, and 3 oxygen atoms).

Sometimes the symbol for an element occurs twice in a formula because of the way the atoms are arranged in the molecule.

The formula for the acetic acid in vinegar is CH_3COOH (1 carbon, 3 hydrogen, 1 carbon, 1 oxygen, 1 oxygen, and 1 hydrogen). The element carbon has the special ability to combine with itself to form long chains. This makes it possible to have many atoms in one molecule. For example, the formula for table sugar is $C_{12}H_{22}O_{11}$ (12 carbon, 22 hydrogen, and 11 oxygen). The elements in a sugar molecule are arranged as shown in the diagram below.

Even though this is a large, complex molecule, it is much too small to be seen with an ordinary microscope. Electron microscopes must be used to get pictures of atoms. Since the ideas about atoms have not been proven conclusively, they are called theories. Some scientific theories (such as evolution) are false because they are contrary to what the Bible teaches. But the theories about atoms have abundant facts to support them, and they do not contradict the Bible.

The marvels of atoms and molecules show the wisdom of God in making the wonderful matter we live and work with every day. How can God, who is so great, make beauty and order so small that we cannot see it? Everywhere we look in creation, we see the glory of God's handiwork. With atoms and molecules, there is a glory in something too small to be seen.

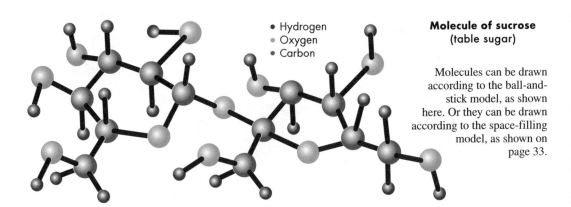

• Hydrogen
• Oxygen
• Carbon

Molecule of sucrose (table sugar)

Molecules can be drawn according to the ball-and-stick model, as shown here. Or they can be drawn according to the space-filling model, as shown on page 33.

———————— Study Exercises: Group H ————————

1. Choose the correct word for each definition.
   a. One or two letters that represent an element.   atom
   b. Smallest particle of a compound.   compound
   c. Basic material that cannot be separated.   element
   d. Smallest particle of an element.   formula
   e. Name for a compound that shows elements in it.   molecule
   f. Material formed by combining two or more elements.   symbol

2. Give the chemical symbol for each element.
   a. carbon   b. chlorine   c. nitrogen   d. sodium

3. Write the number of the correct formula for each compound.
   a. aluminum oxide   (1) NH_4Cl
   b. calcium hydroxide   (2) $Pb(NO_3)$
   c. ammonium chloride   (3) $ZnCO_3$
   d. sodium sulfate   (4) $Ca(OH)_2$
   e. lead nitrate   (5) Al_2O_3
   f. zinc carbonate   (6) Na_2SO_4

4. The formula for the main chemical in sand is SiO_2. What is meant by the small *2* after the O?

5. How can compounds of carbon have high numbers of atoms?

6. Why are the ideas about atoms and molecules called theories?

Group H Activities

1. *Investigating elements.* Scientists announced in 2004 that they had identified a total of 116 elements. Where did the extra 23 come from? What are their names? What properties and uses do they have? Do research to find answers to these questions.

2. *Separating water with electricity.* Mix 4 parts water with 1 part vinegar (acetic acid) to make enough to fill a shallow bowl with the solution. Fill two test tubes with the solution, hold the openings shut with your thumb, and invert them into the bowl of water as shown. Remove about 1 inch of insulation from two pieces of copper wire, and bend the wires so the exposed ends are in the mouths of the test tubes. Connect one of the wires to the positive post and the other to the negative post of a direct current source, such as a battery charger or a car battery. Which wire produces a gas? What is happening at the other wire?

   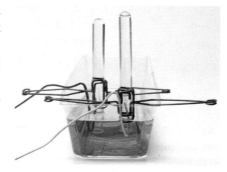

   If the wire you connected to the positive post was stainless steel or (better yet) platinum, you would obtain oxygen at the positive electrode. This process of separating water is called electrolysis of water.

Chemical reactions. In the previous section, you saw that all the different kinds of materials are made of a relatively small number of elements. Compounds are materials made of a combination of different elements. But you cannot make a compound simply by mixing together two elements, such as a little nitrogen and some gold powder. (Neither can you separate a compound by physical methods.) Rather, a compound forms when atoms from two or more elements combine chemically. A. chemical change takes place. Each element gives up its unique properties to form the compound. The properties of the compound are different from those in the elements it contains.

Sometimes compounds react with each other to form new materials. These reactions are valuable for several reasons. One is that they produce important substances not found as raw materials on the earth. For example, sulfuric acid is produced by a series of chemical reactions. This acid is used to make fertilizer, to refine petroleum, and to make batteries for cars and trucks. Many tons of this acid are needed every year.

Much sulfur is found on the earth, but not sulfuric acid. The first step in making sulfuric acid is to burn sulfur. The reaction that takes place in this process is shown by the following symbols.

$$\underset{\text{sulfur}}{S} \quad \underset{\text{plus}}{+} \quad \underset{\text{oxygen}}{O_2} \quad \underset{\text{yields}}{\rightarrow} \quad \underset{\text{sulfur dioxide}}{SO_2}$$

Another reaction is used to change the sulfur dioxide to sulfur trioxide (SO_3), which then reacts with water to form sulfuric acid.

$$\underset{\text{sulfur trioxide}}{SO_3} \quad \underset{\text{plus}}{+} \quad \underset{\text{water}}{H_2O} \quad \underset{\text{yields}}{\rightarrow} \quad \underset{\text{sulfuric acid}}{H_2SO_4}$$

Farmers of past years used a chemical reaction to make lime from limestone. They heated limestone rocks very hot in a kiln.

This caused a reaction in which the limestone rock changed into lime and gave off carbon dioxide.

$$\underset{\substack{\text{calcium carbon-}\\\text{ate (limestone)}}}{CaCO_3} \quad \underset{\text{yields}}{\rightarrow} \quad \underset{\substack{\text{calcium}\\\text{oxide (lime)}}}{CaO} \quad \underset{\text{plus}}{+} \quad \underset{\substack{\text{carbon}\\\text{dioxide}}}{CO_2}$$

The chemical reactions described are just a few of the many thousands that are used to make important chemicals for people's use.

Another value of many chemical reactions is the energy they produce. Coal is a natural form of carbon. When coal burns, the carbon combines with oxygen in the air.

$$\underset{\text{carbon}}{C} \quad \underset{\text{plus}}{+} \quad \underset{\text{oxygen}}{O_2} \quad \underset{\text{yields}}{\rightarrow} \quad \underset{\text{carbon dioxide}}{CO_2}$$

This reaction gives off much heat which in many power plants is used to produce steam that drives turbines to turn electric generators. The heat from burning gasoline and diesel fuel powers millions of vehicles. Your home may be heated by burning some kind of fuel.

There are other useful chemical reactions. Dynamite and gunpowder are explosives that produce great quantities of gas in a fraction of a second. The explosion of dynamite has enough force to blast rocks apart. The intense energy of lightning combines nitrogen and oxygen into nitrogen oxide. Raindrops carry it to the ground where the soil chemically changes it into nitrates, a fertilizer. The chemical reaction in dry cell batteries generates electricity. The reaction of Portland cement with sand and water produces concrete that is as hard as rock. Ammonium hydroxide (NH_4OH), commonly known as ammonia, helps to clean windows by reacting with the film of grease on them.

Some chemical reactions cause problems. Burning can be destructive if it is not controlled. Rusting of iron is a chemical reaction that can ruin tools, bridges, and machinery. The chemicals in hard water react with soap to hinder its cleaning power. Acids formed by bacteria in your mouth react with the hard material of the teeth to cause tooth decay.

This is only an introduction to the fascinating world of chemistry. Even these few details are enough to show that God is great and wise to have made matter with the great variety that it has.

———— Study Exercises: Group I ————

1. A ——— reaction is a process that produces new ———.

2. How is it possible to make chemicals that are not found in nature?

3. Write the number of the "chemical equation" that shows the reaction for
   a. obtaining hydrogen from water.
   b. making lime from limestone.
   c. burning coal.
   d. making sulfuric acid.
   e. removing iron from iron ore.

   (1) $C + O_2 \rightarrow CO_2$
   (2) $Fe_2O_3 + 3CO \rightarrow 2Fe + 3CO_2$
   (3) $Fe + S \rightarrow FeS$
   (4) $2H_2O \rightarrow 2H_2 + O_2$
   (5) $SO_3 + H_2O \rightarrow H_2SO_4$
   (6) $CaCO_3 \rightarrow CaO + CO_2$

4. Tell how the following chemical reactions are useful.
   a. explosion of dynamite
   b. effect of lightning on nitrogen and oxygen
   c. reaction of zinc and ammonium chloride in dry cell batteries
   d. reaction of sand, Portland cement, and water
   e. reaction of ammonium hydroxide with grease

5. Give two examples of chemical reactions that have harmful results.

6. Which statement about God and chemistry is true?
   a. All that chemicals can do is a result of God's wisdom.
   b. Chemicals can do things that are more wonderful than the Creation.
   c. Man is showing God what wonderful things His chemicals can do.
   d. God controls spiritual things but not physical things like chemicals.

Group I Activities

1. *Obtaining a gas from a solid and a liquid.* Baking soda ($NaHCO_3$) is a powdered solid. Vinegar is a liquid that contains acetic acid (CH_3COOH). If you mix these two, a chemical reaction will produce the gas carbon dioxide (CO_2). This is the gas that causes a cake to rise and that makes the holes in bread.

   Put several teaspoons of baking soda in the bottom of a quart or liter jar. Add one-fourth cup of vinegar. Let the reaction fill the jar with carbon dioxide. Carbon dioxide is more dense than air and will stay a long time in the open jar. Lower a lighted match into the jar. What happens? Why?

2. *Using heat to cause a chemical reaction.* Heat some white table sugar in an old teaspoon over a propane or alcohol flame. Continue heating until the sugar turns black. How do you

know a chemical reaction has taken place? What is the black substance left in the spoon? Where did it come from?

3. *Experimenting with hard water.* Put equal amounts of hard water and rain (or distilled water) in two bottles. Add small and equal amounts of scrapings from bar soap, cap the bottles, and shake them vigorously. Notice the difference in the amount of suds and in the cloudiness of the water. You can make the soft water hard very easily. Add a very small amount of a calcium compound, such as calcium chloride (road salt). Now shake the bottle again. How do you know the water has become hard?

Mixtures. Both elements and compounds are pure materials. Each one always has the same properties. And each compound always has the same proportions of the different elements that make up that compound. For example, water (H_2O) always has two hydrogen atoms and one oxygen atom in each molecule.

However, elements or compounds can be mixed together without forming new compounds. The result is simply a mixture; no chemical reaction takes place. A mixture consists of two or more elements or compounds that are not chemically combined. Each ingredient in a mixture retains its unique properties. If a spoonful of sugar is dissolved in a cup of water, the water is still a clear liquid and the sugar retains its sweetness. The water and sugar are a mixture; the whole cupful is sweetened. There are two main kinds of mixtures: suspensions and solutions.

Suspensions. A **suspension** is a mixture formed when very fine particles of one material are suspended (held up) in another material. Smoke is a suspension in which particles of carbon and ash are suspended in air. Soil is a suspension; the different types of soil particles can be separated by putting it in a jar of water, shaking the jar, and letting it stand for a while. Sand and limestone particles will settle to the bottom first, clay will form more layers, and decayed plant material will settle near the top or float in the water.

One characteristic of liquid suspensions is their cloudy appearance. Light that passes through a suspension is scattered by reflection from the particles in suspension. Another characteristic is that the particles in a suspension can be filtered out with fine filters. A suspension of flour in water is very cloudy, but the tiny particles can be quickly removed by pouring the mixture through filter paper.

Milk is an emulsion that contains cream suspended in skim milk. An **emulsion** is a suspension of one liquid in another liquid.

Solutions. A **solution** is a mixture formed when one material spreads molecule by molecule through another material. The process of a solid going into solution with a liquid is called dissolving. The liquid is

Kind of Suspension	Description	Example
Foam Solid foam	Gas in a liquid Gas in a solid	Whipped cream Marshmallow
Liquid aerosol Emulsion Solid emulsion	Liquid in a gas Liquid in a liquid Liquid in a solid	Fog, Mist Cream, Mayonnaise Butter, Cheese
Smoke Sol (sôl) Solid sol	Solid in a gas Solid in a liquid Solid in a solid	Dust in air Jelly Pearls, Opals

Kind of Solution	Example
Gas in a gas Gas in a liquid Gas in a solid	Air Carbonated drink (carbon dioxide in water) Alloy of palladium and hydrogen
Liquid in a liquid Liquid in a solid	Vinegar (acetic acid in water) Rubber cement (benzene in rubber)
Solid in a gas Solid in a liquid Solid in a solid	Perfume (perfume particles in air) Saltwater (salt in water) Steel (carbon in iron)

the *solvent,* and the solid or liquid that becomes dissolved is the *solute* (sol′·yo͞ot). Sugar water is an example of a solution. When solid sugar dissolves in water, it separates into molecules that are distributed among the water molecules.

For a material to dissolve in another material, the two must have something in common. Sugar and water both have hydrogen that bonds and enables the solvent (water) to dissolve the solute (sugar). Water molecules also have a tiny electrical (ionic) charge, the same as atoms in a salt crystal. So water dissolves salt by pulling the crystal apart with its ionic attraction for the sodium chloride in the salt crystal. But water and petroleum products have nothing in common; therefore, water cannot dissolve oil.

Liquid solutions are different from suspensions in that solutions are clear rather than cloudy. Solutions may be colored, but light travels through them without being dispersed. Another difference is that solutions cannot be separated by filtering. However, a solid solute can be separated from a solution by the physical action of evaporating the solvent.

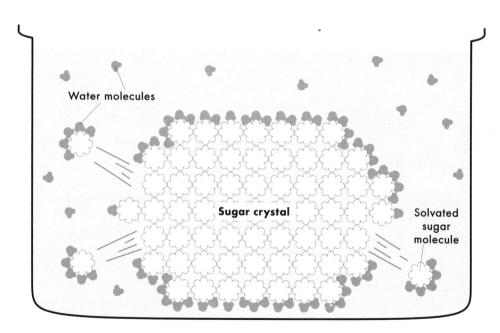

Sugar dissolves in water because the hydrogen in water molecules bonds to the hydrogen in sugar molecules and pulls the sugar crystal apart.

We often think of solutions as liquids, but many gases and solids are also solutions. Air is a solution of nitrogen and oxygen. Divers sometimes carry air packs that are a solution of helium and oxygen. Many *alloys* (mixtures of metals) are solutions. Brass is a mixture of copper and zinc. One common solder is a mixture of tin and lead. Stainless steel is an alloy of iron, chromium, manganese, silicon, and carbon.

Suspensions are cloudy.   Solutions are clear.

Study Exercises: Group J

1. Write *filter* or *chemical reaction*.
   a. A —— is needed to separate the parts of a compound.
   b. A —— can separate the solid and liquid parts of a suspension.
2. How can the different parts of soil be separated?
3. In a solution, the solute is broken down into individual (atoms, molecules, granules) that are distributed through the solvent.
4. Why can water dissolve salt but not oil?
5. How can you tell whether a liquid mixture is a suspension or a solution?
6. How can you separate a solid solute from a solution?
7. a. A mixture of metals is called an ——.
   b. Brass is a mixture of —— and ——.

Group J Activity

Observing a suspension. Put plain water in a quart or liter jar. In a dark room, shine the beam of a flashlight through the water. A laser beam would be even better. Does the water glow from the light passing through it? Add some milk to the water, and again shine the light through it. What evidence do you have that the jar contains a suspension?

Increasing the rate of dissolving. Forming a solution takes time. You know that if you put sugar in water, it takes some time for it all to dissolve. There are three ways to increase the rate of forming a solution.

One way is to grind the solute into small pieces. This allows more surface area of the solid to be exposed to the solvent. Another way is to heat the solvent. Stirring or agitating the mixture is a third way to speed up the process.

If you do not stir a mixture of sugar and water, the sugar will eventually all dissolve, but it will take much longer than if you stir the water.

Sometimes a material will dissolve if it is first changed by a chemical reaction. For example, acid water will dissolve limestone rock by first changing it from insoluble calcium carbonate to soluble calcium bicarbonate. This is the process that causes hard water in limestone regions.

The food you eat must be changed to a solution before your body can use it. We call this process digestion. The process of digestion was wisely provided by God to allow us to eat such insoluble foods as starches, proteins, and fats. These are turned into materials that will dissolve in water so they can be taken into the bloodstream.

Saturated solutions. If you continue adding a solute to a solution, the mixture will eventually reach a point where no more solute will dissolve. At that point, the solution is said to be saturated, and it is useless to add more solute to it. A strong solution of salt is called a brine. If more salt is added to a brine, the salt crystals will eventually settle to the bottom of the container. A saturated or nearly saturated solution of sugar is sometimes called a syrup.

If part of the solvent evaporates from a saturated solution, some of the solute must come out of the solution. The result is crystal formations of the solute. This is a way to grow beautiful crystals of certain chemicals, such as copper sulfate.

Surfactants. Some liquids will not mix. If you put diesel fuel and water in a jar and shake them together, the two liquids will mix, but they will quickly separate. There is no common link between water and diesel fuel whereby the one can suspend or dissolve the other.

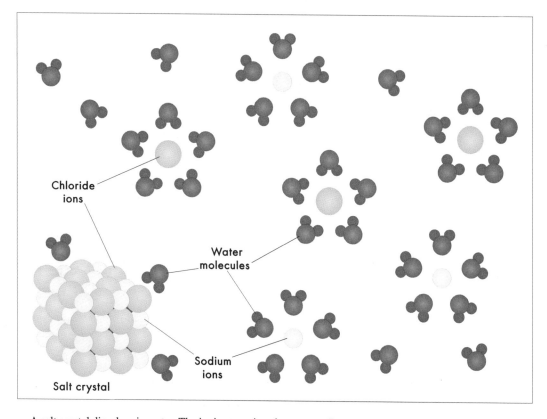

Chloride ions

Water molecules

Sodium ions

Salt crystal

A salt crystal dissolves in water. The ionic attractions between sodium and chloride molecules are weaker than those between water and sodium or water and chloride. Therefore salt crystals will easily dissolve in water. What evidence in this diagram indicates that the solution is not yet saturated?

But if you add a ***surfactant*** (sər·fak′·tənt) before shaking, you will obtain a cloudy mixture (an emulsion) of diesel fuel suspended in water. The word *surfactant* comes from the expression "surface active agent." A surfactant serves as a link between water and particles of oil or grease and allows them to become suspended in water. Soap is a surfactant that helps you to wash oil or grease off your hands. Detergents are soaplike surfactants that clean better than soap in cold and hard water. Detergents are commonly used to wash dishes and clothes.

Surfactants also improve the ability of water to wet things. The surfactant breaks down the surface tension of water, which gives the water surface a "skin." Then the water spreads out on a surface instead of forming beads. Farmers often add a surfactant to spray materials so the spray spreads out and sticks better to the leaves of plants. That is why a surfactant is sometimes called a "sticker spreader."

There is great variety in the state, kind, and form of matter; yet there are certain facts that are true of all matter. All matter was created by God, all matter has weight, and all matter takes up space. God knew that we would need many different materials to live. He gave us plenty of oxygen in the air to breathe. He put an abundance of water on the earth for drinking, washing, and forming solutions. We have many metals for different purposes. This vast array of materials is part of the many good gifts that God has provided.

Left: The lauric acid molecule is a common soap ingredient derived from coconut oil. The CH_3 end of this surfactant molecule will readily bond to petroleum products, which are rich in carbon. The OH end will readily bond to water.

Below: How detergent removes grease.

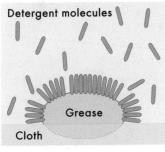

Detergent attaches to grease.

A grease droplet forms.

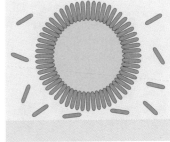

The grease droplet floats away.

———— Study Exercises: Group K ————

1. Give three ways to increase the rate at which a solution is formed.

2. How can an insoluble material like limestone rock be made to dissolve?

3. What term is used for changing food into soluble materials?

4. If you want a saturated solution, you can do which *two* of the following things?
   a. Add more solute.   c. Add more solvent.
   b. Stir the solution.   d. Evaporate some solvent.

5. What name is used for a saturated or nearly saturated solution
   a. of salt?   b. of sugar?

6. a. Oil will mix with water if a ——— is added to the water.
   b. The resulting mixture is called an ———.

7. For what two reasons is it helpful to add a surfactant to agricultural sprays?

Group K Activities

1. *Experimenting with rates of dissolving.* Obtain some rock salt, such as that used in water softeners. Pick out several pieces that are the same size. Crush one with a hammer (or a mortar and pestle if you have one). Using equal quantities of water, note the time it takes to dissolve a lump of rock salt in cold water without stirring in comparison with dissolving an equal amount of powdered salt in warm water while stirring. Design other comparisons involving one or more of the three ways to increase the rate of dissolving.

2. *Growing a copper sulfate crystal.* Make a saturated solution of copper sulfate by adding solute to about one pint (one-half liter) of water until no more will dissolve. It will be best to let the solution stand overnight with extra solute in it to be sure the solution is saturated. If a small crystal of copper sulfate has not appeared, you will need to grow some seed crystals. Do this by putting some saturated solution in a saucer and letting it evaporate for several days. Many small crystals will form. Select one of the largest crystals, tie a fine thread around it, and suspend it from the center of a small stick. Without disturbing the sediment, pour off the saturated copper sulfate solution into a quart or liter jar. Lay the stick across the top of the jar, and wind the thread so that the seed crystal is in the center of the solution. Let the jar stand in a warm, dry place. As the solvent evaporates, the solute will come out of the solution to make the crystal grow.

   Note: If your seed crystal dissolves, it shows that the solution was not quite saturated and you need to add more solute. It will take several days and even weeks to grow a large crystal, but the result is worth waiting for. The crystal will develop angles according to the pattern of the atoms in copper sulfate molecules. Crystals (including snowflakes) are a testimony to the order and beauty with which God designed matter.

3. *Demonstrating the action of a surfactant.* Put about one inch each of water and diesel fuel (or some other oil) into a test tube. Which comes to the top? Why? With your thumb over the mouth of the test tube, shake the two vigorously. Which comes to the top when you let it stand? Add a few drops of dishwashing detergent, and shake again. What has been formed?

   Add a few drops of a surfactant to a test tube of water. With a medicine dropper, let some drops of this surfactant water fall on a piece of cardboard. Also put some drops of plain water on the cardboard. What difference do you see?

Branches of Knowledge About God's World

This unit discusses a few details in the scope of God's great world. Because of the extent of God's vast creation, the study of science is divided into many branches. In the first part of this unit, you read about the laws of energy and motion. This branch of science is called physics. In the section about compounds and reactions, you studied the branch called chemistry. In other units you will study astronomy, biology, and meteorology.

Many branches of science overlap so much that you cannot expect to study one without referring to the others. However, dividing science into branches is a way of organizing things so that the different aspects of God's world can be studied in a systematic way.

Common Branches of Science

Physics is the study of forces and energy. A person who studies physics is called a physicist.

Chemistry is the study of properties of matter and of the changes that produce new materials. A person who studies chemistry is called a chemist.

Biology is the study of life and living things. A person who studies biology is called a biologist.

Botany (bot′·ən·ē), a branch of biology, is the study of plant life. A person who studies botany is called a botanist.

Zoology (zō·ol′·ə·jē), a branch of biology, is the study of animal life. A person who studies zoology is called a zoologist.

Geology is the study of the earth's crust—the outer 25 miles (40 km) of its surface. A person who studies geology is called a geologist.

Meteorology (mē′·tē·ə·rol′·ə·jē) is the study of the earth's atmosphere and weather. A person who studies meteorology is called a meteorologist.

Astronomy is the study of the heavens. A person who studies astronomy is called an astronomer.

————— Study Exercises: Group L —————

Name the branch of science in which you would expect to study each topic named below or suggested in the Bible verses given.

1. How an airplane works
2. What makes it rain
3. How concrete is made
4. The parts of a bird
5. The cause of disease
6. Where to drill a well
7. The phases of the moon
8. The electric motor
9. The parts of a flower
10. Job 14:7–9
11. Job 28:1, 2
12. Job 28:9–11
13. Job 38:4, 5
14. Job 38:22
15. Job 38:31
16. Job 38:34
17. Job 38:41
18. Job 39:13

Unit 1 Review
Review of Vocabulary

Study the list of vocabulary words at the beginning of the unit until you know the meaning and spelling of each term. Then write the missing words in the following exercise without looking back. When you have done the best you can from memory, fill in the others with help from the text. You may change the form of a word, such as from singular to plural, to make it fit in a sentence.

The power to produce changes in God's world is called —1—. Such a change is a —2— if it causes only a different shape or form in matter, but it is a —3— if it results in new materials. Energy in action is called —4—, and stored energy is —5—. There are many different forms of energy, including heat, light, and sound. The tendency of energy to become less concentrated is called —6—.

The tendency of matter to remain in its state of rest or motion is called —7—. When —8— makes an object move in a curved path, an outward force called —9— is produced. A moving object has energy called —10—, which is the product of its —11— and its speed. When a constant force is applied to a moving object, it causes a change in speed known as —12—. The action of a rapidly escaping exhaust causes a forward push called —13—. The study of energy and the motion it produces is called —14—.

Anything that has mass and volume is called —15—. A material is a —16— if it has a definite size and shape. If it has a definite volume but will take the shape of its container, it is called a —17—. A —18— will take both the volume and shape of its container. These last two states of matter are called —19— because the materials flow.

The —20— is the smallest particle of the basic kinds of matter. These basic kinds are called —21—, and they are represented by one or two letters called —22—. When two or more of the basic kinds of matter unite, they form —23—, whose smallest particle is called the —24—. These united materials are represented by —25— that tell what kind of atoms are in them. The study of the reactions that happen between different kinds of matter is called —26—.

Various mixtures of materials can be made. A —27— is a cloudy mixture in which very fine particles of one material are held in another material. A —28— is a clear mixture in which the molecules of the —29— spread evenly through the liquid part of the mixture, which is called the —30—. Oil and water can be mixed to form an —31— if the water contains a —32—. Metals can be mixed to form —33—.

Multiple Choice

1. Which of the following would be in a study of chemistry but not physics?
   a. To push a box forward, you must push backward on the floor with the same force.
   b. If you keep a piece of machinery dry, it will not rust.
   c. If you hang a magnet from the ceiling, one end of it will point north.
   d. If you put a white and a black object in the sun, the black object will get warmer.

2. Since sound is a form of energy, you can be sure that all the following will be true *except*
   a. sound can be transformed into other forms of energy.
   b. sound can produce changes in God's world.
   c. sound has weight and takes up space.
   d. sound is governed by the second law of thermodynamics.

3. If a material goes through a chemical change,
   a. the source of energy is a fuel or some other form of chemical energy.
   b. the material after the change is different from the original material.
   c. heat will be given off.
   d. there will be no movement of the material.

4. Which statement about kinetic energy is *not* true?
   a. A falling rock provides an example of kinetic energy.
   b. A swinging boy has the greatest kinetic energy when he is closest to the ground.
   c. An object that has the inertia of motion also has kinetic energy.
   d. Kinetic energy must be changed to potential energy before it will produce change.

5. According to the first law of thermodynamics,
   a. the sun has less energy today than it had yesterday.
   b. natural laws must have been created by God.
   c. there was a time when God created energy.
   d. entropy increases with the passage of time.

6. According to the second law of thermodynamics,
   a. energy can be transformed from mechanical to heat energy.
   b. energy always moves from a hot area to a cold area.
   c. only a few machines have perpetual motion.
   d. entropy decreases with the passing of time.

7. Which example below does *not* involve inertia?
   a. It takes more energy to ride a bicycle up a hill than down.
   b. Dirt will come out of a rug if you shake the rug.
   c. A heavy hammer will drive nails faster than a light hammer.
   d. Oil and grease sometimes fly off a spinning shaft.

8. Because of centrifugal force,
   a. a car should have brakes.
   b. a car needs several gears.
   c. a curving road should slant inward.
   d. the speed limit needs to be low in towns.

9. Which of the following objects has the greatest momentum?
   a. a stone dropped from a high tower
   b. a ship sailing on the ocean
   c. a truck moving at highway speed
   d. a hard-hit softball

10. The law of action and reaction helps to explain
    a. why water flies out of a wet cloth whirled around.
    b. why a blown-up balloon will fly around when you release it.
    c. why a hammer drives a nail into a board.
    d. why a thrown ball continues to fly through the air.

11. If a material is liquid,
    a. it is an element and takes up space.
    b. it has weight and a definite shape.
    c. it fills its container and can produce changes.
    d. it has inertia and is a fluid.

12. A molecule is to a compound as
    a. a solute is to a solution.
    b. an atom is to an element.
    c. inertia is to centrifugal force.
    d. energy is to motion.

13. Which statement is correct?
    a. The formula for sodium chloride is NaCl.
    b. The formula for oxygen is O.
    c. The symbol for carbon is Ca.
    d. The symbol for water is H_2O.

14. Surfactants are useful for all the following things *except*
    a. forming emulsions.
    b. forming suspensions.
    c. causing insoluble materials to dissolve.
    d. cleaning greasy surfaces.

15. All the following are mixtures *except*
    a. alloys.   b. solutions.   c. compounds.   d. suspensions.

16. Which statement is true?
    a. Sugar forms a suspension in water.
    b. Brass is an alloy of copper and zinc.
    c. Milk is an important solution.
    d. Water is a useful solvent for carbon.

17. A solution will form the fastest if
    a. the water is cooled before adding the sugar.
    b. the water and copper sulfate are mixed in a shallow dish.
    c. the water is filtered before adding the solute.
    d. the salt is put into a jar of water and shaken.

18. A saturated solution can hold no more
    a. solute.   b. surfactant.   c. brine.   d. solvent.

19. The many uses of chemicals are evidence that
    a. men have gone beyond what God had planned.
    b. matter is the product of a wise and loving Creator.
    c. science has given amazing properties to the chemicals.
    d. God has revealed the laws of chemistry to men.

Unit 2
The Body

Working together and adhering to a design plan are important in building a barn. A pile of assorted lumber has little value. But if workers put lumber pieces together by a good plan, the result is a strong, useful barn. Likewise God designed the human body according to a wonderful plan. Every one of its members, from the tiny alveoli in the lungs to the femur bone in the upper leg, has a specific task. Yet all the members need to work together to make one healthy body.

Searching for Truth

"But now hath God set the members every one of them in the body, as it hath pleased him" (1 Corinthians 12:18).

We cannot properly study the human body without first recognizing God as its Creator. God designed the human body according to His pleasure. And what a wonderful design it is! Every one of its members, from the tiny blood vessels in the eye to the large bones in the legs, He fashioned and set in the place He planned for it. Each of its many members He formed precisely for its specific task. Only the foot can walk. Only the ear can hear. Only the eye can see. Yet all the members need each other and work together to make one body, "as it hath pleased him."

Searching for Understanding

1. What is the general organization of the many parts of the body?
2. How did God provide for the body to supply its needs?
3. How do muscles and bones work together to produce movement?
4. What is the purpose of the blood?
5. What provisions did God make to remove body wastes?

Searching for Meaning

alveoli	glottis	organism
antibody	heart	plasma
antigen	immunity	platelet
appendix	integumentary system	red corpuscle
artery	joint	respiratory system
bronchus	kidney	skeletal system
capillary	ligament	spleen
circulatory system	lung	system
dermis	lymph	tendon
diaphragm	lymphatic system	thymus
epidermis	lymph node	tissue
epiglottis	mucous membrane	tonsil
excretory system	muscular system	trachea
extensor	nephron	vein
flexor	organ	white corpuscle

The Organization of the Body

Many Members—One Body

"But now are they many members, yet but one body" (1 Corinthians 12:20).

God created the whole world with beautiful order and mathematical precision, and the human body is the crowning work of this order. Trillions of different body cells work together as one body because of God's infinite wisdom.

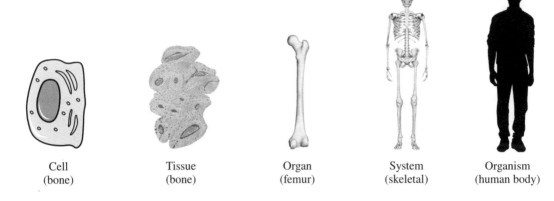

Cell	Tissue	Organ	System	Organism
(bone)	(bone)	(femur)	(skeletal)	(human body)

The human body is so complex that man is still investigating and discovering its organization and function. Medical doctors are still asking questions about the body and how it works, and they are still finding only sketchy answers and ever-changing theories. That is why you cannot expect to learn every detail or the answer to every question about the human body. This unit and the next are only a short discussion that must necessarily omit many important facts. Yet God built the amazing complexity of the body by a fairly simple plan.

From Trillions to One

The smallest unit of the human body is the cell. Though the cell can be divided into smaller pieces, those pieces cannot carry on the functions of life alone. Cells are the building blocks of the human body.

If you tried to count all the cells in your body and could count one cell every second, it would take about thirty-two years to reach one billion cells. Scientists estimate that there are one hundred *trillion* cells in the adult human body. That means it would take you more than three million years to count all the cells in your body! Of course, no one has ever obtained an exact number. Not only are there far too many

cells to count, but they are also far too small. There are about five million cells in a single drop of blood!

Not all the cells in the body are this small. Cells vary in size, in shape, and in the tasks for which God designed them. Yet all cells have two basic parts: the nucleus and the cytoplasm. The nucleus (headquarters of the cell) directs the work that takes place in the cytoplasm of the cell.

All cells are alike in yet another way. They all have the ability to react to a stimulus, to move, and to reproduce themselves. But cells are different because they specialize in certain functions. For example, the long, slender nerve cells specialize in sending messages from one part of the body to another. The flat, short muscle cells contract easily to produce motion. The tough, compact skin cells form a protective covering over the outside of the body.

Cells that specialize in the same function join together to form *tissue*. For instance, nerve cells combine to form nerve tissue. Muscle cells join in groups to make muscle tissue, and skin cells build skin tissue.

Each kind of tissue performs the task for which God especially designed it. But many functions of the body require several of these tasks at once. For this reason, God placed

Systems of the Body

Name	Main Function	Main Organs or Parts
integumentary system*	provides external protection for the body and internal support	skin, hair, mesentery
skeletal system*	provides support for the body and protection for vital organs	skull, ribs, vertebrae, pelvis
muscular system*	provides movement for the body	extensor muscles, flexor muscles
digestive system	prepares food for the body	esophagus, stomach, liver, intestines
respiratory system*	provides oxygen for the body and removes carbon dioxide	trachea, bronchi, lungs
circulatory system*	distributes food, oxygen, and wastes in the body	heart, blood, arteries, veins, capillaries
lymphatic system*	drains body cells and defends against disease	spleen, tonsils, thymus, lymph nodes
excretory system*	removes nitrogen wastes from the body	kidneys, ureter, urinary bladder, urethra
nervous system	provides communication between body parts	brain, spinal cord, nerves
endocrine system	produces various chemicals to regulate the body	pituitary gland, thyroid gland, adrenal gland
reproductive system	provides for propagation of the human family	testes (male), ovaries (female)

*System discussed in this unit

several different kinds of tissues together in units throughout the body.

A unit with various kinds of tissues working together is called an *organ.* Consider several organs that perform specialized tasks for the body. In the eye, blood tissue and nerve tissue work together to provide vision. In the stomach, muscle tissue, skin tissue, and blood tissue combine to store and digest food.

The heart uses muscle tissue nourished by blood tissue to pump blood through the body.

A group of body organs working together to perform a major function is a *system.* For example, the esophagus, stomach, small intestine, and large intestine are organs in the digestive system. Each organ within that system performs its part in helping to prepare or absorb food for the body's use.

All the systems working together form an *organism*. The organism known as the human body has eleven major systems. The table on page 51 lists these eleven systems along with their functions and main organs. Notice that each system does something for the body or for the human family.

The Human Body and the Church

God created the body and made all its different parts work together so well that He used this as an illustration of how Christ's body, the church, should function. "For as the body is one, and hath many members, and all the members of that one body, being many, are one body: so also is Christ" (1 Corinthians 12:12).

Each member of the physical body uses its natural abilities for the good of the whole body. So each member of the church should use his spiritual gift for the good of the whole church. Each member should appreciate the contribution of every other member. All the members should submit to each other and to the Head, Jesus Christ.

Each body system is designed to fulfill its task in the most efficient manner, yet not one of these systems could live without the other systems. Not only does each system function efficiently, but all the different systems are joined together to form an efficient, functioning organism—the human body. "For our comely parts have no need: but God hath tempered the body together, having given more abundant honour to that part which lacked: that there should be no schism in the body; but that the members should have the same care one for another. And whether one member suffer, all the members suffer with it; or one member be honoured, all the members rejoice with it" (1 Corinthians 12:24–26).

King David also recognized God's wonderful design of the human body. Memorize his expression of praise in Psalm 139:14, and consider it often as you study this unit. "I will praise thee; for I am fearfully and wonderfully made: marvellous are thy works; and that my soul knoweth right well."

——— Study Exercises: Group A ———

1. Supply the missing words.
   a. The ——— is the smallest unit of the human body.
   b. These smallest parts group together to form ———.
   c. Several kinds of this material join to form ———.
   d. These units work together in larger groups called ———.
   e. All these larger groups together form the ——— known as the human body.

2. The fact that trillions of cells work together to form one body
   a. proves that God created man in His own image.
   b. shows that the body has developed and improved over many years.
   c. illustrates how God intends believers to work together.
   d. explains why the human body has such a tendency to become ill.

3. Name the kind of specialized cell that does each of the following.
   a. Sends messages from one part of the body to another.
   b. Forms a protective covering for the body.
   c. Contracts to produce motion.

4. The cell is called the smallest part of the human body because
   a. it is smaller than tissue.
   b. there is no way of dividing it into smaller parts.
   c. it is the smallest part that can carry on the functions of life.
   d. the human eye cannot see it without the aid of a microscope.

5. For each boldface word below, write the number of a meaning that applies to the use of that word in the study of the human body. You will not use all the numbers.
   a. **integument**   (1) Absorb with the mind.
   b. **digest**   (2) Change (food) into a usable form.
   c. **respire**   (3) Give off; discharge.
   d. **circulate**   (4) Spread abroad; distribute widely.
   e. **excrete**   (5) Inhale and exhale.
                         (6) Move in a circular path.
                         (7) Natural covering.
                         (8) Organize systematically.

6. Below are four principles about the organization of the human body. Match each with the number of the verse(s) in 1 Corinthians 12, where that principle is mentioned.

Principles	Verse Numbers
a. God has organized the parts of the body.	*14*
b. The body contains many members or parts.	*18*
c. The different parts work for the good of other parts and need each other.	*20*
d. The various members make up one body.	*24, 25*

7. According to 1 Corinthians 12:21, 22, church members are to the church as ——— are to the human body.
   a. cells   b. tissues   c. organs   d. systems

The Integumentary System, Your Body's Protection

Probably no system does as many different things for the body as the *integumentary system* (in·teg′·yŏo·men′·tə·rē). This system includes the two square yards of skin that cover and protect the body. The skin has two layers—epidermis and dermis.

The Epidermis

The *epidermis* is the outer layer of skin. This tough, leathery layer is made mostly of dead cells and is constantly wearing away by friction. The dead cells are replaced by cells from underneath, which begin as living cells in the innermost part of the epidermis. If a certain spot must endure heavy friction, the epidermis develops a thick, tough callus there. Such calluses are quick to form on the hands and feet, and they are welcome to the barefoot boy and the carpenter.

Pigment cells in the epidermis account for the color of the skin. In a dark skin, these cells are numerous and produce large amounts of the skin pigment called melanin (mel′·ə·nin). Some persons called albinos have no pigment cells. When you are exposed to bright sunlight, your pigment cells produce extra melanin to protect your body from the harmful effects of ultraviolet rays. Concentrations of melanin result in freckles.

The Dermis

The *dermis* is the inner layer of skin. This soft layer is thicker than the epidermis and provides most of the functions of the skin.

Fat cells in the dermis provide a protective cushion for the body. They help to pad the sharp angles of the bones. These cells also store energy for the body. If for some reason you would not eat for several days, your body would use these reserves of fat.

Oil glands produce oils that lubricate the surface of the epidermis and make it water resistant. This keeps exterior water from entering the body, and it protects the moist interior tissues from drying out. The oil also helps to destroy bacteria. A blockage in these glands causes the pimples that may appear on your face.

The dermis is laced with tiny *blood vessels* that bring food and oxygen to the skin. These capillaries help to regulate the body temperature. They enlarge when the body becomes too warm, and the increased blood supply releases heat to cool the body. When the body becomes too cool, the same blood vessels constrict to conserve body heat. The blood in the skin capillaries gives light skin a pinkish color.

Nerve endings make the skin a large sense organ. The sense of touch is a very wonderful provision for contact with the outside world. Through the nerve endings in your skin, you perceive the softness of cotton, the smoothness of glass, and the roughness of sandpaper. By touch, pain, and other skin signals, God provided safeguards to protect the body from dangers such as cuts and burns.

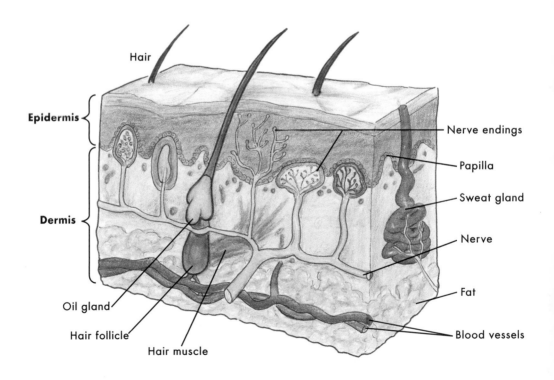

Sweat glands are another major feature of the dermis. The watery secretion of these glands is one method of removing wastes from the body. However, sweat, or perspiration, acts mainly to regulate the body temperature. As your perspiration evaporates and cools you on a warm summer day, you profit greatly from this provision of our wise God.

Exposing the skin to sunlight can be beneficial or harmful. Sunlight helps the body to produce vitamin D; but because of ultraviolet rays, overexposure to sunlight can cause skin cancer. Keeping most of the skin covered and wearing protective headgear promotes modesty as well as physical health.

Other parts of the integumentary system are the *hair, fingernails,* and *toenails,* which contain no nerve cells or blood vessels. Both the hair and nails grow at their bases and soon die as new cells push them outward. Thus you have no feeling in your hair or nails. This makes them an ideal protection for your head and for the ends of your fingers and toes.

The hair grows out of a small bulb-shaped sheath called a hair follicle. A tiny muscle is attached to each follicle. When you are chilly or frightened, these muscles pull your hair erect. The epidermis then rises slightly around the hair, giving you goose bumps.

Some parts of the integumentary system act as connective and supportive tissue inside the body. For example, the mesentery (mez'·ən·ter'·ē) of the abdomen holds the intestines in place. Connective and binding tissues are associated with bones and muscles throughout the body.

Study Exercises: Group B

1. Name the part of the integumentary system that provides each of these.
   a. cleaning and cooling
   b. padding and energy storage
   c. lubrication and waterproofing
   d. sense of touch
   e. temperature control and food

2. A callus is caused by and protects against ———.

3. Which two parts of the dermis perform a similar function?
   a. hair and muscle
   b. callus and epidermis
   c. sweat glands and blood vessels
   d. fat cells and oil glands

4. The parts of the integumentary system composed mostly of dead cells are
   a. the dermis and the epidermis.
   b. the nails and the hair.
   c. the fat and the nerve endings.
   d. the mesentery and the sweat glands.

5. What disease of the skin is caused by overexposure to sunlight?

6. Tell where each of the following is located by writing *epidermis, dermis,* or *inside body.*
   a. oil gland
   b. mesentery
   c. callus
   d. capillaries

7. Write *true* or *false* for each sentence.
   a. The oil glands secrete sweat.
   b. Goose bumps occur when muscles pull on hairs in the skin.
   c. Sweat glands keep dirt from entering the body.

Group B Activities

1. *Testing the sensitivity of your skin.* This activity demonstrates that you can sense greater detail at some places on your skin than at others. With your eyes closed, have a friend touch your skin with the closed, pointed end of a pair of scissors. Repeat this touching with the scissors, each time opening the scissors slightly more than before. Continue until you can just feel that there are two distinct points. Record the distance between the points after testing each of the following areas: back of neck, forehead, cheek, upper lip, palm of hand, back of hand, and tip of finger.

2. *Examining your fingerprints.* Interlocking projections called papillae (pə·pil′·ē) connect the dermis and epidermis of the skin. These papillae appear in rows on the surfaces of the hands and feet, producing prints with unique patterns for every person. For this reason, fingerprints can be used for identification.

   Press a finger or thumb onto a stamp pad with washable ink; then stamp it on a piece of paper. Label your fingerprint; then compare it with those of several classmates. You will find that each print is different, though certain ones have general similarities. Experts place fingerprints in three categories: loops, arches, and whorls. Compare your fingerprint with the general patterns shown, and decide which kind of print it is.

   Press the same finger into the oily region beside your nose. Then place it on a smooth surface, such as a piece of glass. Have other classmates do the same thing, keeping no record of who made each print. Place a dark piece of construction paper behind the fingerprints, and sprinkle baby powder on the fingerprints. Shake off the excess powder; the remainder will stick to the oily portions of your fingerprints. Use your labeled fingerprints to identify the prints on the glass.

   Loop   Arch   Whorl

3. *Checking for freckles.* Melanin is produced by skin cells called melanocytes (mel′·ə·nō·sīts′). Differences in skin color, as in light-skinned and dark-skinned people, are a result of the amount of melanin produced by the melanocytes. Sunlight excites these pigment cells and causes them to produce more melanin to shade the skin. Freckles appear when melanin accumulates in uneven patches in the skin. If you have freckles, look for them before going outdoors for a recess in the sunshine; then check them again after you return. The freckles will be darker after some time in the sun.

   Very fair-skinned people, such as those with red hair, have a type of melanin called phaeomelanin. They are more sensitive to sunlight, and their skin does not tan like people with regular melanin.

The Skeletal System, Your Body's Support

The skeletal system is marvelously designed to provide support and protection for the body. Consider the many ways it is suited for this task.

- It is strong enough to support the weight of the body plus heavy objects that one may carry.

- It is tough enough to withstand sharp blows or falls without breaking.

- It is light enough to be carried about easily.

- It is flexible enough to allow running, turning the head, and a multitude of other body movements.

- It is self-healing to repair fractures (broken bones) so that the injured bone becomes as strong as it was before.

- It is living so that it can grow in length and strength.

The *skeletal system* consists of 206 bones that God designed and fitted together "as it hath pleased him." The bones contain calcium carbonate, calcium phosphate, and other minerals that make them very hard. But a bone is not a dead mass of solid minerals. A bone is alive. A network of tiny canals in the bone carry blood to the living bone cells.

The centers of some large bones are hollow and contain living bone marrow. One important function of marrow, especially in children, is to produce both red and white blood cells. Bones are also covered with a tough, smooth outer membrane called periosteum (per′·ē·os′·tē·əm). The periosteum provides a place for attaching muscles and tendons to bones.

The skeletal system has two main parts. They are the axial skeleton and the appendicular skeleton. The axial skeleton forms the axis of the body, including the skull, spine, ribs, and sternum (breastbone). Plate-shaped bones in the upper part of the skull protect the brain. The jawbone hinges at the lower part of the skull to allow the mouth to open and close. The spine is a column of thirty-three vertebrae; these disk-shaped bones protect the spinal cord.

Twelve ribs form the rib cage, which protects the heart and lungs. Seven of these ribs, called true ribs, attach directly to the sternum with bands of cartilage. Three more ribs, called false ribs, attach to each other and then to the seventh true rib. Two more ribs, called floating ribs, do not connect to the sternum at all.

The appendicular skeleton includes the bones in the appendages—the arms and legs. God placed a remarkably symmetrical number of bones in the limbs. Both the arms and the legs have three long bones each. The humerus (upper bone) of the arm corresponds to the femur (upper bone) of the leg. The femur is the largest bone in the body. In the lower arm are two bones, the radius and the ulna. These correspond to the two bones in the lower leg, the tibia and the fibula.

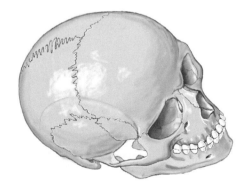

Sutures in the skull are immovable.

The Human Skeleton

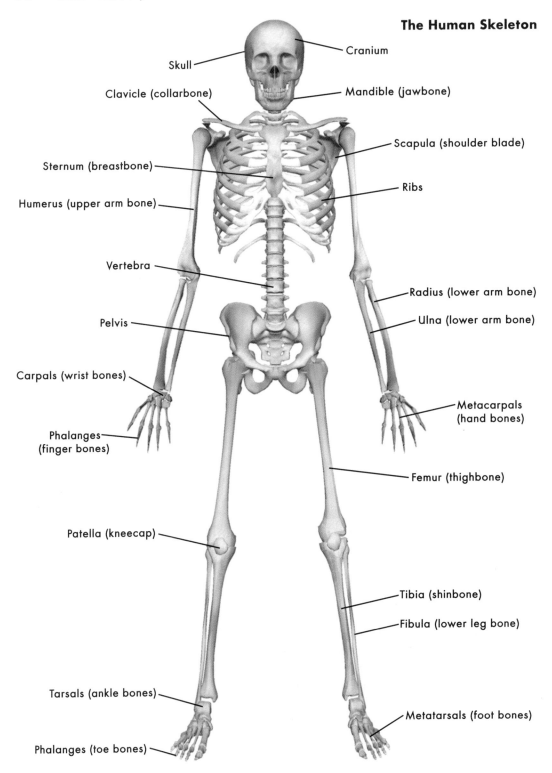

Cranium

Skull

Mandible (jawbone)

Clavicle (collarbone)

Scapula (shoulder blade)

Sternum (breastbone)

Ribs

Humerus (upper arm bone)

Vertebra

Radius (lower arm bone)

Ulna (lower arm bone)

Pelvis

Carpals (wrist bones)

Metacarpals (hand bones)

Phalanges (finger bones)

Femur (thighbone)

Patella (kneecap)

Tibia (shinbone)

Fibula (lower leg bone)

Tarsals (ankle bones)

Metatarsals (foot bones)

Phalanges (toe bones)

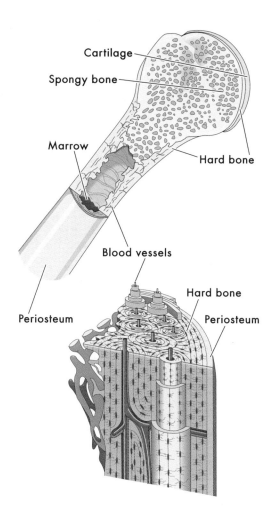

Cartilage

Spongy bone

Marrow

Hard bone

Blood vessels

Hard bone

Periosteum

Periosteum

The wrist has eight bones called carpals, and the ankle has seven bones called tarsals. There are five metacarpals (hand bones) and five metatarsals (foot bones). There are also fourteen finger bones and fourteen toe bones—all called phalanges.

Each arm is connected to the axial skeleton by the shoulder girdle, which has a scapula (shoulder blade) and a clavicle (collarbone) on each side. This arrangement allows you to shrug your shoulders, whereas the legs have a more rigid connection to the axial skeleton at the pelvic girdle.

Skeletal bones are connected at joints. A *joint* is a place where two bones meet. While some joints are nearly immobile, most of them let the bones move. There can be much pressure at a joint, such as when the force of the upper leg bone is transferred to the lower leg bones. To protect the joints, a thin layer of cartilage, or gristle, pads the bone ends. The joints secrete synovial fluid (si·nō′·vē·əl) to lubricate this cartilage.

The joints are further protected by spongy bone at the ends of bones. Spongy bone is simply a branching network of bone surrounding air pockets. This makes the bones both lightweight and springy. Strong bands of cartilage, called *ligaments,* hold the joints firmly in place. This connective tissue keeps bone ends from crashing together to destroy themselves. For instance, without ligaments, the bones at your knee would not meet properly and would strike together with great force when you run.

There are various kinds of mobile joints. The *ball-and-socket joints* of the hips and shoulders provide freedom of movement in many directions. The *hinge joints* of the knees and elbows allow only forward and backward motion. The *pivot joints* in the neck allow a rotating movement. The *saddle joints* allow sliding movement in two directions, much as

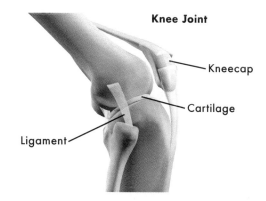

Knee Joint

Kneecap

Cartilage

Ligament

Mobile Joints in the Human Body

Neck vertebra

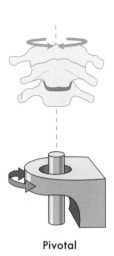

Pivotal

Wrist, Foot

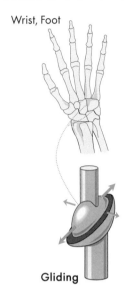

Gliding

Elbow, Knee

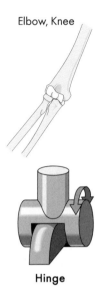

Hinge

Base of thumb

Saddle

Shoulder, Hip

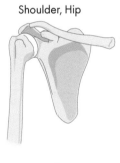

Ball-and-socket

Base of hand

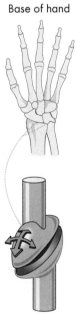

Ellipsoidal

a rider can slide around in a saddle.

The bones in the wrists and ankles provide examples of *gliding joints.* These allow bones to distribute weight evenly as one slides over another. The *ellipsoidal joints* (i·lip′·soid′·əl) occur where the arm bones attach to the wrists and the leg bones attach to the ankles. Allowing movement in two directions, these joints are somewhat like a mushroom inverted in a shallow bowl.

Some joints do not allow the connected bones to move. Most of the twenty-two bones in the skull are locked together with zigzag, immobile joints called sutures. Fibrous connective tissues further strengthen the joints. These bones form the dome-shaped cranium (skull) that supports and protects the brain.

Man has invented many mechanical devices. Skyscrapers prove his success in structural engineering. Kevlar (a synthetic fiber) can withstand bullets. But man continues to improve the design and function of his buildings and machines. Yet the design and function of the skeletal system has not changed since Creation. When God designs something, no one can improve it.

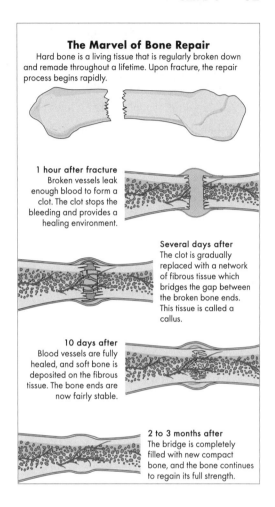

The Marvel of Bone Repair

Hard bone is a living tissue that is regularly broken down and remade throughout a lifetime. Upon fracture, the repair process begins rapidly.

1 hour after fracture
Broken vessels leak enough blood to form a clot. The clot stops the bleeding and provides a healing environment.

Several days after
The clot is gradually replaced with a network of fibrous tissue which bridges the gap between the broken bone ends. This tissue is called a callus.

10 days after
Blood vessels are fully healed, and soft bone is deposited on the fibrous tissue. The bone ends are now fairly stable.

2 to 3 months after
The bridge is completely filled with new compact bone, and the bone continues to regain its full strength.

———————— Study Exercises: Group C ————————

1. Which one of the following is *false*?
   a. The skeleton provides support.
   b. The bones of the skeleton are dead.
   c. The skeleton protects vital organs.
   d. The skeleton allows for motion.

2. Match each fact about bones with the quality or function resulting from that fact.
   a. Bones are hard.
   b. Bones have a network of blood vessels and are covered with an outer membrane.
   c. Some bones are hollow.
   d. Bones contain marrow.
   e. Bone ends contain spongy bone.

   absorption of pressure at joints
   growth and self-healing
   lightness
   production of blood cells
   support and protection

3. Name the part of a joint that provides each of these things.
   a. connection and binding   c. lubrication
   b. padding

4. Explain why the following statement is not completely true.

   A joint is a place that allows bones to move.

5. Label the joints named below, using these words: *ball-and-socket, gliding, hinge, pivot, suture.*
   a. shoulder   d. skull   f. hip
   b. knee   e. neck   g. elbow
   c. ankle

6. Complete the following analogy.

   Rib is to *heart* as *cranium* is to ———.

7. Write the names of the thirteen bones described below. Use the scientific names rather than the common names.
   a. protecting bone for the brain   i. upper arm bone
   b. protecting bones for the lungs   j. bones of the spine
   c. protecting bone for the knee   k. thick bone in the lower leg
   d. thick bone in the lower arm   l. thin bone in the lower leg
   e. thin bone in the lower arm   m. upper leg bone
   f. large, flat bones to which the arms are fastened
   g. large, flat bone to which the legs are fastened
   h. bone that passes between the shoulder blade and the breastbone

Group C Activities

1. *Finding joints in the world around you.* Try to find examples of the seven kinds of joints in the natural world: ball-and-socket, hinge, pivot, saddle, gliding, ellipsoidal, and suture. The universal joints of socket sets and the ball joints of some penholders illustrate the ball-and-socket joint. Door hinges and the hydraulic arms on buckets of trucks and tractors function like the hinge joints of the knees, knuckles, and elbows. Swivel chairs and doorknobs have pivot joints. Few true gliding joints appear in the material world, but some platform rockers and computer-monitor bases have ellipsoidal joints. Sutures appear in the dovetailed corners of some wooden furniture and at the corners of window and door casings.

2. *Comparing the motion of a ball-and-socket joint and a hinge joint.* Use a piece of chalk to draw a circle on the board, keeping your arm straight and pivoting it at your shoulder. Practice until you get a fairly perfect circle. Notice that you can swing your arm nearly 360 degrees at your shoulder.

   Now with your upper arm pointing straight toward the chalkboard and your lower arm hinged upward, prop your elbow in your other hand and try to draw a circle. Notice that the motion of your elbow is much more limited. You cannot draw a complete circle this way.

The Muscular System, Your Body's Means of Motion

The *muscular system* of the body contains specialized cells that can change chemical energy into movement. When stimulated by a nerve impulse, the long strands of muscle cells contract with great force. Many muscle cells work together to make the hundreds of muscles in your body.

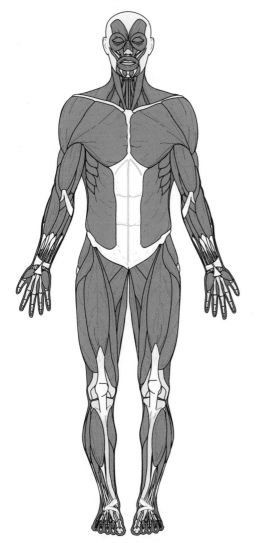

Sixty percent of the body mass is muscle. Notice the long tendons that control the fingers.

There are three types of muscles in the body—skeletal muscles, smooth muscles, and cardiac muscles.

Skeletal muscles attach to the skeleton. They account for most of the muscles in the body, about 700 of them. They also account for about 60 percent of the entire body mass. These are the muscles we mean when we speak of the muscular system. They can also be called voluntary muscles because you can control them by conscious thought.

The fibers of the skeletal muscles are arranged in bundles wrapped in connective tissue. These bundles are further arranged according to the function of the muscle.

The ends of a skeletal muscle are connected to bones of the skeleton by tough connective tissues called *tendons.* A tendon is similar to a ligament: a tendon fastens a muscle to a bone, whereas a ligament connects one bone to another. You can feel the largest tendon in your body just above your heel. This very strong tendon, called the Achilles tendon, lifts your body when you walk, run, or jump.

Muscles can exert force only by contracting. Therefore, muscles cannot push; they can only pull. Since bones need to move both ways, God provided for two-way movement by arranging bones to form levers and by placing muscles in pairs to operate those levers. For example, a muscle called the biceps, on the front of your upper arm, causes your lower arm to bend inward. A muscle called the triceps, on the back of your upper arm, causes your lower arm to straighten.

Muscles that cause a joint to bend (flex) are called *flexors.* Thus the biceps muscles are flexors. Muscles that cause a joint to unbend (extend) are called *extensors.* The triceps muscles are extensors.

You have a number of flexors that allow you to make a fist. You have a number of corresponding extensors that allow you to open your hand. Where are these muscles? God knew that if they were all in the hand itself, the hand would be very bulky. Therefore, the main muscles that operate the hand are in the lower arm. Long tendons run through the wrist and up the front and back of the hand to move the fingers. You can see these tendons work if you watch your wrist and the back of your hand while you move your fingers. The muscular system abounds in such examples of compactness and unity.

Smooth muscles are not fastened to the skeleton. These are called involuntary muscles because they are not governed by conscious thought. Instead, these muscles are controlled by the autonomic (ô′·tə·nom′·ik) nervous system; *autonomic* means "self-ruling."

Smooth muscles move the contents of such organs as the intestines, stomach, and blood vessels. They also cause the hairs in the skin to stand erect.

Cardiac muscles are other involuntary muscles not fastened to the skeleton. These are the muscles of the heart, which pump blood through the body. The autonomic nervous system controls the rate of the heartbeat, but scientists do not fully understand what causes the rhythmic contractions of cardiac muscles.

God created the muscular system to provide motion. This system is designed to operate either with great force or with careful precision, as the situation requires. Lifting concrete blocks requires great force. Writing with a pencil or batting a ball requires controlled and complex muscle movements. Unit 7 contains further discussion of how muscles are controlled.

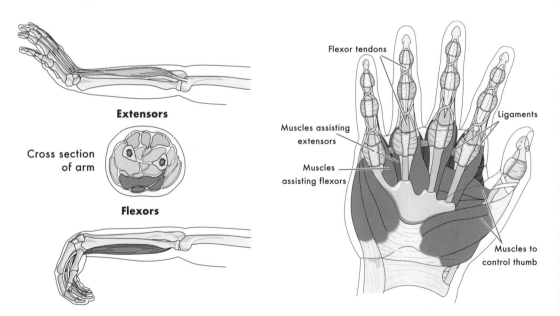

Left: The forearm uses many different muscles to control the wrist and the fingers. The grayed muscles in the cross section are not used for these particular hand movements. *Right:* The palm has additional muscles and tendons to help with finger control.

———— Study Exercises: Group D ————

1. Write *true* or *false* for each statement.
   a. Muscles contract when they are stimulated by nerve impulses.
   b. Tendons connect muscles to bones.
   c. Extensors push, but flexors pull.

2. Why are many skeletal muscles arranged in pairs?

3. Consider the action of the large muscle fastened to the Achilles tendon. Is this a flexor or an extensor?

4. a. Where are the main muscles that operate the fingers?
   b. Why are they located there?

5. Tell whether each description refers to *skeletal, smooth,* or *cardiac* muscles.
   a. Attached to bones in the body.
   b. Controlled by conscious thought.
   c. Move food through the intestines.
   d. Pump blood through the body.
   e. Arranged in pairs.
   f. Attached by tendons.

6. The (voluntary, involuntary) muscles are controlled by conscious thought.

7. The primary function of the muscular system is to
   a. support the body.
   b. protect the body.
   c. produce motion in the body.
   d. provide communication in the body.

Group D Activity

Demonstrating how muscles work in pairs. Get two pieces of wood ½ inch thick, 2 inches wide, and 12 inches long (1 cm by 5 cm by 30 cm). Insert four screw eyes on one piece, putting them on both edges 2 inches (5 cm) from each end. (See "Upper arm" in the diagram.) Insert two screw eyes on the other piece, putting them on opposite edges about an inch from the same end. (See "Lower arm" in the diagram.)

Lay the two pieces on top of each other, and drill a hole through both, ½ inch (1 cm) from one end. Get a bolt that is slightly smaller than the hole you drilled and long enough to go through both pieces and be fastened with a nut on the other end. Fasten the two pieces together as shown in the diagram, and screw the nut on lightly so that the two pieces of wood will hinge. Run a cord along the lower set of screw eyes, and another cord along the upper set of eyes. Tie each cord to the eyes at the ends. (The cords will need to be long enough to let the "elbow" bend.) By pulling on the cords, you can demonstrate the motion of the bones in your arms.

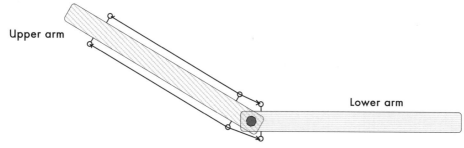

Upper arm

Lower arm

The Respiratory System, Your Body's Source of Oxygen

The human body constantly needs oxygen from the air in order to function. It also needs to release carbon dioxide waste. God created the *respiratory system* to provide both of these.

The process of absorbing oxygen into the bloodstream and releasing carbon dioxide is called respiration (breathing). Oxygen is the only thing you must take from your surroundings at all times. Even while you sleep, God provides for you to continue breathing. Respiration has two parts: inhalation (breathing in) and exhalation (breathing out).

Air enters the body through the nasal passages. These passages have warm, moist mucous membranes that warm the air and filter out dust. Hairs in the nose also trap foreign particles. From the nasal passages, air enters the pharynx (throat). At the bottom of the pharynx, air passes through the *glottis,* an opening at the top of the larynx (voice box). From there, the air goes into the windpipe, or *trachea.*

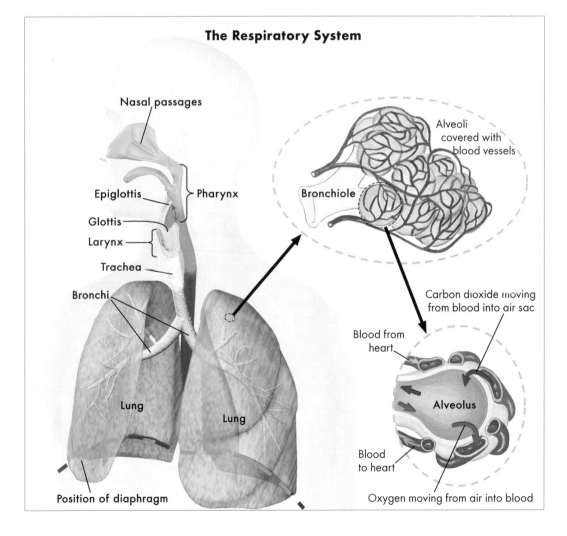

The Respiratory System

Nasal passages

Alveoli covered with blood vessels

Epiglottis

Pharynx

Bronchiole

Glottis

Larynx

Trachea

Bronchi

Carbon dioxide moving from blood into air sac

Blood from heart

Lung

Lung

Alveolus

Blood to heart

Position of diaphragm

Oxygen moving from air into blood

Above the glottis is a trapdoor called the *epiglottis.* As long as this flap is up, air can pass freely into and out of the trachea. Food also passes through the pharynx to enter the body. But the instant you swallow, the epiglottis covers the glottis and causes the food to pass down your esophagus into your stomach. As soon as the food is past, the epiglottis opens and breathing can continue. What a remarkable arrangement!

The trachea and larynx are supported by cartilage, a flexible material similar to that in the nose and the outer ear. This cartilage keeps the trachea open while allowing it to flex with the movement of the neck. When you speak or sing, the vocal cords in your larynx stretch taut to produce sound. When you are not using your voice, the vocal cords are relaxed and open for easy breathing.

The bottom of the trachea divides into two tubes, or *bronchi.* The left bronchus carries air to the left lung, which is divided into two lobes. The right bronchus carries air to the right lung, which is divided into three lobes.

The two main organs of the respiratory system are the lungs. The *lungs* are spongy masses of tubes and air sacs. In them the bronchi divide into smaller and smaller tubes called bronchioles, forming a system much like the branches of a tree. At the end of each branch is a cluster of tiny sacs called *alveoli* (al·vē′·ə·lī′). Here is where the exchange of oxygen and carbon dioxide takes place.

The walls of the alveoli have a network of very fine blood vessels. During inhalation, oxygen enters the blood through the thin walls of the blood vessels. During exhalation, carbon dioxide escapes from the blood and is released into the air. As you breathe in, the inhaled air is rich in oxygen; and as you breathe out, the exhaled air is rich in carbon dioxide.

Why did God make the lungs as a mass of tiny air sacs instead of making them like two balloons? Here again we see the wisdom of God. The oxygen passes through the walls of the blood vessels rather slowly. But because of the many alveoli—300 to 400 million in each lung—the area for exchanging oxygen and carbon dioxide is almost 1,000 square feet (93 m^2). This allows a plentiful amount of oxygen to enter the blood in a short time.

What air pump fills the millions of tiny alveoli? The air pressure of the atmosphere forces air into the lungs. Atmospheric pressure at sea level is about 14.7 pounds per square inch (1.03 kg/cm^2).

Just below the lungs is a large dome-shaped sheet of muscles called the *diaphragm.* When you inhale, this diaphragm contracts and pulls downward. At the same time, the muscles attached to the ribs contract and pull them upward and outward. These two muscular actions enlarge the space of the lung cavity, creating a partial vacuum in your lungs. Atmospheric pressure then forces air in to fill the vacant space.

When you exhale, your diaphragm and rib muscles relax. Your diaphragm rises and your rib cage falls, forcing air out of your lungs. During quiet breathing, you exchange less than a quart of air with each breath; but an adult man can exhale as much as a gallon of air.

While breathing is a natural process, there are certain things you can do to maintain a healthy respiratory system. Running, brisk walking, and hard work are healthful for your lungs because they make you breathe faster. At least once a day, you should exert yourself enough to breathe deeply.

God not only made the body to use oxygen in its life processes; He also provided an efficient respiratory system to supply the needed oxygen. "Marvellous are thy works; and that my soul knoweth right well" (Psalm 139:14).

──────── Study Exercises: Group E ────────

1. Describe two things that the nasal passages do to the air on its way to the lungs.

2. Give the term for each description.
   a. Combined action of breathing in and breathing out.
   b. Action of breathing in.
   c. Action of breathing out.
   d. Gas taken into the body when breathing in.
   e. Gas released by the body when breathing out.
   f. Opening to the larynx.

3. Why do food and water not enter your trachea when you swallow?

4. What is the advantage of lungs containing many small alveoli rather than simply being two large sacs?

5. Which sentence does *not* help to explain how air enters the lungs?
   a. Atmospheric pressure outside the lungs is greater than the pressure inside.
   b. Wavelike muscular contractions pump air down the trachea.
   c. The diaphragm and rib muscles contract.
   d. The space in the lung cavity enlarges.

6. Supply the missing terms in the path of air from outside the body to the tiny sacs in the lungs.

   nasal passages, *(a)* ───, glottis, larynx, *(b)* ───, bronchi, bronchioles, *(c)* ───

7. Which sentence best states the reason that God gave you two lungs?
   a. One lung is for breathing in, and the other is for breathing out.
   b. One lung takes in oxygen, and the other lung gives off carbon dioxide.
   c. Each lung can serve as a spare in case the other stops working.
   d. A large and constant supply of oxygen is important for body cells to function well.

8. Fill in the blanks to show how the respiratory system is associated with two other systems that you have studied.
   a. The ─── is a large, dome-shaped muscle of the muscular system.
   b. The ─── that protect the lungs are part of the skeletal system.

Group E Activities

1. *Testing for carbon dioxide.* You can test for carbon dioxide in exhaled air by bubbling your breath into limewater. The carbon dioxide will turn the limewater milky; this is the test for carbon dioxide.

   To make limewater, add several tablespoons of lime to a quart of water and shake well. Allow the mixture to stand overnight. Pour off the clear limewater without disturbing the sediment. Then use a straw to blow your breath into the limewater. Blow for about half a minute.

   Put some limewater in another container, and use an air pump or the exhaust from a vacuum cleaner to blow regular air into it. The regular air will not turn the limewater milky because it contains little carbon dioxide. It is important that you blow into the limewater with regular air as well as your breath so that you have a control for your experiment.

2. *Measuring your lung capacity.* Pour water into a gallon or half-gallon jug until it is completely full. Hold your hand over the mouth of the jug, and invert it in a large container of water. Insert a flexible tube into the mouth of the jug. Take a deep breath, and then exhale into the tube. The air that you blow into the jug will push an equal volume of water out of the jug. Place your hand over the mouth of the jug, and turn it upright. With a large measuring cup, measure the water needed to refill the jug. That amount is the capacity of your lungs.

The Circulatory System, Your Body's Transportation Network

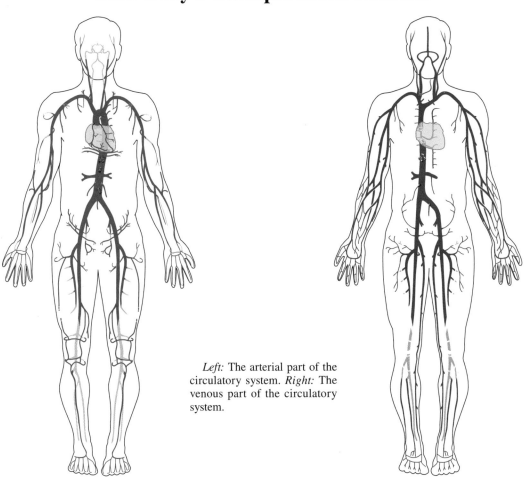

Left: The arterial part of the circulatory system. *Right:* The venous part of the circulatory system.

Although the *circulatory system* is only part of the entire living body, "the life of the flesh is in the blood" (Leviticus 17:11). Every living cell in the human body is sustained by this system. The circulatory system takes food, water, and oxygen to the cells, and it carries waste materials away from the cells. It regulates the body temperature, provides a defense against disease germs, and combats any disease that does develop. Let us take a look at this system.

The Blood Vessels

Blood travels to and from all parts of the body through 60,000 miles (97,000 km) of blood vessels. Laid end to end, all those blood vessels would reach around the earth two and a half times at the equator! Large, strong-walled vessels called *arteries* carry blood away from the heart. These branch out into smaller and smaller arteries until the blood passes through tiny *capillaries.*

Capillaries are so small that it would take ten of them to make the thickness of a hair, but they are so numerous that every living cell of the body is within a hairsbreadth of a capillary. Food, oxygen, and water pass through the thin walls of the capillaries to sustain the cells, and wastes from the cells pass into the bloodstream.

From the capillaries, the blood enters *veins* to travel back to the heart. Veins do not need the strong walls that arteries have, because they carry blood at a lower pressure. Valves inside the veins keep this low-pressure blood from flowing backward.

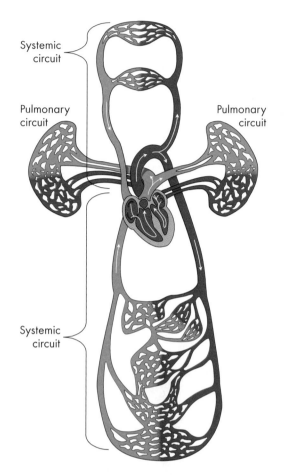

This diagram shows the two circuits of the blood. The systemic circuit moves blood throughout the body, whereas the pulmonary circuit moves blood from the heart to the lungs and back to the heart. Blood vessels carrying oxygen-poor blood are shown in blue, and those carrying oxygen-rich blood are shown in red.

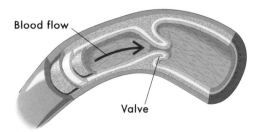

Blood flow

Valve

The Heart

The circulatory system is driven by the heart. The *heart* is a muscular pumping organ about as large as one's fist. The heart pumps about 1,500 gallons of blood a day and continues to do this as long as you live. It builds up enough pressure to shoot a stream of water to a height of 5 or 6 feet (1.5 to 1.8 m).

Blood flows through two separate circuits of arteries and veins. The right side of the heart pumps blood through the pulmonary (pool'·mə·ner'·ē) circuit. As oxygen-poor blood flows into this side, the heart pumps the blood through the pulmonary arteries to the lungs. The carbon dioxide is exchanged for oxygen, and then the oxygen-rich blood returns to the left side of the heart. The left side then pumps the blood through the systemic (si·stem'·ik) circuit, in which the systemic arteries carry oxygen-rich blood throughout the body.

Man has a four-chambered heart. The left atrium (ā'·trē·əm) and the right atrium are the receiving chambers, and the left ventricle and the right ventricle are the pumping chambers. Because of this design, the heart can pump blood through two circulatory paths at the same time.

There are three phases in every heartbeat, as described below.

First phase: The heart muscle relaxes, and oxygen-rich blood flows from the lungs into the left atrium. At the same time, oxygen-poor blood from other parts of the body flows into the right atrium.

Second phase: The upper heart muscle contracts and squeezes blood from both atria into both ventricles.

Third phase: The lower heart muscle contracts. Oxygen-rich blood is pumped from the left ventricle through the aorta to the systemic arteries. At the same time, oxygen-poor blood is pumped from the right ventricle through the pulmonary artery to the lungs.

The heart goes through these three phases over and over as it pumps blood through the body. A system of valves regulates the flow of blood into and out of the ventricles. The sounds of the blood pressing against the closing valves make the *lub-dub* of the heartbeat.

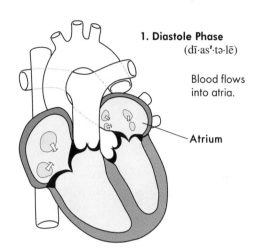

1. Diastole Phase
(dī·as'·tə·lē)

Blood flows into atria.

Atrium

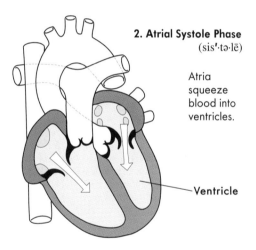

2. Atrial Systole Phase
(sis'·tə·lē)

Atria squeeze blood into ventricles.

Ventricle

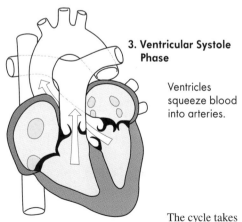

3. Ventricular Systole Phase

Ventricles squeeze blood into arteries.

The cycle takes about four-fifths of a second, but the speed may more than double during vigorous exercise.

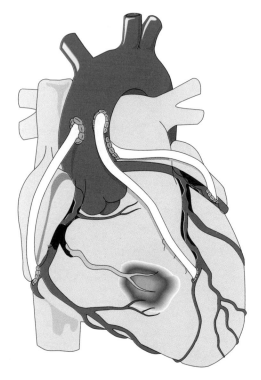

that unites easily with oxygen and carbon dioxide. Thus the red corpuscles are involved in the vital process of respiration.

White corpuscles fight disease germs. Some white corpuscles kill germs by devouring them. Others produce chemicals called *antibodies,* which label and weaken germs. White corpuscles also digest and remove injured and dead body cells, and thus they have a part in the process of healing.

Platelets are involved in the clotting of blood. If blood escapes through the broken tissues of a wound, platelets give off a special chemical that helps to produce a mass of fibers; these form a clot or scab over the wound. Without this provision for blood clotting, even a small wound could be fatal because of the loss of too much blood. God anticipated the problems of physical life and created this solution even before the problem existed.

A heart attack occurs when a heart muscle artery becomes choked with fatty deposits (shown in black), causing some muscles to die (shown in dark gray). The three white tubes are blood vessels placed by bypass surgery to go around the blockages.

The Blood

The blood is the most important part of the circulatory system. The average adult has about ten pints of blood. Though blood looks red, more than half of its total volume consists of *plasma,* a pale yellow liquid. Plasma is water with dissolved food, salt, and other substances from the digestive system. This fluid supplies water and digested food to all the cells of the body.

Besides plasma, blood contains three kinds of living cells: *red corpuscles, white corpuscles,* and *platelets.* The red corpuscles make up nearly 96 percent of the cells. They contain hemoglobin, a reddish iron compound

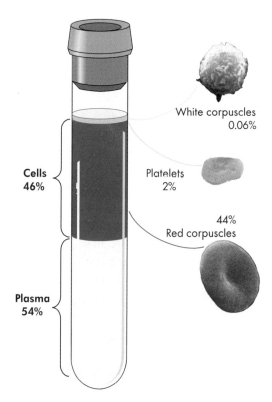

White corpuscles
0.06%

Platelets
2%

44%
Red corpuscles

Cells
46%

Plasma
54%

If a wound is serious, a person may lose so much blood that some of it must be replaced. The patient can receive blood from another person if both have the same type of blood. The four main blood types are labeled A, B, AB, and O, according to the type of *antigens* in the red blood cells. (An antigen is any substance that stimulates the production of antibodies.) Type A blood has antigen A; type B has antigen B; type AB has both kinds of antigens; and type O has neither kind of antigen. Blood of each type may also be either positive or negative, depending on the presence or absence of a substance called the Rh factor.

Doctors must be sure to use a suitable blood type when they give blood to a patient. The accompanying table shows which blood types may be used safely. If type B blood is given to a person with type A, the A-antigens react and thicken the type B blood so that it clogs the arteries and may kill the patient. The same thing happens if a person with type B receives type A. A person with type O can receive only type O, but type O can be given to a person with any of the other blood types.

A study of the blood and its many functions reveals what a marvelous fluid is moving through your arteries and veins. Without it, you would die physically. But the blood of Christ, the purchase price of salvation, is more wonderful yet. Without it, men die spiritually. It is the only means whereby we can be saved from sin, and it is sufficient for all the sins of all men. "The blood of Jesus Christ his Son cleanseth us from all sin" (1 John 1:7).

Compatible Blood Types	
Blood Type of Patient	*Blood Type That May Be Received*
A	A or O
B	B or O
AB	A, B, AB, or O
O	O

———— Study Exercises: Group F ————

1. For each function of the circulatory system, write the number of the other body system to which it is most closely related. Choices may be used more than once. (See the first paragraph of this section and the table "Systems of the Body.")
   a. Taking food and water to the cells.
   b. Defending against disease germs.
   c. Carrying wastes away from the cells. (two answers)
   d. Regulating body temperature.
   e. Taking oxygen to the cells.

   (1) integumentary system
   (2) respiratory system
   (3) digestive system
   (4) excretory system

2. Name the heart chamber that each description refers to. Include *left* or *right* in each answer.
   a. Receives oxygen-rich blood from the lungs.
   b. Receives oxygen-poor blood from the body.
   c. Pumps oxygen-rich blood to the body.
   d. Pumps oxygen-poor blood to the lungs.

3. Tell whether each description refers to an *artery* or a *vein*.
   a. Has strong, heavy walls.
   b. Has valves and thinner walls.

    c. Carries blood from the heart to the body.

    d. Carries blood from the body to the heart.

    e. Carries blood from the heart to the lungs.

    f. Carries blood from the lungs to the heart.

4. The main purpose of the circulatory system is fulfilled in the

    a. arteries.   b. capillaries.   c. veins.   d. heart.

5. Supply the missing terms for the path of blood through the body, starting at the heart and returning again. Choose from the following words.

        body veins   *left ventricle*   *pulmonary vein*
        capillaries   *right ventricle*

    right atrium, *(a)* ———, pulmonary artery, lung, *(b)* ———, left atrium, *(c)* ———, body arteries, *(d)* ———, *(e)* ———, right atrium

6. Name the part of the blood that each description refers to.

    a. Fights disease germs.

    b. Contains dissolved food, minerals, and wastes.

    c. Is involved in the clotting of blood.

    d. Carries oxygen and carbon dioxide.

7. A person with which blood type

    a. can receive any blood type?

    b. can give blood to a person with any blood type?

8. Challenge question: A blood donation is taken from a vein and not an artery. Why is this?

Group F Activities

1. *Taking your pulse.* You can readily detect your pulse on your neck or your wrist. With your palm turned up, place the first two fingertips of your opposite hand on the thumb side of your wrist. You can also place them on the side of your neck just below your chin. Do not use your thumb, because it has a slight pulse of its own that may confuse you.

    Press gently, and wait until you can distinctly feel the beat of your heart. Count the beats for 15 seconds, and then multiply by 4. At rest, a normal adult's heart beats about 70 times per minute. Your heart will likely beat faster because you are younger.

2. *Listening to your heart.* Use a stethoscope to listen to the *lub-dub* of your heartbeat. One complete cycle of contraction and relaxation makes up one heartbeat.

    The ventricles fill up with blood which entered through valves from the atria. When the ventricles contract, the pressure forces those entrance valves shut (the *lub* sound) and opens other valves for the blood to gush out into the arteries. This is the throb you feel when you check your pulse.

    Then the ventricles relax, and the exit valves close (the *dub* sound).

    Blood pressure is measured with two numbers that indicate the ventricular systole and the diastole. The first one, the systole, is higher because the heart muscle is contracting. The second one, the diastole, is lower because the heart muscle has relaxed.

The Lymphatic System, Your Body's Means of Drainage and Defense

No man-made machine, however modern its technology, can work continuously without periodic breakdowns. Considering the intricacy of the body, how can it work nonstop with so few failures? The answer is a remarkable plan of God called the *lymphatic system.* This is a network of vessels and nodes that follow the circulatory system and perform two main functions: drainage and defense. The lymphatic system accounts for about one trillion of the estimated one hundred trillion cells in the human body.

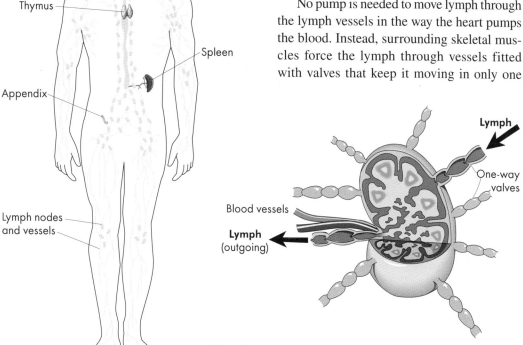

Mucous membranes

Tonsils

Thymus

Spleen

Appendix

Lymph nodes and vessels

The lymphatic system drains the fluid that surrounds body cells. In the previous section, you read that blood plasma brings water and dissolved food to all the cells in the body. Some of this plasma returns to the bloodstream through the capillaries. The remaining fluid is drained away by the lymphatic system.

As with any drainage system, the fluid in the lymphatic system flows in only one direction—away from the cells. Beginning as capillaries with closed ends, the lymph vessels drain into larger and larger vessels until they reach the veins at the base of the neck. The *lymph* (fluid drained from the tissues) contains protein, salt, sugar, and nitrogen wastes that filter out along the way. At the base of the neck, these vessels empty the filtered lymph into the veins of the circulatory system.

No pump is needed to move lymph through the lymph vessels in the way the heart pumps the blood. Instead, surrounding skeletal muscles force the lymph through vessels fitted with valves that keep it moving in only one

Lymph

One-way valves

Blood vessels

Lymph (outgoing)

As lymph passes through a lymph node, the node filters the lymph and supplies it with a certain type of white blood cell.

direction. (Since skeletal muscle are voluntary muscles, physical exercises contributes to a healthy lymphatic system.) An adult's lymphatic system carries between one and two quarts of lymph.

The lymphatic system defends the body against disease. Bacteria and viruses are constantly invading the body through the mouth, the nose, and the skin. Cells in the lymphatic system recognize these invaders and either destroy them immediately or produce other cells to kill them.

The Lymph Organs

The *lymph nodes* are the smallest and most numerous organs of the lymphatic system. These small, bean-shaped capsules of lymph tissue are strung along the lymph vessels like beads on a string. The lymph nodes with their incoming and outgoing vessels are grouped in the largest numbers around the major organs and main arteries of the body. You might think of them as an adopt-a-highway system, keeping our vital organs free of litter, or as a police force, defending them from attack.

The *spleen* is the largest organ of the lymphatic system. About the size of one's fist, the spleen lies at the tip of the left lung, left of the stomach and just below the diaphragm. Its job is to remove old blood cells from circulation and store their hormones and hemoglobin to recycle them in new blood cells. If the bone marrow cannot produce enough red corpuscles, the spleen will help to make them.

The spleen also produces white corpuscles. In fighting disease, the white corpuscles move freely between the vessels of the circulatory and lymphatic systems. As the white corpuscles produce antibodies to recognize and destroy disease germs in the blood, so the spleen produces antibodies to destroy invaders

that enter the body through food, air, and skin contact.

The *thymus* is an organ of the lymphatic system. This small gland, lying between the heart and the sternum, continues to grow until a person reaches his early teens. By adulthood it returns to its original size. The thymus makes lymph cells that recognize and destroy viruses and cancer cells. The lymph cells it produces in childhood will continue to reproduce within the body and supply the person for life.

The *tonsils* are organs of the lymphatic system. Occurring in three pairs, the tonsils lie at the back of the mouth, the sides of the mouth, and the entrance to the larynx. The cells in the tonsils filter out and destroy harmful bacteria and viruses that enter through the mouth and nose. Like the thymus, your tonsils shrink as you reach adulthood.

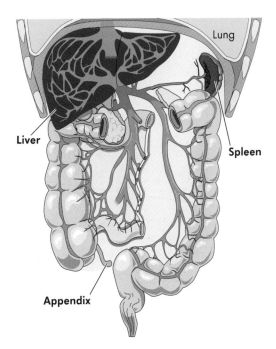

The spleen is connected to veins leading to the liver. The spleen and the liver both service the blood before sending it back to the heart.

The *mucous membranes* of the respiratory and digestive systems function as part of the lymphatic system. These membranes secrete the slippery mucus that filters out and destroys invaders that enter with air and food. You can see this moist lining in your mouth and nose.

The *appendix* is an organ of the lymphatic system. This narrow, dead-end tube connected to the large intestine contains lymph tissue that apparently destroys antigens.

The lymphatic system produces three kinds of cells to destroy invaders. The NK cells (natural killer cells) and the T cells produced by the thymus destroy antigens directly. The B cells, produced in the bone marrow, make antibodies that recognize and destroy specific antigens. This is how your body develops *immunity* to a certain disease, either by having the disease or by receiving a vaccine for it. When you are vaccinated, a small amount of a disease-causing organism is injected into your body. This triggers your body to produce antibodies for protection against later attacks of those disease germs.

Occasionally an antibody will fail to recognize its own body cells and will begin to attack them. When this happens, a kind of allergy such as an autoimmune disease may result.

Not only did God create your body to function well, but He also arranged for it to defend itself against invasion.

────────── Study Exercises: Group G ──────────

1. The two main functions of the lymphatic system are ─── and ───.

2. Supply the missing words in the following sentences, which show that the functions of the lymphatic system are closely related to other systems of the body.
   a. The lymph is pumped by the organs of the ─── system.
   b. Some lymph cells are produced in the marrow of the ─── system.
   c. The lymphatic system removes antigens from the air taken in by the ─── system.
   d. The lymph vessels follow and drain into the vessels of the ─── system.
   e. The lymphatic system removes antigens that enter through the pores of the ─── system.

3. What kind of valves are found in lymph vessels, which are also found in veins?

4. You would expect to find the most lymph nodes
   a. in your legs.   c. around your heart.
   b. in your lower arms.   d. in your feet.

5. The tonsils sometimes become enlarged in childhood. Because of their position, this would most likely cause
   a. irregular heartbeat.   c. muscle spasms of the legs.
   b. difficulty in breathing.   d. high blood pressure.

6. Some lymph cells are destroyers, and others are producers.
   a. What do the destroyer cells destroy?
   b. What do the producer cells produce?

7. The thymus and the tonsils shrink as you reach adulthood. Their main purpose is to recognize and destroy invading bacteria and viruses. This shows that we develop most of our immunity to disease during (childhood, adulthood).

The Excretory System, Your Body's Means of Waste Removal

As the body uses proteins for fuel and growth, it produces a nitrogen waste called urea (yŏŏ·rē′·ə). This waste must be removed from the body to prevent poisoning and death. But since urea is dissolved in the blood, no simple filter can remove it. The blood also contains other dissolved minerals that must be excreted. God created the *excretory system* to do the vital work of removing wastes and maintaining a proper balance of chemicals in the body.

The main organs that remove wastes are the kidneys. The *kidneys* are two bean-shaped organs, each about the size of a fist. They lie against the back wall of the abdomen near the bottom of the rib cage, with the right kidney lying a little lower than the left kidney.

Large arteries and veins carry blood to and from the kidneys. The larger artery entering the kidney divides into smaller arteries that branch through the inner part of the kidney, called the medulla. The arteries continue to divide until they become capillaries in the outer part of the kidney, called the cortex. Here the blood passes through almost a million tiny purifying structures called *nephrons* (nef′·ronz).

In the nephrons, water and dissolved substances from the capillaries filter through the walls of the collecting tubes at the rate of about 48 gallons (180 *l*) per day. God

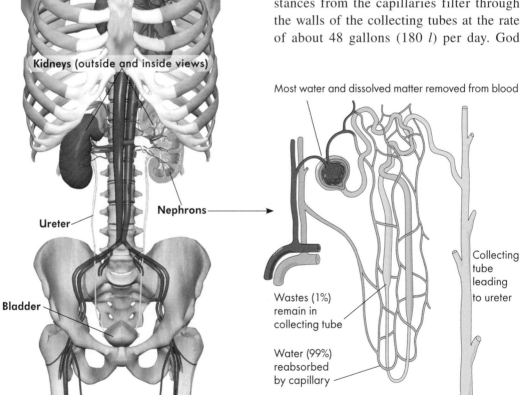

Kidneys (outside and inside views)

Ureter

Nephrons →

Bladder

Most water and dissolved matter removed from blood

Collecting tube leading to ureter

Wastes (1%) remain in collecting tube

Water (99%) reabsorbed by capillary

The Excretory System

designed a marvelous method for reusing most of this water. As the solution continues through the tiny tubes of the nephrons, about 99 percent of the water is reabsorbed into the bloodstream.

The remaining 1 percent of the solution is a concentrated waste product called urine. This solution contains water, urea, and other substances. Urine leaves the kidneys through tubes called ureters (yo͝o·rē′·tərz), which carry it to a collecting sac called the urinary bladder. Here the urine is stored until it passes out of the body through the urethra (yo͝o·rē′·thrə).

A person can live with the use of only one kidney. But if neither kidney is functioning, death is sure to follow. The poisons will build up in the body until they kill the person. A dialysis machine can be used to do the work of poorly functioning kidneys, but such a machine cannot begin to match the convenience and efficiency of the two kidneys God gave you.

The kidneys do their job so quietly and so well that we usually give little thought to them. But we can do certain things that put unnecessary strain on our kidneys. For example,

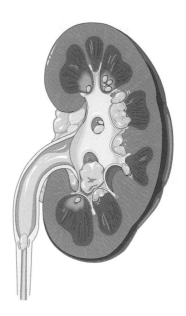

If calcium salts solidify in the kidney, they form kidney stones that can cause blockages in the excretory system. The problem occurs mostly in men, and the stones may be microscopic to an inch in size.

heavy consumption of coffee and other caffeinated beverages may lead to kidney problems. Drinking plenty of pure water every day will promote the health and proper function of our kidneys.

The Unity of the Body

Perhaps in studying the body piece by piece, you may miss the unity of the body. The systems of the body are not only dependent on each other; they are interconnected. For example, every system has the blood vessels of the circulatory system running through it. Muscles cause the hairs of the integumentary system to stand erect. Both the skeletal and the muscular system would be greatly handicapped without each other.

The circulatory system carries the oxygen provided by the respiratory system. The respiratory system depends on the diaphragm, a muscle of the muscular system. The muscular system produces nitrogen wastes that need to be removed by the kidneys. The kidneys in turn need oxygen. So it is throughout the entire body. Each part unselfishly gives what it can to serve the body, and in turn it receives services from the other parts. In fact, the systems are so interrelated that if only one organ of one system fails to function properly, all the systems suffer harm. "And whether one member suffer, all the members suffer with it" (1 Corinthians 12:26).

All the marvels of man's body were planned by the omnipotent Creator. Therefore, it is important that you give careful thought to the way you use your body. You have one short life that is highly dependent on God's continual care. One foolish, thoughtless act could quickly bring your life to an end, or the practice of some unhealthful habit could lead to years of misery and a premature death. Never be careless or thoughtless in the use of your body, thus robbing God of the honor due to His Name. "For ye are bought with a price: therefore glorify God in your body, and in your spirit, which are God's" (1 Corinthians 6:20).

———— Study Exercises: Group H ————

1. What wastes do the kidneys remove from the blood?

2. Which statement correctly summarizes the function of the kidneys?
   a. The kidneys remove much material from the blood and return much to the blood, resulting in much waste.
   b. The kidneys remove little material from the blood and return little to the blood, resulting in little waste.
   c. The kidneys remove much material from the blood and return much to the blood, resulting in little waste.
   d. The kidneys remove much material from the blood and return little to the blood, resulting in much waste.

3. Give the correct name for each description.
   a. Tube that carries blood to a kidney.
   b. Tube that carries blood away from a kidney.
   c. Purifying structures in the kidneys.
   d. Tube that carries wastes away from a kidney.
   e. Organ that stores urine until it leaves the body.

4. Complete these analogies.
   a. Alveoli are to lungs as ——— are to kidneys.
   b. Carbon dioxide is to lungs as ——— is to kidneys.

5. God provided two kidneys for the human body.
   a. From your earlier study in this unit, give one other example of organs that occur in pairs.
   b. What is an important benefit of the fact that God provided these organs in pairs?

6. Tell how the two systems in each pair depend on one another.
   a. integumentary system and muscular system
   b. skeletal system and muscular system
   c. respiratory system and circulatory system
   d. respiratory system and muscular system
   e. lymphatic system and circulatory system
   f. circulatory system and excretory system

Unit 2 Review

Review of Vocabulary

The human body was created by God in a very orderly way. The smallest unit of life in the body is the cell. Many cells of the same kind join to form —1—, of which several kinds join to form —2—. These are arranged into the eleven —3— of the body, each of which performs a major function. The whole body is known as an —4—.

The —5— provides the body with protection from injury and from dehydration. An important part of this system is the skin, with its —6— (outer layer) and —7— (inner layer). The —8— provides firm support for the body, but it has —9— that allow bending and movement. The bones are held together with strong bands called —10—. Movement is provided by the —11—. An organ that causes a joint to bend is called a —12—, whereas one that causes a joint to unbend is called an —13—. Strong —14— join muscles to bones.

The main purpose of the —15— is to take in oxygen and to release carbon dioxide. Air enters the nasal passages and then goes through the —16—, which can be closed by the —17— to prevent food from entering. After the air passes through the —18— (windpipe), it goes through the two —19— and into the tubes and air sacs of the —20—. The air sacs are called —21—. The —22— is a dome-shaped sheet of muscle that causes the chest cavity to become larger and smaller.

The blood of the —23— is pumped through the blood vessels by the —24—. Blood leaves the pumping organ through —25— and returns through —26—. As blood passes through tiny —27—, it provides food and oxygen for body cells. The watery part of the blood is called —28—. Blood also contains three kinds of living cells: —29—, which carry oxygen; —30—, which kill germs; and —31—, which are involved in the clotting of blood.

The —32— provides drainage and defense for the body. Fluid called —33— circulates through the vessels of this system and passes through the —34—, which are strung along the vessels like beads. The largest organ of this system is the —35—, located in the upper left abdomen. The system also includes a small organ above the heart called the —36—, pairs of organs in the throat called the —37—, a small protrusion at the end of the intestines called the —38—, and the —39— that line the nasal passages. The body develops —40— to disease when —41— in the form of foreign invaders stimulate some of these organs to produce —42— that recognize and destroy the invaders.

The —43— removes nitrogen wastes from the blood. Its main organs are the two —44—, which have small purifying structures called —45—. All these systems work together according to God's plan for the welfare of the marvelous organism called the human body.

Multiple Choice

1. Which statement is *false*?
   a. Each organ consists of various tissues, and each tissue consists of various kinds of cells.
   b. Each body system depends on the other systems and contributes to the entire body.
   c. The order in the body is a result of the creating power of God.
   d. The trillions of cells in the body are grouped into larger and larger units that form eleven main body systems.

2. What characteristic of the outer layer of the epidermis makes it wear-resistant?
   a. It is kept moist with perspiration.
   b. It is made of tough, dead cells.
   c. It contains a material similar to the composition of bone.
   d. It contains nerve endings that provide warning of harmful heat, pressure, and friction.

3. Which fact does *not* contribute to the protection and smooth working of the joints?
   a. A layer of cartilage covers the ends of the bones.
   b. Synovial fluid lubricates the joints.
   c. Bones are made of calcium compounds.
   d. The ends of the bones are made of spongy bone.

4. Which of the following is the best comparison?
   a. The femur bone is similar to the carpals.
   b. The shoulder joint is similar to the knee joint.
   c. The pelvis is similar to the shoulder blade.
   d. The tibia is similar to the cranium.

5. The fact that muscles can exert force only by contracting helps to explain
   a. why the muscles of the hand are in the lower arm.
   b. why the muscles are fastened to the bones with tendons.
   c. why the muscles can exert great force or provide carefully regulated motion.
   d. why skeletal muscles are usually found in pairs.

6. The lungs contain many small alveoli instead of one large sac
   a. so that many good sacs would be left if some are damaged.
   b. because the little sacs have more surface area than what one large sac would have.
   c. because the many little sacs hold more air than what one large sac could hold.
   d. because the diaphragm would not allow the atmosphere to push air into a large sac.

7. The greatest concentration of capillaries is found in the
   a. glottis.   c. trachea.
   b. alveoli.   d. bronchi.

8. Compared with the inhaled air, the air exhaled from the lungs contains
   a. more carbon dioxide and less oxygen.
   b. less nitrogen and more oxygen.
   c. more carbon dioxide and less nitrogen.
   d. more oxygen and less carbon dioxide.

9. The ventricles of the heart are directly connected to the
   a. bronchi.   c. veins.
   b. capillaries.   d. arteries.

10. If a person needs a blood transfusion, giving him only plasma would *not*
    a. restore water lost through perspiration.
    b. help the blood to carry food and minerals to the cells of the body.
    c. increase the capacity of the blood to carry oxygen and carbon dioxide.
    d. help the kidneys do their job of waste removal.

11. The work of the circulatory system is most like that of
    a. a telephone company.
    b. a department store.
    c. a trucking company.
    d. a chemical industry.

12. The cells of the lymphatic system function most like
    a. farmers harvesting crops.
    b. servants carrying messages.
    c. dogs chasing cattle.
    d. cats catching mice.

13. The lymphatic system does all the following things *except*
    a. draining excess fluid and putting it into the circulatory system.
    b. destroying foreign bacteria and viruses.
    c. excreting wastes from the body.
    d. producing antibodies to destroy antigens.

14. What is the process by which the kidneys treat wastes in the bloodstream?
    a. They collect the wastes and process them to be reused.
    b. They remove the wastes and replace them with useful substances.
    c. They filter out the wastes and discard the filter.
    d. They remove the wastes and return most of the water to the bloodstream.

15. The operation of the excretory system is most closely involved with the
    a. respiratory system.
    b. circulatory system.
    c. muscular system.
    d. digestive system.

16. Which verse in 1 Corinthians 12 teaches the interdependence of the systems in the body?
    a. verse 14
    b. verse 15
    c. verse 18
    d. verse ~~25~~ 26

17. Which of the following is *false*?
    a. The skeletal system provides support for the body.
    b. Some muscles operate without conscious control.
    c. The pulmonary arteries and veins carry blood to and from the kidneys.
    d. The cells of the hair and fingernails are dead except for the ones at the base.

18. A malfunctioning organ affects
    a. only the surrounding tissues.
    b. only the nearby organs.
    c. only the system of which it is part.
    d. the entire body.

First Quarter Review: Units 1–2

Matching

Write the letter of the word that matches each description.

1. The tendency of matter to remain in its state of rest or motion.
2. A suspension of one liquid in another liquid.
3. The smallest particle of an element.
4. An inward pull that causes an object to move in a curved path.
5. The product of the mass and speed of a moving object.

a. atom
b. centrifugal force
c. centripetal force
d. emulsion
e. inertia
f. molecule
g. momentum
h. solution

6. A tough tissue that connects a muscle to a bone.
7. A tiny blood vessel that brings food and oxygen to cells.
8. The inner layer of the skin.
9. A main organ of the excretory system.
10. The watery part of the blood.

i. capillary
j. dermis
k. heart
l. kidney
m. ligament
n. plasma
o. tendon
p. vein

Completion

Write the correct word to complete each sentence.

11. Energy in action is called —— energy.
12. An agent that allows oil particles to be suspended in water is a ——.
13. Metals can be mixed to form ——.
14. If a material takes both the size and the shape of its container, it is a ——.
15. The study of energy and motion is called ——.
16. Muscle cells join together to form muscle ——.
17. The —— system provides firm support for the body.
18. Resistance to infection by a certain disease is ——.
19. Another name for the windpipe is the ——.
20. The kidneys have small purifying structures called ——.

Multiple Choice

Write the letter of the correct choice.

21. Centrifugal force is caused by
    a. acceleration.　　b. gravity.　　　　c. inertia.　　　　d. entropy.

22. According to the first law of thermodynamics, energy
    a. is always changing in form.
    b. always increases in entropy.
    c. is always potential before it is kinetic.
    d. can never be created.

23. According to the second law of thermodynamics, energy
    a. always becomes less concentrated.
    b. can never be destroyed.
    c. is always changing in form.
    d. always becomes less random.

24. Jet propulsion is based on the law or laws of
    a. inertia.　　　　　　　　c. thermodynamics.
    b. acceleration.　　　　　　d. action and reaction.

25. A solution will form the fastest if
    a. the water is cooled before the sugar is added.
    b. the water is stirred after the sugar is added.
    c. the water and sugar are mixed under pressure.
    d. the water is filtered before the sugar is added.

26. Which is *not* true of the integumentary system?
    a. It includes the skin.　　　　c. It contains cells of only one kind.
    b. It prevents dehydration.　　　d. It gives protection from injury.

27. Muscles exert force only by
    a. expanding.　　b. contracting.　　c. rotating.　　　d. exerting pressure.

28. The ventricles of the heart are directly connected to the
    a. bronchi.　　　b. capillaries.　　c. veins.　　　　d. arteries.

29. The main reason for the many small alveoli in the lungs is
    a. to increase the air capacity.
    b. to let air enter more easily.
    c. to increase the surface area.
    d. to improve the efficiency of the diaphragm.

30. The largest organ of the lymphatic system is the
    a. spleen.　　　b. liver.　　　　c. pancreas.　　　d. lymph gland.

Unit 3
Food and Digestion

The inside walls of the small intestine are lined with microscopic fingerlike projections, called villi, which greatly increase the surface area so that digested food has ample contact with capillaries inside the villi. The exact method by which villi absorb dissolved food is unknown, though a process similar to osmosis seems to be involved for some digested materials. But most of what happens in the villi is the exact opposite of osmosis. It is a mysterious process that has been called "osmosis in reverse."

The absorption of food is one of the wonderful things that God has provided in your body, which thus far remains unknown despite the investigations of science. "O the depth of the riches both of the wisdom and knowledge of God! how unsearchable are his judgments, and his ways past finding out!" (Romans 11:33).

Searching for Truth

"Now the Spirit speaketh expressly, that in the latter times some shall depart from the faith, … commanding to abstain from meats, which God hath created to be received with thanksgiving of them which believe and know the truth. For every creature of God is good, and nothing to be refused, if it be received with thanksgiving: for it is sanctified by the word of God and prayer" (1 Timothy 4:1–5).

The Christian is especially concerned about the care of his body, since it is the temple of the Holy Ghost. One danger in selecting food for our bodies is to refuse what God has provided. The children of Israel loathed the manna and apparently were not very thankful when they ate it. We should never be so ungrateful that we refuse to eat nutritious food that God has created. Our food should rather be "received with thanksgiving."

Searching for Understanding

1. How is food like or unlike the fuel that runs an automobile?
2. In what way is your diet affected by where you live?
3. Why do you need to eat a variety of foods?
4. What is harmful about overeating?
5. What does the body need from food?

Searching for Meaning

alimentary canal	glucose	protein
amino acid	goiter	rickets
calorie	hormone	saliva
carbohydrate	large intestine	scurvy
cellulose	liver	small intestine
colon	macronutrient	starch
deficiency disease	micronutrient	stomach
digestion	mineral	sugar
enzyme	pancreas	villi
esophagus	peristalsis	vitamin
fat		

Food—the Nutrients That Keep You Alive

Perhaps you have an apple as part of your lunch today. Your classmate may eat an orange. On the opposite side of the equator, someone else may eat a mango. Although their lunches may vary, each person is very much alike. Each one must eat food in order to live.

All humans need the same basic nutrients so that their bodies function as God designed. Nutrients are special chemicals that provide three things for the body: (1) fuel for energy, (2) materials for growth, and (3) chemicals to change food into materials that the body can use. A wide variety of foods can supply the basic nutrients.

Most food cannot be absorbed directly into the bloodstream; the food must first be digested. You will study digestion later in this unit, but first you will read about the foods that the body needs and the various things that food provides for the body.

The human body cannot always use a material just because it contains needed chemicals. Gasoline is an excellent fuel, but the body was not made to use the energy in gasoline. Hair is high in the proteins that are used to build body cells, but the body cannot use hair as food. God created termites with the ability to use wood as food, but humans could not receive much

nourishment by eating wood.

God has provided a great variety of food that is suitable for man to eat. "Every moving thing that liveth shall be meat for you; even as the green herb have I given you all things" (Genesis 9:3). Plants and animals together provide a broad array of food for man. Since the time of Christ, God has given man much liberty in the choice of his food. "For every creature of God is good, and nothing to be refused, if it be received with thanksgiving" (1 Timothy 4:4).

It is good and proper for the Christian to study the needs of his body and then to eat the foods that will help to make his body a sound, healthy temple for the Holy Spirit.

Macronutrients—Carbohydrates, Fats, and Proteins

Food provides two categories of nutrients: macronutrients and micronutrients. *Macro* means "large." *Macronutrients* are the nutrients that the body needs in large amounts every day: carbohydrates, fats, and proteins. They form the greatest part of your diet.

Carbohydrates and Fats Provide Energy

Ninety-five percent of the food that you eat provides energy. Just as an automobile needs fuel to provide energy for its functions, the cells of your body must have fuel. Before gasoline can power an automobile, it must combine with oxygen to release its energy. This process is called oxidation or burning. A similar thing happens in the cells of the body. Blood carries oxygen from the lungs to the various cells, and there the oxygen combines

with fuel foods to release energy.

Carbohydrates are the main fuel food. The word *carbohydrates* may remind you of the carbon in coal, an important fuel. Carbon in carbohydrates is the same as carbon in coal, but the molecules of coal are too complex to digest. Most carbohydrates come from plants. Through the process of photosynthesis, plants take carbon from the carbon dioxide in air and combine it with water to form carbohydrates.

Water (H_2O) has two atoms of hydrogen for every atom of oxygen. A *carbohydrate* is any carbon compound that has this same ratio of hydrogen to oxygen. Body cells break down carbohydrate molecules into carbon dioxide and water, releasing energy in the process. This is the energy that operates muscles and other body processes. The carbon dioxide then goes to the lungs and returns to the air, and the water joins other fluids in the body.

Simple carbohydrates are called sugars. *Sugar* is what gives fruit and many other foods their sweet taste. There are many different kinds of sugar. Scientists divide them into two groups called monosaccharides and disaccharides. These names explain what is simple about simple carbohydrates. *Mono* means "one," *di* means "two," and *saccharide* (sak′·ə·rīd′) means "sugar unit of carbon, hydrogen, and oxygen." Monosaccharides have one sugar unit per molecule, and disaccharides have two. Only monosaccharides, the simplest sugars, can be absorbed into the bloodstream. That means the body must break all carbohydrates down into this simple form before it can use them.

Monosaccharides (1 sugar unit per molecule)	**Disaccharides** (2 sugar units per molecule)
Glucose Nearly all plants	*Maltose (glucose + glucose)* Sprouting grain
Fructose Fruits and saps	*Sucrose (glucose + fructose)* Sugar beets, sugar cane
Galactose Seldom found except in lactose	*Lactose (glucose + galactose)* Milk

Foods rich in simple carbohydrates

Two simple sugar molecules can combine to form a more complex sugar molecule.

The most important and most available one-unit sugar is *glucose* (glōō′·kōs′). Its chemical formula, $C_6H_{12}O_6$, shows the 2:1 ratio of hydrogen to oxygen that is characteristic of carbohydrates. Many fruits contain glucose. The body of a healthy person maintains a constant level of glucose in the blood at all times. Another one-unit sugar is fructose, which comes from fruits as well as from honey.

Sucrose, the ordinary table sugar, is made from sugar cane or sugar beets. This is a disaccharide, a sugar with two sugar units. Sucrose has the formula $C_{12}H_{22}O_{11}$. Other disaccharides include maltose, found in sprouting grain, and lactose, found in milk. Some of these sugars are less sweet, but each disaccharide must be broken down to monosaccharides before the body can use it.

Complex carbohydrates include starches and some other foods. Molecules of *starches* are very large, which is why scientists call them polysaccharides. They have so many sugar units that an estimate of their chemical symbol might be $C_{1200}H_{2000}O_{1000}$. Despite the size of the molecule, the ratio of its hydrogen and oxygen atoms is 2:1, the same as in sugars. The body must break each starch molecule into glucose molecules before it can use the starch. This means that starches are harder to digest than sugars. However, starches are a valuable part of a healthful diet.

Starches abound in root plants such as potatoes; in grains such as corn, wheat, rice, and oats; and in legumes such as peas, beans, and peanuts. Various methods are used to prepare these products for eating. Peas and beans are often cooked. Other starchy products are ground into flour and then baked to make bread, breakfast cereals, noodles, and cakes. The processes of grinding, cooking, and baking help to make starch easier for the body to use.

Cellulose (sel′·yə·lōs′) is another complex carbohydrate, or polysaccharide. It is the strong supporting membrane that surrounds plant cells, which is also called roughage or fiber. However, because of the way these molecules are put together, the body has no way of breaking them down into sugar. In other words, humans cannot digest cellulose.

However, cellulose is very important in the diet. The mass of indigestible material helps to move food through the intestines. This promotes the regular removal of solid wastes from the body. Good sources of cellulose include leafy vegetables (such as lettuce, celery, and cabbage) and pulpy fruits (such as oranges, peaches, and plums).

Fats are another important fuel food. Like carbohydrates, *fats* are compounds containing carbon, hydrogen, and oxygen, but their formulas reveal a different composition. The formula for the main fat of beef tallow is $(C_{17}H_{35}COO)_3C_3H_5$. Because of these large molecules, fats are much harder to digest than other foods. Each fat molecule must be broken into small glucose molecules before the body can use it. This takes a long while, but breaking apart this complicated molecule releases a

Peas

Egg
noodles

Flour

Bread

Vegetable macaroni

Pinto beans

Brown rice

BLACK
BEANS

BLACK BEANS

Foods rich in complex carbohydrates

is that certain vitamins will dissolve only in fat. Without fat to carry those nutrients to the cells, the body would not be able to use fat-soluble vitamins. Fats also form a large part of brain cells and nerve cells. Finally, fat in the diet helps to build up a store of body fat to provide insulation against cold and padding to protect vital organs.

Many common foods contain fats and oils. Solid fats appear in animal meats, such as beef and pork, poultry, and fish. (Note that *meat* refers to the muscles of mammals; poultry and fish are grouped separately.) Animal products are also rich in fats. Milk contains butterfat, which can be churned into butter. Eggs contain about 12 percent fat. Because animal fats contain more hydrogen, they are generally solids. These are called saturated fats.

large amount of energy for the body.

For this reason, fats are a much more concentrated form of energy than either sugars or starches. In fact, they provide more than twice as much energy than an equal weight of carbohydrates.

If the body takes in more food than it can use, it stores the excess as fat. Diets that are high in fat also encourage the accumulation of fat and fatty deposits in the blood vessels. These deposits can build up until they close an artery that supplies blood to the muscular tissue of the heart. The result is a heart attack. Thousands of people die every year as a result of heart attacks and diet-related heart disease.

The body does need some fat. One reason

Liquid fats or oils appear mostly in plants, including soybeans, peanuts, coconut, and corn. These plant oils are used in cooking. Margarines and vegetable shortenings are made of hydrogenated vegetable oil. This means hydrogen has been added to the liquid (unsaturated) fats to make them solid. Liquid fats are also found in a few animals such as fish. Although liquid fats provide just as much energy as solid fats, they are less likely to form fatty deposits in the blood vessels.

Foods rich in fats

———————— Study Exercises: Group A ————————

1. List three things that food provides for the body.

2. List the three chemical elements found in all carbohydrates.

3. A sugar molecule that has 14 atoms of hydrogen would have ——— atoms of oxygen.

4. Why are fats and starches harder to digest than sugars?

5. Cellulose is a carbohydrate that
   a. the body cannot break down into simple sugars.
   b. is not useful to the body.
   c. the body must break down into simple sugars before using.

6. The quickest source of energy is
   a. sugars.   b. starches.   c. fats.

7. The most energy per gram of food comes from
   a. sugars.   b. starches.   c. fats.

8. Give two reasons why you should have some fat in your diet.

9. Write the letter of the food combination that is most likely to contribute to heart disease if eaten in excess.
   a. whole-grain bread with margarine and jelly
   b. corn pancakes fried in vegetable oil
   c. bacon with eggs fried in butter

Group A Activities

1. *Testing for starch in foods.* Set out six small plastic or glass containers, and place 2 table-spoons of warm water in each of them. Stir in 1 tablespoon of flour, cornstarch, oatmeal, rice, tapioca, and sugar, putting each of these in a different cup. Set out four more cups and put a piece of bread, a raw apple slice, a raw potato slice, and some popped popcorn in them. Next put two drops of tincture of iodine onto the item in each cup. Some items will instantly turn blue-black. This color indicates that there is starch in those foods.

   The foods that contain no starch will stay the same color with perhaps a touch of the yellow-brown color of iodine. You can tell which ones have the most starch by which ones turn the darkest. Test other foods. The ones that turn blue belong to what group of foods?

   Iodine will stain, and it is poisonous. So be careful not to spill iodine on your clothes or other surfaces, and do not eat any food that has been touched by iodine.

2. *Testing for fats in foods.* From a brown paper bag, cut a rectangle measuring 3 inches by 4 inches. Use a pencil to divide the paper into twelve 1-inch squares. In the squares, write the following labels: peanut butter, cheese, milk, mayonnaise, banana, nut, margarine, bologna, honey, apple, bacon, and chocolate chip. Rub a small amount of each one of these foods on the respective squares. Use a paper towel to rub off any excess food. Lay the paper in the sun to dry for ten minutes.

   Now hold the paper up to the light. A translucent spot indicates the presence of fat in that food. You will be able to see the most light through the spots from foods with the most fat. Which foods are those? Which foods do not contain any fat? Test other foods for fat.

Proteins Provide Materials for Growth and Maintenance

Like a machine, the body constantly needs food for energy; but it is more marvelous than any machine built by men. With the proper food, the body can grow larger and repair itself by producing new cells and replacing old ones.

The body uses *proteins* to make proto-plasm, the building material for new cells. Like carbohydrates and fats, proteins contain carbon, oxygen, and hydrogen. But all protein molecules have a very important additional element—nitrogen. Not all proteins are alike.

Some contain minerals besides the nitrogen, carbon, oxygen, and hydrogen.

The body needs nitrogen as it is found in proteins to build living protoplasm for its cells. Proteins are very large, complex molecules made from smaller units called *amino acids* (ə·mē′·nō). The body breaks down these large protein molecules into amino acids through the process of digestion. Then it puts the amino acids together again to build the particular protein molecules it needs to make body cells.

There are about twenty different amino acids. Eleven of these are called nonessential amino acids because the body can make them itself; it does not need to get them from food. The other nine amino acids are called essential because the body cannot manufacture them; it must obtain them from food. The twenty amino acids are the building blocks that include the different proteins taken into the body along with the different proteins made by the body.

Proteins are classed in six groups according to the work they perform in the body. One group of proteins, called *enzymes,* speeds up chemical reactions in the body. For instance, one enzyme speeds the rate at which hemoglobin in the blood exchanges oxygen for carbon dioxide. Without this enzyme, hemoglobin could not work fast enough to supply the body's need of oxygen.

Another group of proteins transports materials through the body. Hemoglobin itself is a protein that transports oxygen and carbon

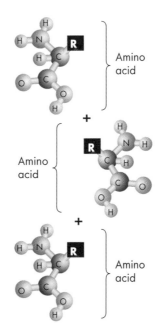

A protein molecule is a chain made of many smaller amino acid molecules.

dioxide. Structural proteins form the fibers of skin, hair, cartilage, and blood clots. Yet another group of proteins enables muscle cells to produce motion.

A *hormone* is a kind of protein that regulates cell activities, controlling such things as how fast the body grows or uses food. Hormones are very important in the endocrine system. You will read more about them in chapter 7.

The antibodies produced by the lymphatic system are another group of proteins. Remember that antibodies protect the body against disease by destroying foreign invaders.

The human body needs proteins daily. As young children grow taller and stronger, they especially need proteins to build new cells. Even after a person reaches maturity, his body still needs protein to make protoplasm. A certain amount of growth continues in the hair and nails, but an adult uses most of the protein in his diet to replace worn-out cells. The cells of the active lining in the small intestine die and need to be replaced every two days or so. Healing of wounds and internal injuries also requires protein to build body tissue. The body makes millions of new cells every day to grow, to replace dead cells, or to repair damaged tissue.

Many foods are rich in proteins. These include animal meats, such as beef and pork, poultry, and fish, as well as animal products, such as eggs, milk, and cheese. Proteins from these foods are called complete because they

Foods rich in proteins

provide all nine of the essential amino acids. Legumes, nuts, grains, and some vegetables also provide protein, but they are called incomplete proteins because they do not supply all nine essential amino acids.

Complete proteins can be obtained by combining incomplete proteins, such as nuts or grains, with legumes. For example, a dish of rice and beans or of corn and beans includes each of the nine amino acids. So does a peanut butter sandwich. Many people in poor countries cannot afford expensive animal proteins, such as meat and milk. But God wisely provided that these countries often have a ready supply of two incomplete proteins, which together make an affordable complete protein.

Water—an Essential for Life

Although water is not a nutrient, the body needs large amounts of it every day. You can live longer without food than without water. About two-thirds of a person's weight comes from water, with about 60 percent of that water inside the cells and 40 percent around the cells.

Foods that contain much water

The body can take water directly from the digestive tract without any special processing.

Water is important to the body in a number of ways. First, water helps food and food wastes to slide through the digestive tract. Water also dissolves nutrients and chemicals in the digestive process. Water in the blood helps to carry digested nutrients to the cells, and it helps to carry cell wastes away. The kidneys use much water in removing these wastes from the body. Water fills the cells and the spaces around them. Without water, many chemical reactions in the body could not take place. Water lubricates the joints of the body. By evaporating in perspiration, water helps to regulate the body temperature.

Every day the body loses water that must be replaced. Many foods contain more than 50 percent water. Soups and milk as well as melons and other juicy fruits supply large amounts of water. However, you do not get enough water from your food alone. You should drink eight to ten glasses of water daily. The amount of water you need will vary with your activity and the weather. At rest, you lose about 40 ounces (1.13 kg) of water per day and must replace that same amount. When you

An adult should drink about 9 glasses of water every day.

work hard on a hot summer day, you may need twice as much water.

How good it is that God made something as essential as water to be so plentiful on the earth! For good health, the body needs plenty of fresh, pure water every day.

———————— Study Exercises: Group B ————————

1. If the body is compared to an automobile, sitting down to a meal would be like taking the car to a service station. Complete these comparisons with *carbohydrates* or *proteins.*
   a. Filling up the fuel tank is like eating ———.
   b. Getting new tires is like eating ———.

2. Proteins are different from carbohydrates because they include the element ——— in addition to the elements ———, ———, and ———.

3. The body's use of proteins is like
   a. reading a book and telling someone else what you read.
   b. buying a flat of flowers and dividing them out to fill your flower bed.
   c. buying used cars, tearing them apart, making some new parts, and using all these parts to build new cars.

4. For each word, give the specialized name of the protein which does that work.
   a. accelerate   b. regulate   c. defend

5. True or false? After a person is fully grown, he still needs a considerable amount of protein.

6. Which food combination would provide a complete protein with the least fat?
   a. barbecued chicken   c. peas and carrots
   b. baked beans and corn bread

7. Explain how water is important to each of the following body systems.
   a. digestive system   d. skeletal system
   b. circulatory system   e. integumentary system
   c. excretory system

8. Which person will need to drink the most water?
   a. A secretary working in an office with a temperature of 70°F (21°C).
   b. A boy digging a ditch when the temperature is 90°F (32°C).
   c. A man shoveling snow when the temperature is 25°F (–4°C).
   d. A child reading a book in a tree house with a temperature of 80°F (27°C).

Group B Activities

1. *Testing for protein in foods.* **Note: This test requires great caution.** Place a hard-boiled egg white in a small glass dish. Add a few drops of nitric acid. The egg white should turn yellow, which shows that there is protein in the egg white. Also test some white cheese for protein.

   Nitric acid is very corrosive, so you should take special care not to get any on your skin or clothes. If the acid touches your skin, the skin will turn yellow because it contains protein.

2. *Measuring the water content of an apple.* Slice an apple into thin strips. Weigh the strips to the nearest gram or tenth of an ounce. Place them in an oven, and bake them at the lowest heat setting for about twelve hours or until the slices are dry and shriveled. Remove the strips from the oven, and weigh them again. Find the difference between the original weight and the weight after drying. Divide the difference by the original weight, and change the result to a percent. This will tell you what percent of the apple was water.

Micronutrients—Vitamins and Minerals

Micro means "small." *Micronutrients* are the vitamins and minerals that the body needs in very small amounts every day. Together the micronutrients make up less than a teaspoon of your daily diet, but without them you could not live.

Vitamins are naturally occurring organic substances that enable macronutrients and minerals to do their work for the body. Without

vitamins, your other foods would be of no benefit to you. The *vit* part of *vitamins* means that these nutrients are vital—necessary for life.

Scientists have known for centuries that eating certain things (poisons) will make a person sick. But they failed to realize that *not* eating certain things will also make a person sick. For example, they did not understand that sailors became sick (and many died) from a

Water-soluble Vitamins			
	Food Sources	**Functions**	**Deficiency Disorders**
B_1 *(Thiamine)*	Organ meats, nuts, sunflower seeds, whole grains, legumes	Helps digestive enzymes release energy from food; aids nervous, circulatory, and muscular systems	Beriberi, neuritis, edema, mental depression, muscle weakness
B_2 *(Riboflavin)*	Yogurt, spinach, beets, milk, almonds, liver, eggs, whole grains	Carbohydrate metabolism; tissue growth and repair; helps digestive enzymes; maintains immunity	Eye irritation and light sensitivity; cracked, inflamed skin and lips
B_3 *(Niacin)*	Peanuts, poultry, fish, organ meats, mushrooms, dried yeast, legumes	Helps digestive enzymes; promotes elastic skin and healthy nerves; aids hormone production and DNA repair	Pellagra (skin sores, nervous and digestive disorders, mental decline)
B_5 *(Pantothenic acid)*	Egg yolks, dairy, whole grains, legumes, meats	Stimulates metabolism and growth by aiding chemical reactions	(rare)
B_6 *(Pyridoxine)*	Whole grains, green leafy vegetables, bananas, fish, poultry	Needed for 60 different enzymes; helps most body functions; vital for overcoming 100 disorders	Skin disorders, irritability, kidney stones, weak muscles, insomnia, depression
B_{12} *(Cobalamin)*	Red meats, eggs, dairy products, brewer's yeast	Red blood cell and DNA production; protects nerves; works with folic acid	Pernicious anemia, nervous disorders, numbness
C (Ascorbic acid)	Citrus fruits, dark leafy greens, tomatoes, peppers, cabbage, guavas	Serves as an antioxidant; assists enzymes; builds connective tissue; resists colds and infections	Scurvy (bleeding gums, loose teeth, muscle soreness, bruise easily)
Folic acid (a B vitamin)	Whole-wheat foods, dark leafy greens, legumes, potatoes, nuts, peanuts	Protein and DNA metabolism, aids red blood cell production, healing processes, and cell reproduction	Anemia, diarrhea, poor digestion and circulation, hair loss, depression

lack of vitamin C. Such a sickness is called a *deficiency disease* because it develops through the deficiency (lack) of a certain vitamin or mineral.

There are two kinds of vitamins: those that dissolve in water, and those that dissolve in fat.

Water-soluble vitamins are absorbed directly into the bloodstream. They can move easily into and out of the body cells. For this reason, you are not likely to receive too many of these vitamins; your kidneys will remove any excess. When vegetables are cooked in water, some of the vitamins leach out of the vegetables and into the water. So it is good to save this water and use it in another dish, such as a casserole.

Why are the B vitamins classified as they are? When scientists first studied vitamins, they thought there was only one B vitamin. But more careful analysis showed that what had appeared as one substance was really eight vitamins, each essential for the normal functioning of the body. The B vitamins work with a number of different enzymes that aid the digestive process in various ways. God actually put substances in food that help the body to digest food.

Note that vitamin C is an antioxidant. Cells use oxygen to release energy from food and to carry out their normal functions. As they use oxygen, they convert it to a form that would damage body cells. An antioxidant is a substance that binds with this oxygen and keeps it from harming the cells. This is one more proof of the marvelous way that God cares for His creation even when we are hardly aware of it.

Lack of vitamin C causes *scurvy,* a common affliction of sailors long ago. Their gums and other soft tissues swelled and bled, and their teeth became loose. They also developed painful muscles and low energy, and eventually they died. In the late 1700s, sailors learned to prevent scurvy by taking citrus fruits along to eat on their voyages.

Fat-soluble vitamins are carried by fats to the body cells. Because they do not dissolve in water, excess fat-soluble vitamins cannot be removed by the kidneys. Therefore, fat-soluble vitamins can build up in body cells and cause various kinds of poisoning. This generally happens only when a person takes far too many vitamin pills. For this reason, you should not

Fat-soluble Vitamins			
	Food Sources	**Functions**	**Deficiency Disorders**
A *(Retinol)*	Orange, yellow, or green vegetables and fruits; dairy; fish-liver oil; liver; eggs	Promotes good vision; aids immunity and cell growth; serves as an antioxidant	Night blindness, scaly and dry skin, bad gums, frequent infections
D	Vitamin D milk, egg yolks, liver, fatty fish, cod-liver oil, sunlight	Helps the body to use calcium and other minerals that build strong bones; supports thyroid	Pyorrhea, tooth decay, bone disorders, rickets (bones soften and deform)
E	Green leafy vegetables, whole grains, liver, eggs, nuts, seeds, flaxseed oil	Serves as an antioxidant, dilates blood vessels	Blurred vision, poor body and eye coordination, hemolytic anemia
K	Dark leafy greens, cabbage, asparagus, beef liver, made by intestinal bacteria	Essential in blood clotting; helps to form proteins for bone building; helps liver and nerve function	Uncontrolled bleeding, poor blood clotting

supplement your diet with fat-soluble vitamins in excess of the recommended daily allowance (RDA). Also, vitamin pills should be kept out of the reach of young children.

Children with *rickets* fail to grow properly, and they develop bowed legs because their bones are too soft to support their bodies. Milk is commonly fortified with vitamin D to prevent rickets. However, getting too much vitamin D can result in kidney stones.

We marvel as we consider how God first created these life-supporting substances in the plants and animals that we eat. Then He created our bodies in such a way that they depend on small amounts of these vitamins to function properly.

Minerals			
	Food Sources	**Functions**	**Deficiency Disorders**
Calcium	Dairy products, small fish, dried legumes, raw green vegetables, sesame seeds	Builds strong bones and teeth; regulates muscle action; aids nerve cells and blood clotting	Muscle spasms, rickets, brittle bones, irritability, tooth decay, insomnia
Phosphorus	Egg yolks, legumes, nuts, corn, seeds, dairy, meats, fish, whole grains	Builds strong bones and teeth; combines with fat so that it can travel in the bloodstream	(rare) Weakness, poor growth, pain in bones
Sodium	Table salt, olive, celery, turnip	Electrolyte balance; aids digestion and nerve functions; regulates heartbeat	(rare) Muscle weakness, poor appetite, confusion
Chlorine	Table salt, fresh fruits and vegetables	Electrolyte balance; makes up part of hydrochloric acid in stomach	(rare) Indigestion, hair and tooth loss, poor fluid levels
Potassium	Leafy vegetables, potatoes, bananas, cantaloupe, meats	Electrolyte and pH balance; supports muscles and kidneys	Edema, heart failure, muscle weakness, cramps
Magnesium	Whole grains; wheat bran; green, leafy vegetables; peanuts; almonds; peaches	Builds protein; activates enzymes; aids blood clotting; aids nerve and muscle functions;	Twitching or cramping muscles; insomnia, cancer, heart disease, depression
Sulfur	Eggs, meats, cabbage, dried beans, onion family, kale	Builds protein and amino acids; disinfects blood; antibiotic	(rare) Dull hair and brittle nails
Iron	Raisins, prunes, egg yolks, whole grains, leafy greens, legumes, black molasses	Aids hemoglobin in carrying oxygen; helps make muscle cells and various enzymes	Anemia (headaches, paleness, fatigue) dizziness, low immunity
Iodine	Leafy greens, garlic, citrus, egg yolks, saltwater fish, iodized salt	Major component of thyroid hormone that regulates most functions of the body	Enlarged thyroid (goiter), cretinism, deaf-mutism, obesity, low pulse
Copper	(Present in most diets; abundant in iron-rich foods)	Works with iron in similar functions; needed for RNA and new blood cells	(rare) Slow pulse, indigestion, weak heart
Zinc	Pecans, split peas, Brazil nuts, whole grains, oats, rye, lima beans, buckwheat, lean meat	Component of 200 enzymes and many hormones; needed for DNA; promotes immunity	(rare), Growth failure, delayed wound healing, mental disturbances
Selenium	(Present in most diets), vegetables, Brazil nuts, cereals	Powerful antioxidant; works with vitamin E; slows aging	(rare), Premature aging, muscle and liver decline

The vitamins that your body needs come from organic substances, just as do the macronutrients. The four main elements needed by the human body are carbon, hydrogen, and oxygen (supplied by carbohydrates and fats) and nitrogen (added by proteins). These four elements make up about 96 percent of your body weight. The remaining 4 percent consists of twenty-two elements known as *minerals*.

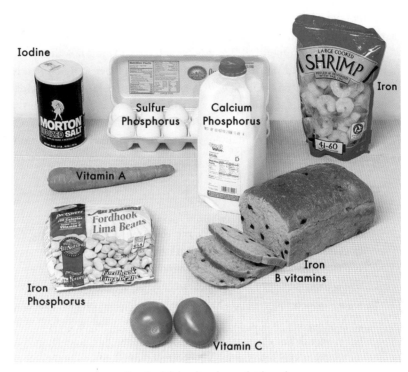

Foods rich in vitamins and minerals

Minerals are not organic compounds produced by living things (as vitamins are), and they are not broken down by the body. You receive many minerals by eating plants that take the minerals from the earth, along with meat and other products from animals that eat plants. Some minerals come in the form of substances added directly to the diet, such as table salt.

The two categories of minerals are major minerals and minor minerals, classed according to the amounts needed by the human body. The table on page 100 gives the seven major

Citrus fruit

minerals (shown in boldface) and five of the sixteen minor, or trace, minerals.

Another minor mineral is *fluoride,* which is important to healthy teeth. The remaining ten are *manganese, chromium, molybdenum, arsenic, nickel, silicon, boron, cobalt, vanadium,* and *tin.* Some of these minerals would be toxic in large doses, but small amounts are essential for good health.

When lack of iodine causes *goiter,* the thyroid becomes larger so that it can capture any available iodine in the body. The thyroid may grow so much that it cuts off the air supply to the lungs. Problems may also result from an excess of a certain mineral. For example, too much sodium can cause high blood pressure. This is why nutritionists recommend that most people reduce their intake of salt.

—————— Study Exercises: Group C ——————

1. Vitamins are different from minerals in that
   a. vitamins come from plants, but minerals come from animals.
   b. vitamins are water-soluble, but minerals are fat-soluble.
   c. vitamins are needed in larger amounts than minerals.
   d. vitamins are produced by plants and animals, but minerals are inorganic compounds.

2. Name the vitamin that is probably lacking in the diet of a person who
   a. cannot see well at night and does not like liver or carrots.
   b. works in an office all day, does not enjoy eggs, and often has broken bones.
   c. dislikes spinach and red beets, lacks immunity to illnesses, and has red, cracked skin.
   d. catches colds easily and does not like citrus fruits.

3. Which food would supply the vitamin needed to combat each deficiency disease?
   a. beriberi (pumpkin pie or whole-wheat bread)
   b. scurvy (orange juice or macaroni and cheese)
   c. rickets (spinach salad or egg and cheese sandwich)
   d. pellagra (turkey-mushroom casserole or fresh tomato sandwich)

4. a. Vitamins C and E both serve as ———.
   b. What important service do these substances perform for the body?

5. How does vitamin D help a mineral to do its job?

6. Write *true* or *false* for each statement.
   a. A person who has muscle spasms probably would benefit from drinking more milk.
   b. A person who has anemia would benefit from eating more citrus fruit.
   c. A person who has goiter would benefit from eating more fish.

7. Phosphorus makes it possible for ——— to travel in the bloodstream.

8. To be sure you have enough vitamins and minerals in your diet, you should
   a. take a vitamin supplement daily.
   b. eat moderately of a wide variety of foods.
   c. eat what you enjoy.
   d. carefully compute the amount of each vitamin and mineral in your daily diet.

Group C Activities

1. *Testing foods for vitamin C.* Get four commercially prepared fruit or vegetable juices. Three should be good sources of vitamin C, and one should not be. Read the labels to see. Orange juice and tomato juice are high in vitamin C.

   Set out 4 clear glass or plastic cups. Put 2 tablespoons of water in each cup, add 1 tablespoon of cornstarch, and stir until smooth. Next add 2 drops of tincture of iodine to each cup. (Add 4 drops if you use regular antiseptic iodine.) The cornstarch will turn dark blue, showing that starch is present.

   Stir 1 teaspoon of each juice into a separate cup of blue cornstarch. The color of the cornstarch will change according to the amount of vitamin C. If the blue color leaves, the juice is high in vitamin C. If the color does not change much, the juice does not have much vitamin C.

2. *Testing foods for iron.* Get four commercially prepared fruit juices with varying amounts of iron. (Read the labels.) Prune juice and pineapple juice are good sources of iron; cranberry juice and white grape juice contain some iron; apple juice has little iron.

   Prepare a cup of hot tea, making it three times stronger than normal. Allow the tea to sit for an hour. Set out 4 clear glass or plastic cups, and put 4 tablespoons of each fruit juice in each cup. Stir ¼ cup of tea into the fruit juice in each cup. Let the mixtures sit for 20 minutes.

   Gently lift the cups, and look for dark particles at the bottoms. These particles are the result of iron in the juices combining with chemicals in the tea. Check the cups every hour. The juices that first form dark particles contain the most iron. The juices that form few or no dark particles have little or no iron.

3. *Extracting iron from cereal.* Get a stand mixer of the Kitchen Aid design. Tape a piece of sheet magnet to one side of the beater, with the magnetic side out. (Trim the magnet to the shape of the beater.) Insert the beater into the mixer. Pour 3 cups of water into the mixer bowl.

   Crush 2 cups of vitamin-enriched cereal that includes 100% of the recommended daily allowance of iron. Turn the mixer on low speed, and pour the cereal slowly into the mixer bowl. Cover the bowl with a splash guard, or drape a cloth over the mixer. Let the mixer stir at low speed for 30 minutes.

   Stop the mixer, and remove the beater carefully. You should see small particles of iron collected on the magnet.

A Healthful Diet

The food that you eat from day to day is your diet. For you to remain healthy, your diet needs to include sufficient nutrients from each of the groups discussed in the first part of this unit. But how can you be sure you are eating the right foods? This question was much harder to answer in the days before canning and freezing. Today we can eat a great variety of fruits, meats, and vegetables all year round. The three keys to healthy eating are variety, moderation, and balance.

A healthful diet requires variety. God

provided many different kinds of food for us to eat. Although a desire for too much variety can lead to expensive living, you should learn to enjoy nearly all foods. By eating a wide variety of foods, you are more likely to consume some of each of the almost 50 nutrients that your body needs. Eating a little of every food that is served to you shows gratitude both to the cook and to God.

A healthful diet requires moderation. Eating in moderation means not eating too much or too little food. Either extreme (too much or too little) can lead to serious health problems.

The energy in foods is measured by a unit called the *calorie.* One calorie is the amount of heat needed to raise 1 gram of water 1 degree Celsius. The calorie used to measure food energy is actually the kilocalorie, which is 1,000 times as large. Boys twelve to fourteen years old should consume about 2,800 of these calories per day, and girls of that age need about 2,400 calories per day.

You use this energy in three main ways. First of all, you need energy to fuel your vital processes while your body is at rest. Scientists call this the basal metabolism (mi·tab′·ə·liz′·əm); it includes the chemical processes in all the basic functions of life, such as respiration, digestion, blood circulation, and immune responses. The basal metabolism is the general process of simply staying alive.

To find how many calories you need for basal metabolism, first find your body weight in kilograms. (Divide your weight in pounds by 2.2.) Then multiply the result by 24. For example, a 100-pound person weighs about 45 kilograms, and $45 \times 24 = 1,080$. This means that a 100-pound person needs 1,080 calories daily for his body to function at rest.

The second way you use energy is to work or play physically. The more strenuous your activity, the more calories you need. A secretary or an office worker will need fewer calories than a carpenter. Cleaning out calf pens or hoeing weeds requires more calories than

Selected Foods and Calorie Values

Food Item	Calorie Value
Fruits	
Apple (1 large)	70
Banana (1 large)	85
Cantaloupe (½)	40
Orange (1 large)	70
Raisins (1 cup)	460
Strawberries (1 cup)	55
Watermelon (2 lb. slice)	120
Vegetables	
Broccoli (1 cup)	60
Carrots, cooked (1 cup)	45
Celery stalk (8 inch)	5
Sweet corn (5-inch ear)	65
Cucumber (1-inch slice)	5
Onions (1 cup)	80
Green peas (1 cup)	110
Pepper, green (1 raw)	15
Potato, baked (5 oz.)	90
Potatoes, French-fried (2 oz.)	200

Food Item	Calorie Value
Dairy Products	
Butter or margarine (1 T.)	100
Cottage cheese (1 oz.)	30
Ice cream (1 cup)	295
Milk, whole (1 cup)	165
Yogurt (1 cup)	120
Meats	
Chicken, fried (½ breast, 3⅓ oz.)	154
Beef, roast (3 oz.)	265
Beef stew (1 cup)	250
Egg (1)	80
Frankfurter (1)	155
Hamburger, on bun (¼ lb.)	416
Ocean perch, fried (3 oz.)	195
Seeds and Nuts	
Almonds, shelled (1 cup)	850
Peanuts, shelled (1 cup)	840
Peanut butter (1 T.)	90

Food Item	Calorie Value
Breads and Cakes	
White bread (1 slice)	62
Angel food cake (2 inch section)	110
Pancake (4-inch diameter)	60
Pumpkin pie (4-inch section)	265
Cheesecake (4 oz.)	215
Boston cream pie (3.3 oz.)	329
Doughnut, sugar (1.8 oz.)	233
Miscellaneous	
Tomato catsup (1 T.)	15
Salad dressing, French (1 T.)	90
Mayonnaise (1 T.)	110
Honey (1 T.)	60
Sugar (1 T.)	50
Dill pickle (1 large)	15
Chicken soup (1 cup)	75
Fudge (1 oz.)	115

riding a tractor or folding clothes. On average, a person your age needs about 1,200 calories per day for work. These calories plus those required for basal metabolism come to about 2,280 calories per day.

The third way you use energy is to digest and absorb your food. Normally this is about 10 percent of all the other calories you need. Divide 2,280 by 10, and add the result to the former total: 228 + 2,280 = 2,508. This is the approximate number of calories that nutritionists recommend for young teenagers.

Your calorie needs change as you grow older. The larger your body mass, the more calories you need. Boys need more energy than girls because their metabolism is faster. In other words, boys use food at a faster rate. Metabolic rate also varies from one person to another. Some persons simply need more food to sustain their bodies. We cannot change this, neither can we change our basic

body build. Some persons are more inclined to store fat than others. Such a person can control his weight by not consuming excess calories, but he should not become displeased with himself. Neither should you tease others for being underweight or overweight. As a person reaches his upper teens and is no longer growing as fast, he needs fewer calories. This decrease will continue throughout life.

Your body stores unneeded energy as body fat. Some body fat is desirable, but an excess is dangerous to health. The extra weight of too much body fat strains the heart and other organs of the body. Overweight due to undisciplined eating is a serious health problem in the United States.

God gave you an appetite for food so that you eat enough to keep healthy, but appetite must be kept in control like all other bodily desires. *Surfeiting* and *gluttonous* are two words the Bible uses to mean overeating.

Food Guide Pyramid

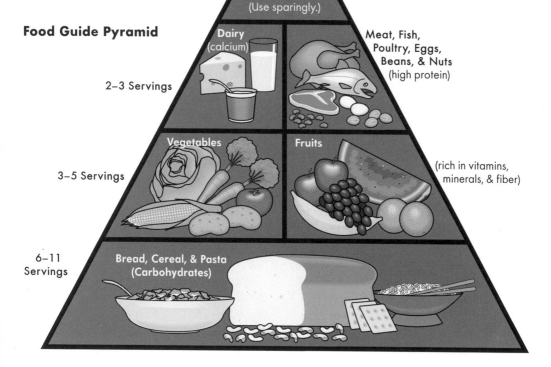

OILS
FATS
SWEETS
(Use sparingly.)

Dairy
(calcium)

2–3 Servings

Meat, Fish,
Poultry, Eggs,
Beans, & Nuts
(high protein)

Vegetables

3–5 Servings

Fruits

(rich in vitamins,
minerals, & fiber)

6–11
Servings

Bread, Cereal, & Pasta
(Carbohydrates)

"And take heed to yourselves, lest at any time your hearts be overcharged with surfeiting, and drunkenness, and cares of this life, and so that day come upon you unawares" (Luke 21:34).

When we think of the thousands of hungry and starving people in the world, it should encourage us to be moderate in our eating. Despite the many products that claim to make you lose weight while continuing to overeat, there is only one sound way to lose weight—eat less and exercise more. However, you should not try to lose weight too fast. Make sure you truly need to lose weight before attempting it, and then aim to lose only one pound per week. If you lose weight faster than this, you frequently lose muscle. Your body receives the signal that you are starving and slows its metabolism. Then you stop losing as fast. It is much better to lose weight slowly. It is better yet to always eat moderately and maintain a healthy body weight.

Nutritionists use a figure known as the body mass index to determine whether a person is overweight. This figure compares a person's height with his weight. To determine your body mass index, multiply your weight in pounds by 700. Divide that figure by the square of your height in inches. A body mass index between 20 and 24 is considered healthy. A person with a body mass index of 25 or above is probably overweight, and one with a body mass index below 20 is probably underweight. (*Note:* This method is designed for persons of average build. It is not accurate for someone with an unusually large or small frame.)

Eating too little can be just as harmful as overeating. Sometimes young people, especially girls, think they are too fat. Their weight may actually be desirable or even low. But to attain an even lower weight, these people refuse to eat much at all. Such unhealthy eating starves the body of vital nutrients, and it can lead to organ damage and even death. It is not necessary to be constantly concerned about calorie counting and excessive exercise. Relax and eat moderate, nourishing meals. If you are not overweight, accept the fact that you may be larger than someone else your age. God did not intend for everyone to be the same size.

A healthful diet requires balance. A balanced diet is one that provides the proper amounts of the nutrients you need every day. These nutrients are carbohydrates, fats, proteins, vitamins, and minerals. Along with those nutrients, you also need cellulose and plenty of water. Nutritionists say that 55 percent of your calories should come from carbohydrates. These are an important fuel food because they do not cause fatty deposits in blood vessels as fats do. Carbohydrates are the only nutrients that your central nervous system can use. A diet low in carbohydrates cannot supply sufficient nutrients to your nerves and muscles.

Nutritionists have developed a pyramid as a guide to healthful eating. You should eat the most of the type of food at the bottom of the pyramid, and successively less of each type toward the top. Note that carbohydrates form the base of the pyramid. You should have 6 to 11 servings of complex carbohydrates every day. Many cakes, cookies, and pies contain some starch in the form of flour, but they have an overabundance of sugar and fat. However, sugars do not cause you to gain weight as quickly as fats do. A moderate amount of sugar in the diet is not known to be harmful.

Vegetables and fruits come next on the pyramid. You should eat 3 to 5 servings of vegetables and 2 to 4 servings of fruits every day. Many fruits and vegetables are high in fiber and rich in vitamins and minerals. The important thing is that you eat a wide variety of fruits and vegetables so that you receive the full range of vitamins and minerals, especially vitamins A and C.

The third level of the pyramid includes dairy products, such as milk, butter, cheese, yogurt, custard, and ice cream. You should have 2 to 3 servings of these products per day to supply the calcium you need. Growing children especially need plenty of milk for bone formation. Many dairy products include large amounts of fats. For this reason, milk itself should make up the largest part of your servings from this group.

Also on the third level of the pyramid are meat, fish, poultry, (dried) beans, and eggs. These foods are rich in protein, and they also include a fair amount of fats. You should eat 2 to 3 servings of these foods per day. Protein should make up between 10 and 15 percent of the calories in your diet. Although growing children need protein-rich diets for growth, most Americans eat too much protein. Since the nitrogen in proteins is the main waste excreted by the kidneys, excess protein puts needless stress on your kidneys as well as your liver. Also, high-protein foods are often more expensive than foods that are lower on the pyramid.

Fats and sweets come at the very peak of the Food Guide Pyramid and should be eaten sparingly. You definitely need some fat in your diet so that fat-soluble vitamins can travel to your cells. However, your body is able to make most of the fats that it needs. No more than 30 percent of your calories should come from fat. One gram of fat yields at least nine calories, more than twice the amount supplied by equal weights of other nutrients.

Nutrition-fact labels describe the nutrients in a food. Since the early 1990s, most packaged foods come with labels similar to the one shown below. This label is for a can of grits. It gives the size of one serving and the number of servings in the container. Then it gives the total calories per serving and the number of those calories that come from fat. One serving of this cereal provides 130 calories, of which 5 calories come from fat. Since the fat calories are less than one-tenth of the total, this cereal is a low-fat food.

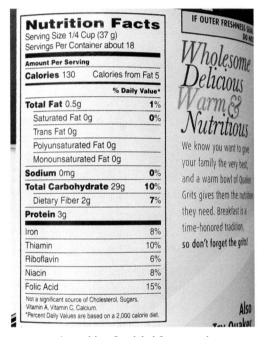

A nutrition-fact label for a cereal

Other items on nutrition-fact labels are the amounts of protein, carbohydrates, and important vitamins and minerals. The amounts of cholesterol and sodium are given for the benefit of people who are limiting their intake of these.

Attention to the various aspects of nutrition should make us wise and cautious. Our bodies are the temples of the Holy Spirit. But we should not become overanxious. Every gift of God is good if we receive it in moderation and with thanksgiving.

For the best nutrition, you should be relaxed and cheerful at mealtime. You will not get the full benefit from a meal if you are rushed or emotionally upset when you eat. Mealtimes should be happy family activities. The prayer before a meal is more than a Christian custom; it is a privilege to acknowledge God's gracious provision for your body. "Whether therefore ye eat, or drink, or whatsoever ye do, do all to the glory of God" (1 Corinthians 10:31).

———— Study Exercises: Group D ————

1. a. How many calories will a 165-pound person need for his basal metabolism?
    b. About how many additional calories will he need for his daily work? (Use the same figure as in the text example.)
    c. How many calories will he need for digesting and absorbing his food? (Calculate from the total of your answers to *a* and *b*.)
    d. How many calories will he need altogether in one day?

2. Which person will need the most calories?
    a. A five-year-old boy riding a bicycle.
    b. A seventeen-year-old boy digging a ditch by hand.
    c. A forty-year-old woman picking peas.
    d. An eighteen-year-old girl washing windows.

3. What happens to the calories that a person consumes beyond his needs?

4. a. Use the formula in the text to find the body mass index of a person with a weight of 126 pounds and a height of 63 inches. Answer to the nearest whole number.
    b. Is this person underweight, of desirable weight, or overweight?

5. Tell why each of the following is not a good way to control or reduce body weight.
    a. eating few carbohydrates
    b. choosing fruits and vegetables with the lowest natural calories
    c. eating few dairy products

6. The greatest number of calories is provided by 100 grams of (carbohydrates, protein, sugar, fat).

7. The best way to maintain a healthy body is to
    a. carefully compute the calories and nutrients in your daily diet.
    b. eat moderately of foods from all the food groups, and exercise regularly.
    c. eat a low-calorie diet and exercise rigorously.

8. The following meals are not balanced. Tell which food group is not represented in each: *vegetables, fruits, meats, grain products,* or *dairy products.*
    a. macaroni, carrots, roast beef, egg custard
    b. orange juice, bologna sandwich with cheese, hard-boiled egg
    c. glass of milk, green beans, fish, banana
    d. biscuit, chicken, apple, celery
    e. tomato juice, cereal with milk and blueberries

Group D Activities

1. *Computing your body mass index.* Use the formula in the text to calculate your body mass index according to your height and weight. If you would like to lose weight by eating less, remember to plan a healthful diet. You should not try to lose more than one pound per week.

2. *Collecting nutrition-fact labels.* With your mother's permission, collect nutrition-fact labels from food packages in her kitchen and pantry. See how large a collection you can make. Classify the labels according to the macronutrients: *carbohydrates, fats,* and *proteins.* Decide whether the nutritional value of each food is high, medium, or low, according to the percentages of important nutrients. Decide which ones you should eat sparingly.

3. *Finding what nutrients and calories you consume in one day.* Make a list of all the foods you eat in one day. Find a chart that lists the nutritional values of most foods. Record the amount of each nutrient in each food that you ate. Find the total for the day, and compare this result with a Recommended Daily Allowance chart.

The Digestive System

Only simple sugars, vitamins, minerals, and alcohol can be absorbed into the body without being changed. All other foods must be changed physically or chemically before they can be fuel or building blocks for the body. The process of changing food so that the body can use it is called **digestion.** Through digestion, carbohydrates are changed into simple sugars, proteins are changed into amino acids, and fats are changed into fatty acids.

The entire process of digestion is one of the masterpieces of God's handiwork. By eating a meal, you start a complex series of events that may last for hours. What you eat, how you eat, and how you feel can affect the process; but much of the digestive process takes place without your being aware of it. The passage by which food goes through the body is called the **alimentary canal** (al'·ə·men'·tə·rē).

Digestion begins in the mouth. The first part of the alimentary canal is the mouth, where the teeth start the digestive process by physically changing food. The teeth cut and break food into very small pieces, thus increasing its surface area so that all parts of the food are exposed to the chemical action of digestive juices.

Normally an adult has thirty-two teeth, arranged in four symmetrical sets of eight teeth each. Two sets in the upper jaw mirror two sets in the lower jaw. The teeth on the right side mirror those on the left. First the

Position of Teeth

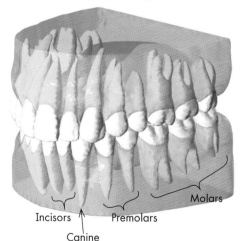

Incisors Premolars Molars

Canine

food meets the incisors, a group of eight chisel-shaped teeth that cut the food. Next are the four canine teeth, commonly called eyeteeth, whose sharp points grasp and tear food. Eight premolars follow; these round teeth have somewhat flat tops for grinding food. Last are twelve teeth called molars, which have squarish tops that also grind food. The four molars farthest in the back, called wisdom teeth, do not appear until early adulthood.

An adult's teeth, called the permanent teeth, are preceded by a set called the milk teeth or baby teeth. These begin to appear within the first year of a child's life and usually are all lost and replaced by sixteen years of age. But if you lose a permanent tooth, you will not grow another one. Therefore, it is highly important that you establish good eating and teeth-brushing habits during these years when God is providing you with a new set of teeth. The following principles will help you to maintain good teeth.

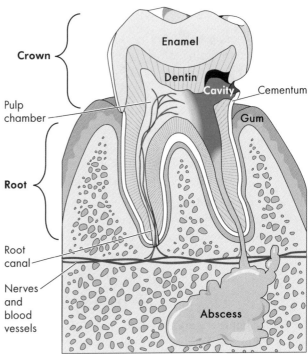

Left side: Parts of a healthy molar. *Right side:* An untreated cavity can lead to an infected tooth.

1. *Do not use your teeth as a pliers or nutcracker.* This may crack or chip the outer coating of enamel. Although tooth enamel is the hardest substance in the body, only a thin shell covers the top of each tooth. If you damage the enamel, you will expose the softer dentine underneath, and your tooth will decay. Your body cannot replace enamel.

2. *Eat sweet foods sparingly.* Sugar feeds the acid-producing bacteria that grow on teeth. This acid reacts with the calcium carbonate in the teeth to cause decay. You should avoid large amounts of candy, soft drinks, and other overly sweet foods. When you eat sweets, you should promptly remove the sugar by brushing your teeth or at least rinsing your mouth with water.

3. *Clean your teeth regularly.* The ideal is to brush your teeth after every meal and avoid snacks between meals; this will almost guarantee that you will not develop cavities. But since it is usually practical to brush your teeth only in the morning and evening, some other method can be used at noon. Carrot sticks, celery, and apples are sometimes called "nature's toothbrushes." Eating these crunchy foods after lunch will scrub your teeth and leave them almost free of harmful food deposits.

Flossing your teeth regularly will do much to prevent cavities below the gum line. A further step is to use an antiseptic mouthwash, which helps to kill germs that cause tooth decay and those that cause bad breath.

4. *Visit your dentist regularly.* Despite your best efforts to keep your teeth clean, a cavity may begin. Once a cavity has exposed the dentine, decay may ruin a tooth. You should visit your dentist at least once and perhaps twice each year. He can check the condition of your teeth and promptly fill any small cavities.

Saliva changes starch into sugar. In the chewing process, the tongue moves food around the mouth and mixes saliva with it. *Saliva* is the watery secretion of three pairs of salivary glands in the mouth, and it contains an enzyme that breaks down starch molecules into smaller sugar molecules.

As chewing continues, the food becomes a soft mass and moves into the pharynx. Swallowing pushes the food into the *esophagus,* a tube surrounded by muscles and extending to the stomach. These muscles move food into the stomach by contracting in a wavelike motion called *peristalsis.*

─────────── Study Exercises: Group E ───────────

1. Food travels through the body by way of the ——.

2. Your teeth change food (physically, chemically) so that digestive juices and enzymes can more easily change it (physically, chemically).

3. Name the teeth that do each of the following.
   a. cutting food
   b. grasping and tearing food
   c. final grinding of food

4. Improper practices can harm your teeth.
   a. What improper practice could harm your teeth physically?
   b. What improper practice in eating could harm your teeth chemically?

5. To maintain healthy teeth, you should brush them every —— and visit your dentist every ——.

6. A chemical change takes place in the mouth when an enzyme in —— is mixed with food.

7. Which kind of food begins to digest chemically in the mouth: carbohydrates, proteins, or fats?

8. Peristalsis explains why
   a. starches are changed into sugars.
   b. saliva must be mixed with food.
   c. astronauts can swallow in zero gravity.
   d. food must be chewed well.

Group E Activity

Tasting the changing of starch into sugar. Get a saltine cracker, and chew it well. At first it will taste salty and starchy; but as you mix the cracker with saliva, you will notice a sweet flavor. This happens because an enzyme in saliva (called ptyalin) changes starch molecules into sugar.

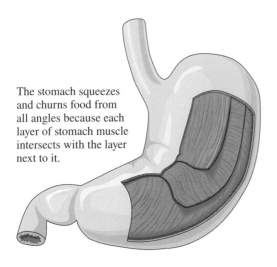

The stomach squeezes and churns food from all angles because each layer of stomach muscle intersects with the layer next to it.

If unchecked, an ulcer can eventually eat all the way through the stomach wall.

Digestion continues in the stomach. The *stomach* is a hollow organ about 9 or 10 inches long (23–25 cm). It is a complex of muscles and glands that performs three main functions.

Storing food. The stomach makes it possible to eat several pints of food in one meal. Soon afterward, the food is released, a small amount at a time, into the small intestine. Usually after a well-balanced meal, all the food is out of the stomach in about four hours. But if the meal was high in fats, the food may remain in the stomach as long as six hours.

Producing fluids to digest food. One of these digestive juices contains an enzyme that breaks proteins into amino acids. Another is hydrochloric acid, which curdles milk. This acid would eat a hole in the stomach wall

if the wall were not protected by a layer of mucus. A hormone produced in the stomach regulates the acidity, or pH level, inside the stomach, keeping it between 1.5 and 2.5. This is a very acidic environment, which is too harsh for most germs.

In spite of the protective mucus, sores known as stomach ulcers sometimes develop on the inside wall of the stomach. Some kinds of stomach ulcers are more common in people who are tense and anxious. A calm faith in God is a safeguard against much of the tension that can worsen an ulcer problem.

Mixing food with digestive juices. Layers of muscles in the stomach walls work in rhythmic contractions, churning the contents of the stomach into a soupy semiliquid. These contractions may actually start before you eat, causing the growling sounds that you hear when you are hungry. When the food is sufficiently mixed and ready to move on, the pyloric valve at the base of the stomach opens and lets the food enter the small intestine.

Final digestion and absorption occur in the small intestine. The *small intestine* is small in diameter but not in length. It is about 22 feet long (6.7 m) and has three main parts: the duodenum, the jejunum, and the ileum.

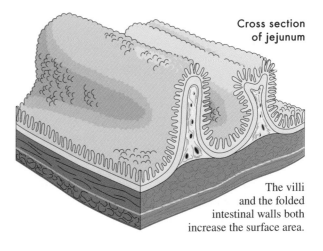

Cross section of jejunum

The villi and the folded intestinal walls both increase the surface area.

The Digestive System

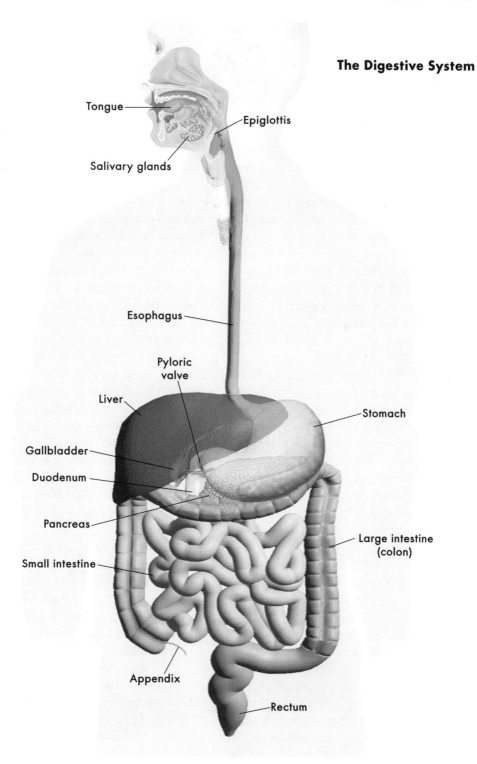

Tongue

Epiglottis

Salivary glands

Esophagus

Pyloric
valve

Liver

Stomach

Gallbladder

Duodenum

Pancreas

Large intestine
(colon)

Small intestine

Appendix

Rectum

The duodenum. The first part of the small intestine, the duodenum (doo′·ə·dē′·nəm), is about 10 inches long (25 cm). It curves around the *pancreas,* a digestive organ lying beneath and behind the stomach. Here most of the chemical digestion is completed. Hormones from the small intestine trigger the pancreas to release pancreatic enzymes. These enzymes finish the digestion of starch (which began in the mouth) and of protein (which began in the stomach). Fats begin to break down into fatty acids. Another pancreatic enzyme turns the acidic food from the stomach into a base.

Hormones also signal the liver to produce bile. Acting much like detergent on grease, bile breaks down fat particles so that pancreatic enzymes can digest them more easily. Extra bile goes into the gallbladder, a small organ beneath the liver, where it is stored until it is needed.

The *liver,* the largest organ in the body, weighs about 3 pounds (1.4 kg) and lies just beneath the diaphragm. It has two lobes, the larger one hanging over the duodenum to the right and the smaller one hanging over the stomach to the left. This large organ performs a variety of functions. Besides producing bile, it makes and stores vitamin A, builds blood-clotting proteins, stores a reserve of sugar for the blood, and filters excess nutrients and poisons from the blood. Some of these toxins are converted to dissolve in water; then the kidneys discharge them. Other toxins remain with the bile; but when the bile leaves the body, so do the toxins.

The jejunum. The next 8 feet (2.4 m) of small intestine is the jejunum (jə·joo′·nəm). Its inside walls are lined with fingerlike projections, called *villi* (vil′·ī), which greatly increase the surface area so that digested food has ample contact with capillaries inside the villi. The jejunum is lubricated with mucus so

that peristalsis can easily move the digested food along. Blood vessels carry nutrients from the villi to the liver. From the liver they go to the heart, which pumps them to all parts of the body.

But how is dissolved food absorbed by the villi? The exact physical or chemical method is unknown, though a process similar to osmosis seems to be involved. By osmosis, dissolved materials in a more concentrated solution pass through a dividing membrane into a less concentrated solution so that the two solutions become equally concentrated. Osmosis does take place in the absorbing of some digested materials, but most of what happens in the villi is the exact opposite of osmosis. It is a mysterious process that has been called "osmosis in reverse."

That makes a nice name, but it does not answer the question *how.* The absorption of food is one of the wonderful things that God has provided to work in your body every day, which thus far remains unknown despite the investigations of science. "O the depth of the riches both of the wisdom and knowledge of God! how unsearchable are his judgments, and his ways past finding out! . . . For of him, and through him, and to him, are all things: to whom be glory for ever. Amen" (Romans 11:33, 36).

The ileum. The third and last part of the small intestine is the ileum (il′·ē·əm), measuring 12 feet long (3.6 m). Here the remaining food particles pass into the bloodstream, and blood vessels absorb bile and return it to the liver. Now all that is left is water and any indigestible materials, especially cellulose. Peristalsis moves this into the large intestine.

Water is removed in the large intestine. The *large intestine* has a diameter of about 2½ inches (6 cm) and a length of about 5 feet (1.5 m). The main part of the large intestine is the *colon.* It starts at the lower right side of the abdomen, ascends to the liver, transverses in front and below the stomach, and descends on the left side of the abdomen. Water from the undigested material is absorbed through the walls of the large intestine and returns to the cells. Bacteria in the colon break down any remaining nutrients and change them to vitamin K—the vitamin essential in blood clotting. Peristalsis moves the solid undigested food to the end of the large intestine, where it is stored until it is passed out of the body.

Near the junction of the small intestine with the large intestine is the appendix, a small fingerlike projection that you studied in connection with the lymphatic system. Sometimes appendicitis develops, which requires surgery to remove the infected appendix.

Promoting Good Digestion

Knowing how food travels through the alimentary canal is useful in helping the digestive system to work as efficiently as possible. You will put needless stress on your digestive system if you rush through a meal or exercise strenuously afterward. It is also unwise to overload the stomach with fats, since they do not begin to digest until they reach the duodenum—and even then they digest slowly. Often you learn from past mistakes and avoid foods and eating habits that cause digestive discomfort.

Worry and tension affect the responses of the muscles and glands. The cheerful, wholesome life of the Christian does much to help the body maintain good health. A balanced diet and moderate eating are also part of God's will for the care of your body.

—————— Study Exercises: Group F ——————

1. Without a stomach, you would probably need to
   a. exclude fats from your diet.
   b. eat oftener and less at a time.
   c. mix your food before putting it into your mouth.

2. In the stomach, digestion begins on
   a. carbohydrates and fats.
   b. proteins and dairy products.
   c. carbohydrates and proteins.
   d. sugars and fats.

3. The pH, or acid level, in the stomach is between 1.5 and 2.5. What would probably happen if the fluids in the stomach were
   a. more acidic?
   b. less acidic?

4. Write *ileum, duodenum,* or *jejunum* for each description.
   a. Most of the chemical digestion is completed.
   b. Most of the digested food is absorbed into the bloodstream.
   c. Remaining food particles are absorbed into the bloodstream.

5. In the duodenum,
   a. food enters as an acid and leaves in a neutralized form.
   b. food enters in a neutral form and leaves as a base.
   c. food enters as an acid and leaves as a base.

6. Without the liver, pancreatic enzymes could not
   a. finish digesting starches and proteins.
   b. digest fats as easily.
   c. store a reserve of sugar for the blood.

7. Since the absorption of food through the villi is called osmosis in reverse, apparently there is a greater concentration of dissolved materials in the (bloodstream, intestine) than in the (bloodstream, intestine).

8. As applied to the two main parts of the intestines, the words *small* and *large* refer to their (length, diameter) and not their (length, diameter).

9. From the matter remaining after food is digested, the body reclaims two materials that were added to aid digestion.
   a. From the ileum, —— returns to the ——.
   b. From the large intestine, —— returns to the body cells.

Group F Activities

1. *Observing enzyme action on protein.* Pour ¼ cup of cold water into a small bowl. Sprinkle 1 tablespoon of unflavored gelatin over the cold water, and let it soften. Meanwhile heat 1 cup of water just to a boil. Remove the water from the heat, and stir in the softened gelatin until it dissolves. Stir in three ice cubes until they dissolve. Divide the gelatin mixture evenly into four small containers.

   Get the following fresh fruits, and chop them fine: kiwi, pineapple, and apple. (Do not use canned pineapple.) Stir the chopped fruits into three of the containers of gelatin. Do not put any fruit in the fourth container. Refrigerate all four containers. An hour or two later, the plain gelatin and the apple mixture should be firm. The kiwi and pineapple mixtures will not gel because those fruits contain an enzyme that breaks down the protein in gelatin.

   Now put a bit of kiwi and pineapple juice on the gelatin in the plain container. The fruit will turn the gelatin to a liquid. Enzymes in the stomach break down protein in a similar way.

2. *Observing how bile prepares fats for digestion.* Place ½ cup of water in a small jar. Add 2 drops of food coloring, cap it with a lid, and shake it. Add ¼ cup of salad oil, and shake it again. The oil will not mix with the water but will stay in separate droplets. Let the mixture sit, and the oil will soon separate from the water and form a layer on top of the colored water.

   Now add about a tablespoon of liquid dish detergent, and shake the jar again. This time the oil will break up into smaller droplets that are suspended in the water. If you let the jar sit, the liquids will eventually separate again, but not nearly as quickly or completely. In the digestive system, bile acts like soap to break down fats so that other digestive juices can digest them more easily.

Unit 3 Review

Review of Vocabulary

Various foods are essential to the health of the body. The foods needed in large quantities are the —1—. Most of the energy for the body is provided by —2—, which contain hydrogen and oxygen in the same ratio as water. These foods are placed in two groups, of which the first is the simple carbohydrates known as —3—. For digestion, all food must be broken down into simple sugars. The most important one-unit sugar is —4—.

The complex carbohydrates include —5—, which are found in roots, grains, and legumes, as well as indigestible plant fiber known as —6—. Another fuel food is —7—, which is harder to digest and contributes to heart problems. The amount of energy that a fuel food provides is measured in —8—.

The body also needs large quantities of —9— for the growth and maintenance of cells. These are made of smaller units called —10—, and they include —11— to speed up chemical reactions and —12— to regulate cell activities.

The —13— are the materials that the body needs in small quantities. These include organic substances called —14— and natural elements called —15—. If any of these are lacking, the body may develop a —16—. A lack of vitamin C causes —17—; a lack of vitamin D causes a condition of soft bones known as —18—; and a lack of iodine causes —19—, in which the thyroid becomes enlarged.

The process of preparing food for absorption into the bloodstream is called —20—. It occurs in the —21—, the passage by which food goes through the body. Digestion begins in the mouth, where the teeth grind the food and an enzyme in —22— converts starch into sugar. After food leaves the mouth, a muscular action called —23— moves it down the —24— and into the —25—. There, further digestion takes place before the food enters the —26—. In its first part, the duodenum, digestion continues with the aid of enzymes from the —27—, the largest organ of the body, and from the —28—, a long, narrow organ just beneath the stomach. Then the digested food enters the jejunum, where it is absorbed into the bloodstream by many fingerlike —29—. The remaining materials go into the —30— and travel through its main part, called the —31—, before passing out of the body.

Multiple Choice

1. Which of these groups contains a material that the body *cannot* use as fuel?
   a. fats, starches, sugars
   b. sugars, fats, carbohydrates
   c. sugars, starches, carbohydrates
   d. starches, proteins, cellulose

2. Sugar provides quick energy because
   a. sugar yields more energy per gram than any other fuel food.
   b. sugar contains three common elements—carbon, hydrogen, and oxygen.
   c. sugar has a good flavor and is easy to eat.
   d. sugar is absorbed directly into the bloodstream.

3. Iron is an important mineral because
   a. it helps the thyroid to function properly.
   b. it helps the hemoglobin in the blood to carry oxygen.
   c. it serves as an antioxidant.
   d. it makes muscles tough and strong.

4. Water is necessary for the proper function of
   a. the digestive and circulatory systems.
   b. the excretory and respiratory systems.
   c. the skeletal and integumentary systems.
   d. all systems of the body.

5. If you did not eat any animal products, your diet could easily be lacking in
   a. carbohydrates.   c. important minerals.
   b. complete proteins.   d. sufficient water.

6. If a person has sore gums and painful muscles, he should eat more
   a. meat.   c. oranges.
   b. eggs.   d. whole-wheat cereal.

7. Milk and dairy products are good sources of
   a. water, protein, calcium, and phosphorus.
   b. fat, calcium, iron, and sugar.
   c. water, fat, calcium, and vitamin C.
   d. water, protein, sugar, and iron.

8. Which statement about a deficiency disease is *false*?
   a. A lack of vitamin B_1 (thiamine) causes beriberi.
   b. A lack of iodine causes goiter.
   c. A lack of vitamin D causes scurvy.
   d. A lack of niacin causes pellagra.

9. A nutrient that speeds up chemical processes is called
   a. an antioxidant.   c. an enzyme.
   b. a vitamin.   d. a hormone.

10. A vitamin is
    a. a medicine that helps the body fight certain deficiency diseases.
    b. an organic substance that the body needs in small amounts so that it can use nutrients.
    c. a mineral that growing children need because it aids the proper development of various parts of the body.
    d. a nutrient that the body uses to build cells.

11. Which of the following meals is *not* balanced?
    a. green beans, fish, yogurt, cake, peaches
    b. rice, carrots, beef, cottage cheese, fruit salad
    c. hamburger, cheese, lettuce, cookies, strawberries
    d. egg, orange, cereal, milk, honey

12. To maintain a healthy body weight, you should
    a. exercise frequently and eat moderately.
    b. eat many vegetables but few carbohydrates.
    c. eat much protein but little starch.
    d. eat few calories but take vitamin and mineral supplements.

13. Which list names the teeth in the right order from the front of the mouth to the back?
    a. incisors, canines, premolars, molars
    b. canines, incisors, molars, premolars
    c. premolars, incisors, canines, molars
    d. incisors, molars, premolars, canines

14. The motion that causes food to pass through the alimentary canal is called
    a. digestion.   c. absorption.
    b. peristalsis.   d. osmosis.

15. Enzymes are added to food in the
    a. stomach, liver, and small intestine.
    b. mouth, stomach, and small intestine.
    c. mouth, esophagus, and large intestine.
    d. stomach, pancreas, and large intestine.

16. The main function of the digestive system occurs in the
    a. mouth.   c. small intestine.
    b. stomach.   d. large intestine.

17. The liver does all the following things for the body *except*
    a. turns food into a base.   c. stores sugar reserves.
    b. produces clotting proteins.   d. filters out poisons and extra nutrients.

18. Digested food enters the bloodstream through the villi by
    a. the process of osmosis only.
    b. a process similar to the way a sponge soaks up water.
    c. a process that is not fully understood by man.
    d. the process of dissolving minerals in water.

19. Both the liver and the large intestine are involved in (choose two)
    a. producing enzymes.
    b. producing vitamins for the body.
    c. clotting blood.
    d. absorbing food into the bloodstream.
    e. storing food wastes for excretion.

20. A Christian's faith helps his body to use food properly because
    a. trust in God reduces anxiety that hinders digestion.
    b. he offers thanks to God before eating his meal.
    c. God is the Creator of the digestive system as well as the entire body.
    d. he knows that God created all food to be eaten with thanksgiving.

Unit 4
The Starry Heavens

"Whoso is wise, and will observe these things, even they shall understand the lovingkindness of the LORD" (Psalm 107:43). The Palomar Observatory in California has a large reflecting telescope with a mirror 200 inches across (508 cm). Its detailed images have helped scientists learn more about the structure of distant galaxies.

Searching for Truth

"When I consider thy heavens, the work of thy fingers, the moon and the stars, which thou hast ordained; what is man, that thou art mindful of him? and the son of man, that thou visitest him?" (Psalm 8:3, 4).

David was the king of Israel, the chosen of God, and a mighty warrior. Yet David felt very humble when he thought about the moon and the stars. You do not need a telescope to be impressed with God's handiwork in the heavens. The heavens are so vast, so beautiful, and so orderly in comparison with our feeble efforts to make things, that we wonder why God would think about us. He not only thought about us, but He also visited us through His Son. God is worthy of our praise!

Searching for Understanding

1. Why does the moon appear to change its shape?
2. How could you tell time by the sun, moon, and stars?
3. What is the difference between a star and a planet?
4. How can a person become familiar with the names and locations of the stars?
5. What celestial wonders can you see through a telescope?

Searching for Meaning

astrology	light-year	planetarium
astronomy	lunar eclipse	retrograde motion
aurora australis	magnitude	right ascension
aurora borealis	meteor	satellite
autumnal equinox	meteorite	solar eclipse
comet	nebula	summer solstice
constellation	nova	vernal equinox
declination	observatory	winter solstice
ecliptic	phase	zodiac
galaxy	planet	

The Wonder of the Stars

A view of the stars from a hilltop on a cool, clear night is a breathtaking experience. Against the darkness of empty space, the stars stand out like sparkling diamonds in the sky.

A casual observance of the night sky may lead you to think there is no order among the stars. Some stars are bright; some are dim. In some areas there are many bright stars; in other areas there are few. Sometimes you see a group of stars in one position; at another time those stars are in a different position. They look as disorderly as rice sprinkled on a dark tablecloth. But the more time scientists spend in studying the heavens, the more marvels they discover.

In this science unit, you will learn about the order and majesty of the stars. The study of the heavenly bodies is called *astronomy.*

God's Reasons for Stars

God has a purpose for everything He does. Why did God create all the thousands and millions of stars? Could we not live just as well without them? The stars are actually very important to us, and God was very wise in creating them. The Bible gives four reasons for God's creation of the stars.

1. *The stars were created to provide measures and divisions of time.* "And God said, Let there be lights in the firmament of the heaven to divide the day from the night; and let them be for signs, and for seasons, and for days, and years" (Genesis 1:14).

Our clocks and calendars are set by the stars. The day, the seasons, and the year are based on the movements of the earth in relation to the sun. The stars are a giant time dial on the celestial clock. That giant clock is the handiwork of God.

2. *The stars were created to provide light and heat for the earth.* "And let them be for lights in the firmament of the heaven to give light upon the earth: and it was so. And God made two great lights; the greater light to rule the day, and the lesser light to rule the night: he made the stars also" (Genesis 1:15, 16).

During the day the sun is ruling. It brilliantly illuminates one side of the earth and warms us with its rays. But at night the moon is the queen of the sky. That lesser light is bright enough to help us walk about, but not so bright that it disturbs our sleep. The stars, too, give light to guide us. Sailors as well as land travelers can find their way by the position of the stars.

3. *The stars were created to teach us God's glory and ability.* "The heavens declare the glory of God; and the firmament sheweth his handywork" (Psalm 19:1).

If God had made only a dozen stars, we would wonder at the greatness of their Creator.

But with billions of stars, we are overcome with awe. The Bible says that God can do anything. The stars are visible evidence of that fact. God would not have needed to make so many stars, but their existence is like a signature of God to inspire us to worship and trust Him.

4. *The stars were created to give us a sense of humility and unworthiness before our great God.* "When I consider thy heavens, the work of thy fingers, the moon and the stars, which thou hast ordained; what is man, that thou art mindful of him? and the son of man, that thou visitest him?" (Psalm 8:3, 4).

Men are prone to be proud and to glory in their accomplishments. But men do not have anything that God did not give to them. God—not man—should receive all glory. A reverent view of the stars can help man see himself as he really is: not very big in relation to God. Then man can repent of his pride and turn to God with praise and gratitude.

God's Glory in Distance and Number of Stars

The psalmist said, "For as the heaven is high above the earth, so great is his mercy toward them that fear him" (Psalm 103:11). How high is the heaven? Since no one can find out by taking a journey to the stars, some other means must be used to measure the distance.

Scientists have estimated the distance by observing how the stars appear to change position as they are viewed from different angles. This is similar to the changes you see when you move from one place in a room to another. From one point of view, for example, Mary may appear to be standing very close to Susan. From another point of view, the girls appear to be standing two feet apart. The girls did not change position, but your change of position caused a shift in where they appear in relation to each other. This shift is called parallax (par'·ə·laks'). The farther away the girls are, the less the parallax will be.

The diameter of the earth's orbit is 186 million miles (300 million km). Since the earth is that distance from where it was half a year ago, one could expect to observe some parallax in viewing the stars. Copernicus (1473–1543) was the first to look for parallax in the stars, but he could find none. No one else could find parallax either until 1838. That was not because there was no parallax, but because their instruments were not accurate enough to detect it.

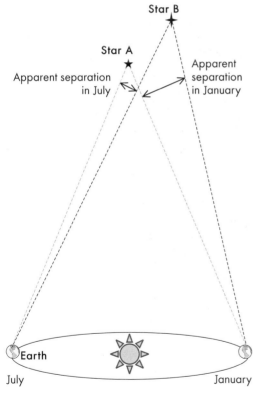

Star B

Star A

Apparent separation in July

Apparent separation in January

Earth

July   January

This diagram shows the principle of parallax, but it is not to scale. If the diameter of the earth's orbit were reduced to 1 inch, the nearest star would be over 2 miles away!

The fact that there is such extremely small parallax leads to the conclusion that the stars are exceedingly far away. With the use of parallax, astronomers were eventually able to measure the distance to some of the nearest stars. These distances are so great that a unit of length called the light-year is used. A *light-year* is the distance that a light ray travels in one year, or nearly 6 trillion miles (nearly 9.5 trillion km).

The nearest star other than the sun, named Proxima Centauri (sen'·tôr'·ē), is 4.2 light-years away. That star is too faint to be seen with the unaided eye. Close by is the bright star Alpha Centauri, about 4.3 light-years away. Both of these stars are so far south that they cannot be seen from countries north of the equator. The closest star visible in the Northern Hemisphere is Sirius (sir'·ē·əs), 8.6 light-years away. Sirius is the brightest star in the entire sky.

Most of the stars are so far away that they show no parallax at all. That means some of them are hundreds of light-years away. Such distances stagger the mind. The starry heavens are very high indeed!

How many stars are there? The word of God came to Jeremiah, saying, "As the host of heaven cannot be numbered . . . so will I multiply the seed of David" (Jeremiah 33:22). With the best modern telescopes for seeing celestial objects, men still do not know the total number of stars.

People on earth can see about 12,000 stars with the unaided eye—6,000 from the Northern Hemisphere and 6,000 from the Southern Hemisphere. Only about 3,000 stars can be seen in the night sky at one time; the other 3,000 are in the daytime sky. A telescope will reveal many more thousands of stars, along with "clouds" in outer space. Some of these are masses of glowing gas. Other "clouds" are families of stars so distant that they look like tiny patches in a telescope. Such a star family is called a *galaxy.* (See the cover photo of this book.)

Our solar system and all the stars we can see are part of the Milky Way. This is a flattened spiral galaxy with a diameter of about 100,000 light-years and an average thickness of about 2,000 light-years. The sun is a star in one of the spiral arms, about 30,000 light-years from the center of the galaxy. Orbiting the sun is the earth, with its equator tilted about 63° in relation to the plane of the galaxy as a whole. The concentration of stars in our galaxy forms the band of light that we call the Milky Way.

The Milky Way is just one of billions of galaxies in the universe. Some of the smaller galaxies have fewer than a billion stars, but others have over a trillion stars. A billion galaxies times a billion stars in each galaxy would make 1 quintillion stars (1 followed by 18 zeros). And that is just an estimate. Only God knows how many stars there really are.

"It is he [God] that sitteth upon the circle of the earth, and the inhabitants thereof are as grasshoppers; that stretcheth out the heavens as a curtain. . . . Lift up your eyes on high, and behold who hath created these things, that bringeth out their host by number: he calleth them all by names by the greatness of his might" (Isaiah 40:22, 26).

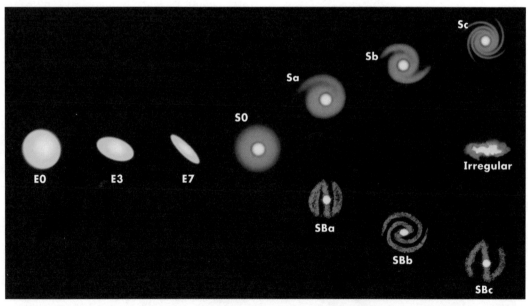

Stars inside a galaxy orbit around a common center. Galaxies are divided into four main groups: elliptical (type E), spiral (type S), barred spiral (type SB), and irregular. The Milky Way is a galaxy of type SBc.

————— Study Exercises: Group A —————

1. The branch of science that deals with the stars is called ———.

2. Name three divisions of time that are set by the sun and the stars.

3. For each reference, write one of the four reasons that God created the stars.
   a. Isaiah 13:10   b. Psalm 136:5   c. Job 38:31–33   d. Joshua 10:12–14

4. For many years, men were not able to observe parallax in the stars because
   a. only a few stars show parallax.
   b. they had not examined enough stars.
   c. the parallax of even the closest stars is very small.
   d. they had not yet discovered the stars that show parallax.

5. What is the name and size of the unit for measuring the distance to a star?

6. Why do many of the stars not show parallax?

7. About how many stars are visible to the unaided eye at one time?

8. Which of these discoveries did *not* help men understand that the total number of stars is extremely great?
   a. The Milky Way is a galaxy.
   b. Some of the "clouds" among stars are galaxies.
   c. The galaxies are families of stars.
   d. Astronomical distances separate the galaxies.
   e. There are billions of galaxies.

9. What do the distances and number of the stars show us about God?

Group A Activities

1. *Observing parallax.* Even the separation between your two eyes is enough to produce parallax with objects that are close to you. Hold your two forefingers up in front of your nose, with one behind the other and about 8 inches apart (20 cm). Close one eye, and observe which finger appears to the right of the other. Then close the other eye, and again observe. Do this alternate blinking while looking at objects 6 feet away (2 m). Do it with objects 20 feet away (6 m).

   Try to observe parallax with poles or trees at a distance. Can you still do it? If objects are too far away, the distance between your eyes will be too small for parallax to be noticeable. In the case of stars, most of them are so far away that even the diameter of the earth's orbit is not enough to cause an observable parallax.

2. *Calculating the distance to the nearest star.* The speed needed to escape the gravitation of the earth is about 24,000 miles per hour (39,000 km/h). At that speed, how long would it take to get to Proxima Centauri, the nearest star? Why do you think men will never try to reach that star?

3. *Counting the stars.* Cut an 8½-inch square from a piece of typing paper. Bring two opposite edges together to form a tube, and tape the edges to each other with no overlap of paper. This will be your counting tube. Point the tube at various spots around the night sky, each time counting the stars you can see through the tube at that spot. When you have taken a number of sample counts, average your results. You can see about $\frac{1}{161}$ of the whole sky through such a tube; therefore, multiply your average count by 161 to obtain an estimate for the whole sky.

The Northern Sky at Night

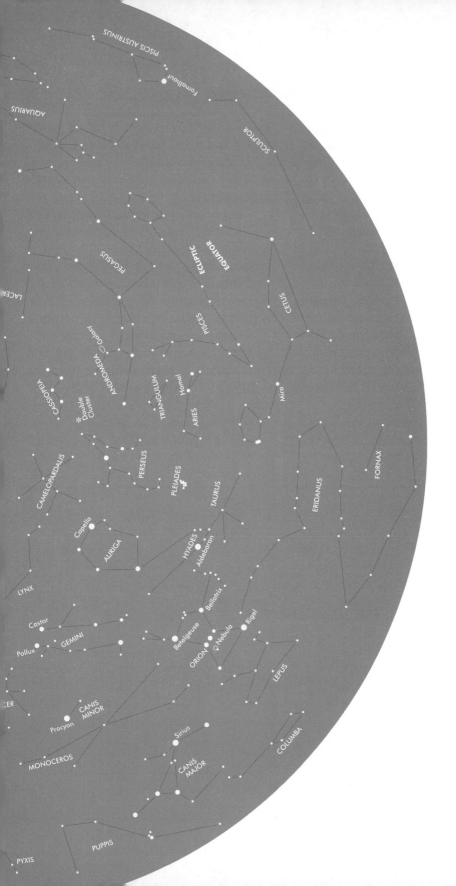

Learning to Identify Stars

Only God knows the number and names of all the stars. Astronomers have assigned names to many of the stars visible to the unaided eye. You can find these names on a detailed star map.

One difficult part about studying the stars is knowing how to start. Just going out and gazing at the night sky is inspiring, but it will do little in becoming familiar with individual stars and the patterns they form. From ancient times, the stars have been associated in groups called constellations. A *constellation* is a group of stars that forms a picture in the sky.

Today the sky is divided into 88 sections (somewhat like the states of the United States), each containing a star picture. Each section is called a constellation and is named for the star picture that it contains. So the word *constellation* refers both to the star picture and to the sky region containing that picture.

Dividing the sky into constellations in this way is useful for identifying the location of other objects besides the star pictures, such as comets, unnamed stars, or the center of a meteor shower.

The constellations themselves are marvels of God's handiwork. "By his spirit he hath garnished the heavens; his hand hath formed the crooked serpent" (Job 26:13). To "garnish" is to decorate; indeed, the heavens are lavishly decorated with stars. The "crooked serpent" is probably the constellation called Serpens.

The star pictures have names like Cygnus, Leo, and Scorpius. These names generally come from Latin, but they have English meanings that you can learn. The Bible usually refers to a constellation or particular star by its ancient name. For example, one of the first and most beautiful constellations that you will locate is Orion, the Hunter.

Selected Constellations

Latin Name	English Name	Latin Name	English Name
Andromeda (an·drom′·i·də)	Chained Lady	Hydra (hī′·drə)	Water Serpent
Aquarius (ə·kwâr′·ē·əs)	Water Carrier	Leo (lē′·ō)	Lion
Aries (âr′·ēz)	Ram	Leo Minor (lē′·ō mī′·nər)	Little Lion
Cancer (kan′·sər)	Crab	Libra (lē′·brə)	Balance
Canis Major (kā′·nis·mā′·jər)	Great Dog	Lyra (lī′·rə)	Lyre
Canis Minor (kā′·nis·mī′·nər)	Little Dog	Orion (ō·rī′·ən)	Hunter
Capricorn (kap′·ri·kôrn′)	Goat	Pegasus (peg′·ə·səs)	Winged Horse
Cassiopeia (kas′·ē·ə·pē′·ə)	Queen	Perseus (pûr′·sē·əs)	Deliverer of Andromeda
Centaurus (sen·tôr′·əs)	Monster	Pisces (pī′·sēz)	Fishes
Cepheus (sē′·fyüs′)	King	Sagittarius (saj′·i·târ′·ē·əs)	Archer
Cetus (sē′·təs)	Whale	Scorpius (skôr′·pē·əs)	Scorpion
Crux (kruks)	Southern Cross	Serpens (sûr′·pənz)	Serpent
Cygnus (sig′·nəs)	Swan	Taurus (tôr′·əs)	Bull
Draco (drā′·kō)	Dragon	Ursa Major (ûr′·sə mā′·jər)	Great Bear
Gemini (jem′·ə·nī′)	Twins	Ursa Minor (ûr′·sə mī′·nər)	Little Bear
Hercules (hûr′·kyə·lēz′)	Giant	Virgo (vûr′·gō)	Virgin

Do not think that by studying constellations in a book, you will be able to walk out at night and say, "There is Cygnus" or "There's Leo Minor." You must go out and actually look at the stars to learn their positions. The best method is for someone who already knows the constellations to point them out. The beam of a strong flashlight makes an excellent pointer for showing stars and constellations to others.

A very helpful tool in star study is a star map. The best ones are larger maps like those available from Rod and Staff Publishers. These star maps come with a mask that lets you find which constellations are visible at any time of the night on any day of the year.

Begin your journey through the stars by identifying one prominent constellation both in the sky and on the star map. A good place to start would be a constellation at the zenith (zē′·nith), the point directly overhead. The zenith will be in the center of the masked star guide. From that zenith constellation, locate and name the constellations around it.

Become acquainted with only four or five constellations at a time. You will be surprised at how quickly you can find your way around in the sky. Soon the heavens will seem very orderly, with Cassiopeia always next to Cepheus, and Pisces always below Andromeda.

You will also want to learn the names of some especially bright stars. Sirius in the constellation Canis Major is the brightest star in the sky. Arcturus (ärk·tŏor′·əs) in Boötes (bō·ō′·tēz) is the fourth brightest star. Arcturus is estimated to be twenty-seven times larger than the sun. The sixth brightest star is Capella (kə·pel′·ə) in Auriga (ô·rī′·gə).

Your eyes need about ten minutes to adjust to the darkness. The pupils dilate almost immediately, but it takes longer for the chemicals in the retina to adjust to low levels of light. If you need to look at a star map or reference book, cover the end of a flashlight with red glass or plastic. Red light will not hinder your seeing ability at night as much as other colors will.

Rich blessings lie in store as the world of stars opens up to you. After a day of hard work, night comes and with it the stars. These silent messengers of God's greatness are visible for all to see and enjoy. Determine as you study this unit that the stars will cease to be a meaningless scattering of lights in the sky. God created the stars for a purpose. They are in the heavens for your benefit.

——— Study Exercises: Group B ———

1. Cygnus, Orion, and Leo Minor are names of ———.

2. Arcturus, Capella, and Sirius are names of ———.

3. Which three of these are good rules to follow in learning the star names and positions?
   a. Be able to identify as many constellations as you can the first night.
   b. Obtain a star map, and learn how to use it.
   c. First identify constellations near the horizon.
   d. Read much about the stars before you try to locate them.
   e. First learn only a few bright constellations.
   f. Have someone point out all the constellations before you start on your own.
   g. Have someone who knows the stars show you the location of a few constellations.

4. The stars and constellations directly overhead are near the —— of the sky.

5. Why will star viewing improve after you have been in the darkness for a while?

6. What color of light will least interfere with night seeing?

7. Use a star map to find the names of the prominent constellations shown below.

a.   b.   c.   d.   e.

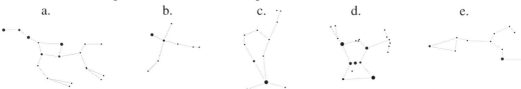

Star Study Aids

Star guides. A good star guide can tell you many things about the heavens. As with any map, it will be most helpful to you if you take a little time to learn how to use it.

The brightness of the stars is shown by dots of different sizes. The biggest dots are for the brightest stars. The brightness of a star is called its ***magnitude.*** As the magnitude number of a star increases, the star brightness decreases. So a magnitude 2 star would be brighter than a magnitude 3 star. The faintest stars visible to the unaided eye are about magnitude 6.

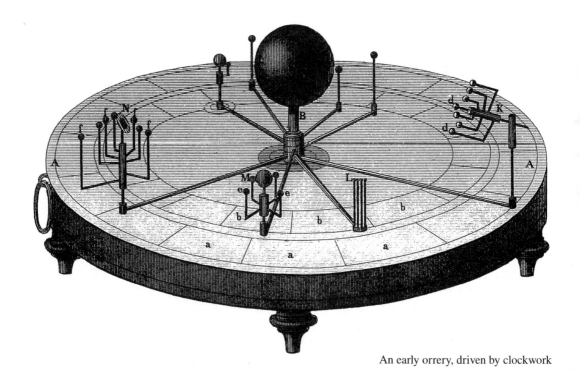

An early orrery, driven by clockwork

The lines that connect stars show the main patterns of the constellations. Some of the patterns look like the figures suggested by the constellation names. Orion looks something like a hunter wearing a belt and holding a bow in his left hand. The three stars of Orion's belt were significant even in Job's day because God asked Job, "Canst thou . . . loose the bands of Orion?" (Job 38:31).

Serpens looks like a serpent and is probably the constellation referred to in the clause, "His hand hath formed the crooked serpent" (Job 26:13). Even Leo looks like a lion with its curved mane over its head, its front leg bent at Regulus, and its tail at Denebola (də·neb´·ə·lə). You will more easily remember the constellations if you can see the figures suggested by their names.

Two kinds of lines form a grid for identifying the positions of stars. The circular lines are lines of *declination* (dek´·lə·nā´·shən). They correspond to the latitude lines on an earth map and are labeled in degrees from the equator. The equator of the sky (celestial equator) is directly above the equator of the earth. The distance of stars north of the

equator is stated in positive degrees of declination, and that of stars south of the equator is given as negative degrees declination. Polaris is near +90° declination, and the point directly above the South Pole is –90° declination.

The straight lines going to the center of a star map are called lines of *right ascension.* They correspond to the longitude lines on an earth map and are labeled in hours east of the

The Ten Brightest Stars			
Star	**Constellation**	**Apparent Magnitude**	**Distance (light-years)**
Sirius	Canis Major	–1.44	9
Canopus	Carina	–0.62	314
Alpha Centauri	Centaurus	–0.27	4.4
Arcturus	Boötes	–0.05	37
Vega	Lyra	0.03	25
Capella	Auriga	0.08	42
Rigel	Orion	0.18	773
Procyon	Canis Minor	0.40	11
Achernar	Eridanus	0.45	144
Betelgeuse	Orion	0.45*	427

The magnitude scale was devised by Ptolemy, an astronomer who lived between A.D. 100 and 200. He divided the stars visible to the unaided eye into six categories, from the brightest (first magnitude) to the faintest (sixth magnitude). With modern instruments, astronomers can measure the brightness of stars much more accurately than Ptolemy could; but they still use his system. A magnitude of 1.0 is 100 times brighter than 5.0. Objects brighter than 0.0 are given a negative value; for example, the magnitude of Sirius (the brightest star) is –1.46.

There are two reasons that stars appear to differ in brightness: (1) stars give out different amounts of light, and (2) stars are located at greatly varying distances from the earth. Therefore, brightness is often stated as "apparent magnitude"—how bright a given star appears from the earth.

*variable star

vernal equinox (the point where the sun crosses the celestial equator when spring begins). The beginning line passes through the constellation Pegasus and just to the right of Cassiopeia, which extends across the first and second hours of right ascension. The meeting point for the lines of right ascension is directly above the North Pole of the earth. Polaris (pə·lar′·is), the North Star, is very near that point. With right ascension and declination, the position of any star or other sky object can be stated very accurately.

Another important line on the star guide is the ecliptic. The *ecliptic* is the path of the sun through the stars. You can find where the sun is on any day of the year by finding that day on the outside circle of dates and drawing a line from there to the North Pole. The sun will be at the place where your line crosses the ecliptic. For example, on August 25, the sun is passing Regulus.

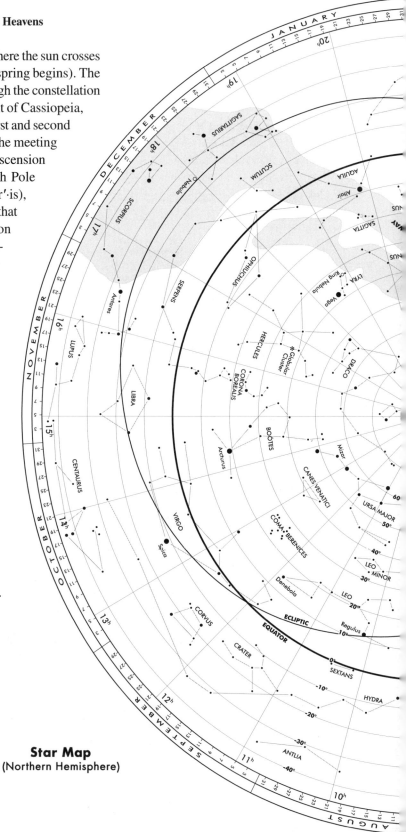

Star Map
(Northern Hemisphere)

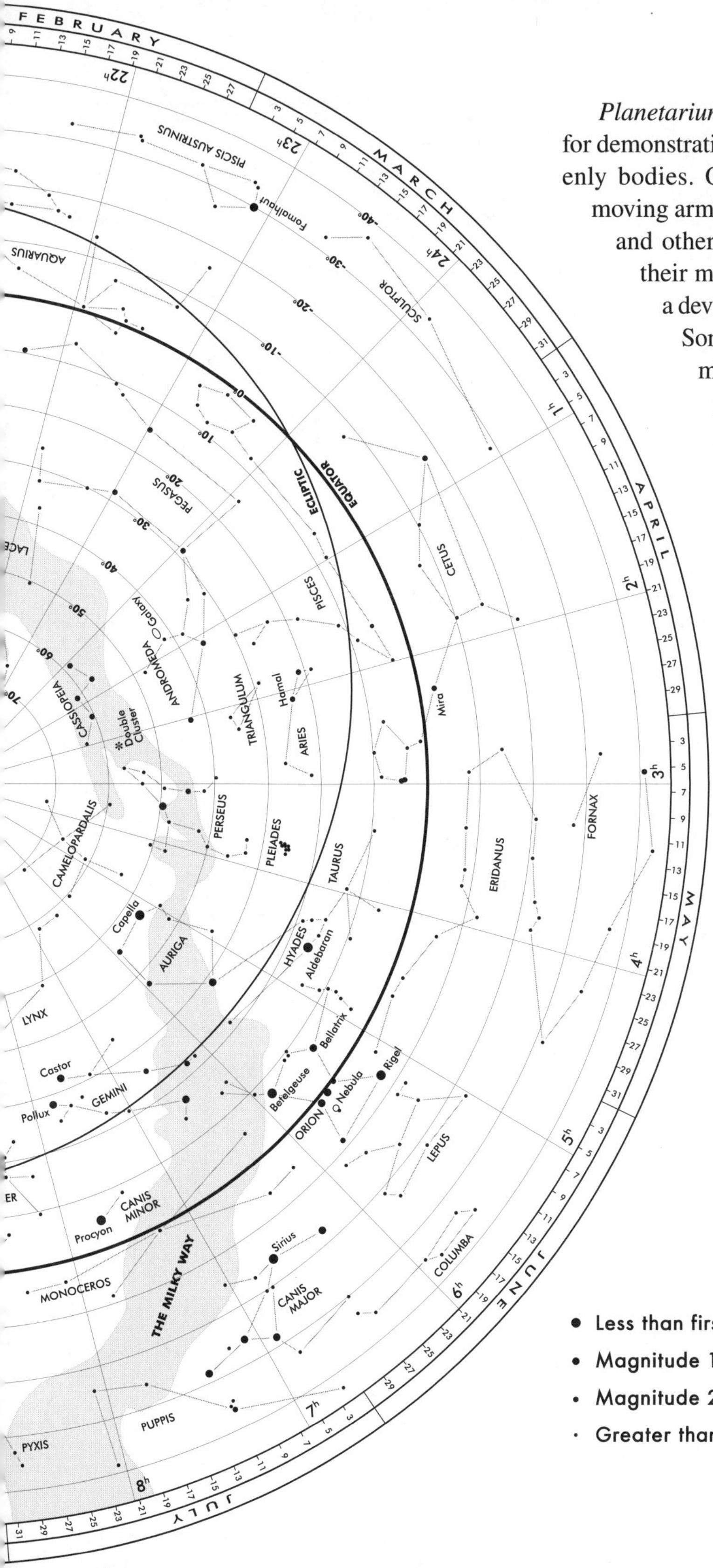

Planetariums. A **planetarium** is a device for demonstrating the movements of the heavenly bodies. One kind of planetarium has moving arms on which models of the earth and other planets are attached to show their movement around the sun. Such a device is called an orrery (ôr´·ə·rē). Some complicated orreries include models of many or all of the planets and their moons.

A star-projector planetarium reproduces the starry sky by projecting tiny points of light onto the domed ceiling of a darkened room. Such a device is very useful in studying the constellations and the principles of astronomy. A simple star-projector planetarium can be made by putting a flashlight inside a can with holes in it. Directions with templates for making such a planetarium can be found in the book *Discovering God's Stars,* available from Rod and Staff Publishers. That book has many other helps for studying and appreciating the stars.

- Less than first magnitude
- Magnitude 1 to 2
- Magnitude 2 to 3.4
- Greater than 3.4 magnitude

Stargazing With Binoculars

The size of a binocular is given by stating both the magnifying power and the size of the objective lens. For example, a 7×35 binocular will magnify an image 7 times, and its objective lens is 35 mm in diameter.

The magnifying power is one of the most important features to consider in selecting a pair of binoculars. Low power is good for wide-field views of the Milky Way, but higher power is better for observing the moon and planets. Keep in mind that as the magnifying power increases, so does the weight of the binocular. Most people can hold 7-power binoculars with minimal fatigue, but a 10-power pair may need additional support for extended periods of use.

Most manufacturers offer zoom binoculars, which allow the user to quickly double or even triple the magnifying power. Zoom models require far more complex optical systems than fixed-power binoculars. More optical elements increase the risk of imperfections and inferior performance. Image brightness also suffers greatly at the high end of the magnification range, with the result that faint objects may vanish.

The diameter of the objective lens should be matched to your eye and your observation site. For most people, the pupil of the eye dilates to about 7 mm in total darkness. If your location suffers from light pollution, your pupils may expand only to 5 or 6 mm. To be perfectly matched for nighttime sky watching, the cone of light leaving the binocular eyepiece, known as the exit pupil, should be the same diameter as the diameter of the pupil of the eye. The size of the exit pupil is determined by dividing the diameter of the objective lens by the power of the binocular. For example, a 7×50 pair has a 7.1-mm exit pupil, while a 10×50 has a 5-mm exit pupil. The larger diameter objectives will pick up more light pollution along with more starlight. This is one application where bigger is not always better.

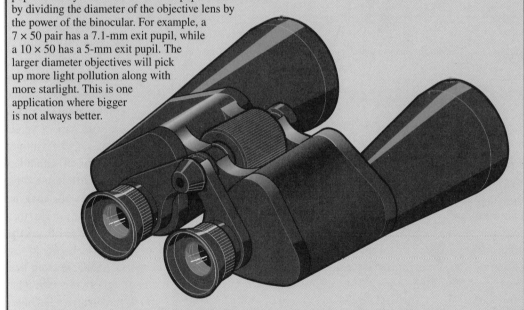

Comparison of Binoculars

Size	Comments
6×30	Small and lightweight. Low power limits usefulness.
7×35	Easily held by hand. Offers excellent wide-field views. Exit pupil limits usefulness in dark sky.
7×50	Easily held by hand. Best choice overall. Larger objective may cause more problems with light pollution.
10×50	Good choice for users who want higher magnification. Reveals many details on the moon. May require a tripod.

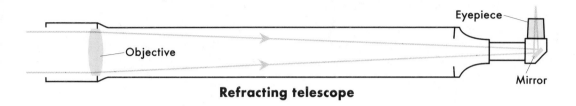

Refracting telescope

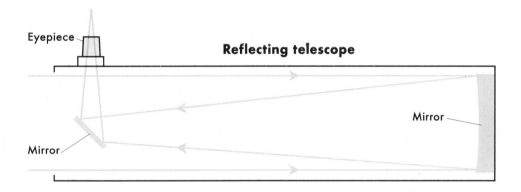

Reflecting telescope

Optical aids. A pair of binoculars or a telescope is very useful for studying the moon, the star clusters, and the Milky Way. The better the optical aid, the more you can see. There are two kinds of telescopes: refractors that collect light with a large lens, and reflectors that collect light with a curved mirror. The light-gathering lens or mirror is called the objective.

If you buy a telescope, be aware of one thing: The diameter of the objective is more important than the magnifying power. A telescope may have high-powered lenses; but if the objective is too small, it will not gather enough light for satisfactory viewing. This is why telescopes are rated by the size of the objective (such as 3-inch or 6-inch) rather than by magnifying power.

You also need a high-quality eyepiece. The word *achromatic* is a sign of better-quality optics. Another important thing is a sturdy mount. The telescope magnifies whatever you see, but it also magnifies any movement of the instrument. This means that even a slight vibration will cause the view of a star to dance about.

An *observatory* is a building with a large telescope and other equipment for observing heavenly bodies. Such a structure is often built on top of a mountain to reduce the amount of air that starlight must pass through. The condition of the atmosphere greatly affects the ability to see clearly through telescopes. Atmospheric layers of unequal temperature cause distortion. Another hindrance to good seeing conditions is bright lights that illuminate the night sky (sometimes called light pollution). The top of a mountain largely avoids these problems. However, many amateur astronomers and educational institutions have observatories in populated areas.

—————— Study Exercises: Group C ——————

1. Give the vocabulary words that match these definitions.
   a. Distance of a star from the celestial equator, measured in degrees.
   b. Device for demonstrating the movements of heavenly bodies.
   c. Measure of the brightness of a star.
   d. Distance of a star east of the vernal equinox, measured in hours.
   e. Building with a large telescope and other equipment for studying stars.
   f. Path of the sun through the starry sky.

2. The heads of the Gemini Twins are bright stars. Castor has a magnitude of 1.58, and Pollux has a magnitude of 1.16. Which of these two stars is brighter?

3. Find the declination of the star Fomalhaut (fō′·məl·hôt′) in the constellation Piscis Austrinus, near the edge of the star map.

4. The star Vega in the constellation Lyra is about halfway between what two hours of right ascension?

5. Near what bright star would the sun be on October 13?

6. A planetarium with models of planets and moons moving about on rotating arms is called an ———.

7. A star-projector planetarium is a useful tool for studying the ——— and for demonstrating the principles of astronomy.

8. A refracting telescope collects light with an objective ———, and a reflecting telescope collects light with an objective ———.

9. Which of the following is the *least* important feature of a telescope?
   a. a wide objective
   b. a stable mount
   c. high power or magnification
   d. an achromatic eyepiece

10. What would be a solution to the problem of light pollution in viewing the stars?

Group C Activities

1. *Identifying stars and constellations.* Learn the names and locations of the stars and constellations. Determine to become familiar with God's handiwork in the sky. The pronunciations of the names are found in the book *Discovering God's Stars,* as well as the meanings of the constellation names.

2. *Finding the brightest stars.* There are only eleven stars on the star map of the Northern Hemisphere with a magnitude less than 1. Can you find all of them?

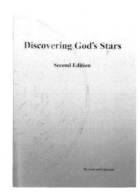

Discovering God's Stars

Second Edition

Revised and Enlarged

"The Sun to Rule by Day"

The sun is our main source of energy. It is the closest star to the planet Earth, even though the distance to the sun is an average of 93 million miles (150 million km). The sun provides almost all the heat and light on the earth. If the sun were suddenly to stop shining, it would be only a matter of days until the earth became so cold that life would begin to die. Even if every home and factory owner operated his heating plant at full power, the cold would still come. It would be like trying to heat an apartment building with a candle flame. Imagine how many light bulbs would be needed to light one-half of the earth as bright as day. Remember, you would need to light the oceans too. You can easily see that the sun is a tremendous source of heat and energy.

God created the energy of the sun in the form of tons and tons of hydrogen under such tremendous heat and pressure that the hydrogen atoms unite to form helium atoms. This is no ordinary chemical reaction, such as burning coal. It is an atomic reaction known as nuclear fusion, in which one element is changed into another and much energy is released. Briefly stated, the sun is a giant atomic reactor that produces tremendous amounts of heat and light.

The surface of the sun has many areas of violent activity called sun storms, which produce solar flares and sunspots. Solar flares

The Solar System				
	Diameter in Relation to Earth	Distance From Sun in Astronomical Units*	Time for One Orbit	Satellites
Sun	109			
Mercury	0.38	0.39	88 days	0
Venus	0.95	0.72	225 days	0
Earth	1	1	365.24 days	1
Mars	0.53	1.52	687 days	2
Asteroids (millions) *and 1 dwarf planet* (Ceres = diameter 605 miles [975 km])				
Jupiter	11.2	5.2	11.9 years	63
Saturn	9.5	9.54	29.5 years	47
Uranus	4.0	19.2	84.0 years	27
Neptune	3.88	30.0	165.0 years	13
Kuiper belt objects (billions; large ones include Pluto and other dwarf planets)				
Comets (orbits vary from a few years to thousands of years)				

*1 astronomical unit = about 93 million miles (150 million km)

The time for one orbit is measured by Earth days and Earth years.

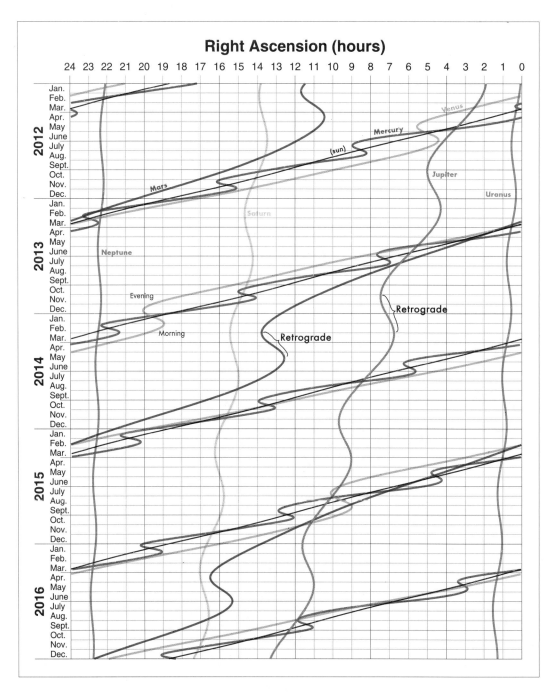

The paths of the planets among the stars

are great "flames" that erupt from the surface of the sun. They cause some areas of the sun to look extra bright, though these are usually not large enough to be seen with a small telescope.

Sunspots are produced by sun storms that cause some areas to be cooler and to look darker than the rest of the sun. These may appear as dots or as large, irregular patches with fuzzy edges. They may change in size and shape from day to day and are often clustered together. Sunspots can easily be seen on a reflected image of the sun.

The number of sun storms increases and decreases in 11-year cycles. Sunspot numbers increase and decrease accordingly, but almost always a few spots are visible. You can see sunspots by doing the activity on page 141.

The sun is the gravitational center of the solar system. The solar system is the sun and all the heavenly bodies that are held in orbit by its gravity. Large bodies orbiting the sun are called *planets,* and smaller bodies are called dwarf planets. Small bodies orbiting the planets are called *satellites.* The moon is the only satellite of the earth, but some of the planets have many satellites. Besides the planets, small objects called asteroids travel around the sun between the orbits of Mars and Jupiter. Other small objects called comets trace very elongated orbits around the sun.

The sun is so huge that its gravity easily holds all the planets in their orbits. Actually, the sun contains over 99 percent of all the matter in the solar system. It would take 109 earths side by side to equal the sun's diameter. The sun has 333,000 times as much mass as the earth, and its gravity is 28 times stronger than that of the earth.

The planets do not have fixed positions in the sky as the stars do. Jude was probably thinking of the planets when he referred to evil

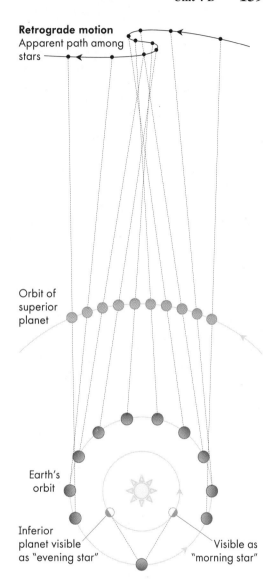

men as "wandering stars, to whom is reserved the blackness of darkness for ever" (Jude 13). Actually, the planets follow very regular elliptical orbits; most of them are almost perfect circles. But because they are much closer to us than the stars are, and because of their orbital motion, the planets appear to wander.

Mercury and Venus are called inferior planets because their orbits are inside the orbit of the earth. This makes them appear to wander first to one side of the sun as "morning stars" and then to the other side as "evening stars." In ancient times, the rising of bright Venus was taken as a sign that morning would soon come.

Mars, Jupiter, and Saturn are superior planets, with orbits outside the orbit of the earth. These planets travel more slowly than the earth. As the earth passes them, they appear to move backward among the stars. This is called *retrograde motion.* The planet chart shows that retrograde motion is not some slight backward movement that is hard to detect. Retrograde motion can be as much as 1 hour of right ascension extending over four months.

Retrograde motion is evident in all the superior planets, but it is most easily seen in Mars, Jupiter, and Saturn. To observe retrograde motion, you will need to note the position of a planet from week to week in relation to the stars. Usually the superior planets travel eastward through the stars, but during retrograde they move westward for several months before moving eastward again.

Of all the bodies in the solar system, only the sun produces its own light. The moon and planets are visible because they reflect the light of the sun. Sometimes a planet is close enough to the earth to make it brighter than Sirius, the brightest star. To the unaided eye, the planets look like stars. But in a telescope, the planets show up as disks and not just points of light as do the magnified stars.

Because Mercury and Venus pass between the earth and the sun, they have phases similar to those of the moon. On rare occasions they pass directly between the earth and the sun, and then they look like dots traveling across the face of the sun. The movement of a heavenly body across the face of a larger body is called a transit. Only the planets Mercury and Venus can have transits across the sun.

If you look at Jupiter with a large telescope, you will see bands across its face. Four of its satellites are visible in even a small telescope. They change position from hour to hour as they orbit Jupiter. Saturn with its broad rings is the highlight of telescope observation. How can such a beautiful and marvelous object exist! Each of the planets shows God's creative ability in its own special way.

─────────── Study Exercises: Group D ───────────

1. The sun provides two main benefits for us. Tell what the sun contains that gives us each of these.
   a. heat and light   b. gravity to hold the earth in orbit

2. Which of these is the best explanation of how the sun produces its energy?
   a. The sun is a huge atomic reactor in which atoms unite to form larger atoms.
   b. The sun is a huge furnace that supports burning at a rapid rate.
   c. The sun is a huge ball of a hot, glowing metal that gives off heat and light.
   d. The great pressure and motion on the sun creates great heat by friction.

3. What are two kinds of violent storms on the surface of the sun?

4. Large bodies that orbit the sun are called ———.

5. Bodies that orbit planets are called ———.

6. Use the solar system table to find each of these.
   a. The largest planet.
   b. The names of two dwarf planets.
   c. The planet whose diameter and distance from the sun are almost ten times as great as those of the earth.
   d. The planet that orbits in the least time.
   e. The planet that is almost the same size as the earth.
   f. The planet that is about 140 million miles away from the sun.
   g. The very cold planets that are about equal in size.

7. In Revelation 22:16, Jesus is called "the bright and morning star." Which planet does this probably refer to?

8. Make each statement true by changing one word.
   a. The planets appear to be stationary among the stars.
   b. The planets outside the earth's orbit are morning and evening "stars."
   c. During retrograde motion, the planets move eastward through the stars.
   d. Mercury is one of the planets that show retrograde motion.

9. a. Use the graph of planet position to find the planet that shows retrograde motion in April 2014.
   b. At what right ascension is that planet near at the end of April 2014?

10. During a ——— of Venus, you could see a small dot moving across the face of the sun.

11. With a small telescope, you can see the four largest ——— of Jupiter and the ——— of Saturn.

12. The order and beauty of the solar system are the handiwork of ———.

Group D Activities

1. *Viewing sunspots.* You can observe sunspots by using a binocular or a simple telescope. **Never view the sun by looking through a binocular or a telescope.** The bright, magnified light of the sun can blind you instantly. Rather, project the sun's image onto a piece of white paper as shown here.

   You will need to move the paper back and forth to focus the image of the sun on it. If there is a large sunspot, you should be able to see a dark central area called the umbra and a surrounding gray area called the penumbra.

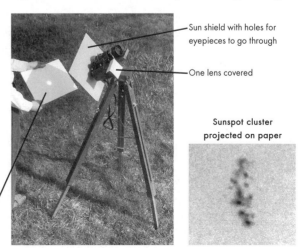

Sun shield with holes for eyepieces to go through

One lens covered

Sunspot cluster projected on paper

White paper with the sun's image projected on it. The farther from the eyepiece, the larger the image.

2. *Finding the diameter of the sun.* Take a mirror outside, and set it exactly 200 feet (2,400 inches) from a flat outside wall so that the light of the sun reflects from the mirror onto the wall. Cover the mirror with a paper having a 1-inch circle cut out of it; this will cause the reflection on the wall to be an image of the sun. Now measure the diameter of the sun's reflection. Is it between 20 and 25 inches? Calculate the diameter of the sun by using the following proportion.

$$\frac{\text{diameter of sun (miles)}}{\text{distance to sun (miles)}} \quad \frac{d}{93,000,000} = \frac{}{2,400} \quad \frac{\text{diameter of image (inches)}}{\text{distance to wall (inches)}}$$

Find the sun's diameter in an encyclopedia. How close did you get to the official value?

3. *Making a model of the solar system.* Get a good idea of the relative sizes and distances in the solar system by making a scale model. The illustration below has the sizes to make the planets. The sun should be 9⅞ inches in diameter, which is very nearly the size of a basket-ball. Search for the right size pinhead or bead for each planet. Perhaps you can carve Jupiter and Saturn out of wood or Styrofoam. Then cut Saturn in half and sandwich the halves over a 2-inch disk of clear, thin plastic to form the rings. Paint the rings on the plastic. Paint all the planets with their colors and markings according to photos in an encyclopedia. Fasten the planets on pins for easy mounting.

Have classmates hold the different planets to show the size in relation to the distances. Of course, you will need to go far away to place Jupiter and Saturn, and it may not be practical to put Uranus and Neptune in your distance model. How amazingly small the planets are in relation to their distances from the sun! The gravity of the sun holds the planets in orbit through millions of miles of space.

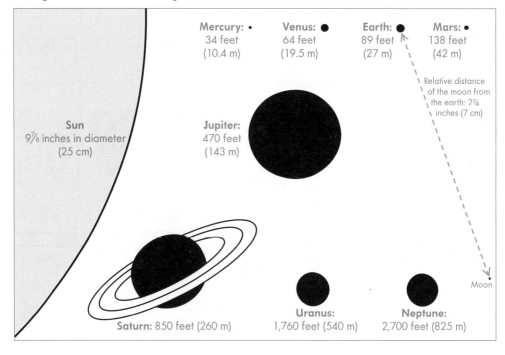

"For Seasons, and for Days, and Years"

Modern calendars and clocks are so common and accurate that few people know how to tell time by the stars. However, our major divisions of time—the day and the year—are still set by the motions of the earth in relation to the sun and stars.

The day is set by the rotation of the earth. Every 24 hours the earth rotates one time on its axis. The half of the earth facing the sun is having daylight, and the other half is having night. Since the earth is rotating eastward, the sun appears to rise in the east and set in the west. This rotation also makes the stars appear to move from east to west. The stars move as fast across the sky as the sun moves. You can observe this movement if you note the position of a constellation at one time and then see where it is an hour later. The motion of the stars was used in ancient days to tell time during the night. People could tell the time of night much as you can estimate the time of day by the position of the sun.

Some ancient clocks showed the time of day by the shadow that a rod cast when the sun shone on it. Such a timepiece is called a sundial, and the rod is called a gnomon (nō´·mon´). King Hezekiah had a sundial that God used to give him a sign. "Behold, I will bring again the shadow of the degrees, which is gone down in the sun dial of Ahaz, ten degrees backward. So the sun returned ten degrees, by which degrees it was gone down" (Isaiah 38:8). Of course, it was a great miracle that the shadow moved backward.

God controls the movements of the earth, and thus He controls the length of the day. On one occasion, God made a day longer than 24 hours by causing the sun and the moon to stand still (Joshua 10:12–14). That was another miracle of God's power and wisdom. It is also part of God's wisdom that normal days are no longer than 24 hours. God knows what our physical bodies need, and He made the day the right length for our welfare.

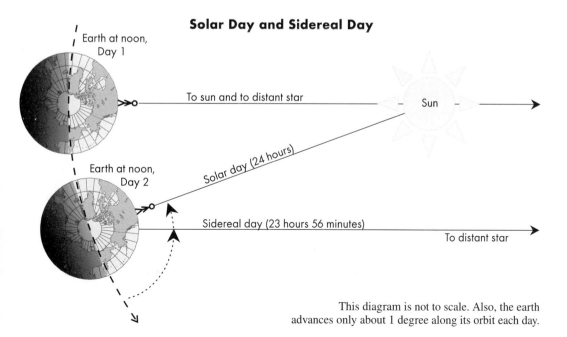

Solar Day and Sidereal Day

Earth at noon, Day 1

To sun and to distant star

Sun

Solar day (24 hours)

Earth at noon, Day 2

Sidereal day (23 hours 56 minutes)

To distant star

This diagram is not to scale. Also, the earth advances only about 1 degree along its orbit each day.

The time of day according to the position of the sun is called sun time or solar time. This time changes gradually as one moves east or west. The amount of change is 1 minute for every 17¼ miles (27.8 km) at the equator. If solar time at one equatorial city is exactly 12:00, 17¼ miles to the west would be 11:59 and 17¼ miles to the east would be 12:01.

But it is confusing for each city to have a slightly different time. To have the time in large areas uniform yet reasonably close to solar time, the earth was divided into standard time zones in 1884. Each time zone has a width of 15 degrees, or 1,037 miles (1,669 km) at the equator. The main part of the United States has four time zones: Eastern, Central, Mountain, and Pacific Standard Time. Thus when the time is 5:00 P.M. in New York City, it is 2:00 P.M. in San Francisco—three hours behind because San Francisco is three time zones to the west.

The system of standard time zones was established for convenience. Many boundaries between time zones do not follow longitude lines, but rather they zigzag to keep from unnecessarily dividing states, countries, and large population centers. It is true that people on the edge of a time zone may be using time that is half an hour off solar time, but this is no great handicap. Most people do not watch the position of the sun closely enough to know the difference.

Solar day and sidereal day. The length of time from noon one day (when the sun is at its highest point overhead) to noon the next day is called a solar day. But remember that while the earth rotates on its axis, it is also revolving around the sun. So at noon each day, the earth is at a slightly different place in its orbit. Instead of rotating 360 degrees from one noon to the next, the earth rotates about 361 degrees. This takes exactly 24 hours, the length of a solar day.

The time needed for the earth to rotate 360 degrees is measured by the stars instead of the sun, and it is about 4 minutes less than a solar day. So a day measured in terms of the stars is about 23 hours 56 minutes long. This is called a sidereal day (sī·dir′·ē·əl); *sidereal* means "according to the stars." It takes one sidereal day for a star that is directly overhead one night to be directly overhead again the next night. But each night, the solar time when that star is directly overhead will change by about 4 minutes, until finally the star is overhead in the daytime.

Finding location by the stars. Navigators can calculate their position by using an instrument called a sextant to find the angle from the horizon to the sun or a star. A ship's latitude is equal to the angle of the North Star above the horizon. For example, if Polaris is 30 degrees above the horizon, the observer is at 30 degrees north latitude.

Polaris

Light from Polaris

0°

45°

90°

North Pole (90°)

45°

Tropic of Cancer

Equator   0°

A stargazer can find his latitude by measuring how far Polaris is above the horizon.

Calculating longitude is more difficult. First, solar time is found by measuring the angle of a certain star and using a set of tables to find the ship time. This solar time is then compared with that shown by an accurate clock set to the time at Greenwich, England—the location of the prime meridian (0 degrees longitude). The difference between the solar time and the Greenwich time tells how far around the earth the observer is located.

For example, suppose a ship's solar time is 6:00 A.M. when Greenwich time is 9:00 A.M.— three hours behind. Since each hour is equal to 15 degrees longitude, the ship is located at 3 × 15, or 45, degrees west longitude. If the ship time is ahead of Greenwich time, its location will be east longitude. Before the age of radio and satellite guidance, such star reckoning was very important for safe sea travel.

—————— Study Exercises: Group E ——————

1. Why do the sun, moon, and stars appear to be revolving around the earth?

2. In which direction do the sun, moon, and stars appear to move?

3. What indicates the time of day on a sundial?

4. Suppose city A is 500 miles west of city B, and both cities are at the equator.
   a. The solar time at city A will be (ahead of, behind) the solar time at city B.
   b. The difference in solar time will be ——— minutes, to the nearest minute. (Remember that there is a time difference of 1 minute for every 17¼ miles.)

5. Why are modern clocks set by standard time and not solar time?

6. Suppose you live in the Mountain Standard Time Zone and want to call a friend in the Eastern Standard Time Zone so that he receives the call at 5 P.M. where he lives. When should you call according to your time?

7. a. The length of the ——— day is set in relation to the sun. It is exactly ——— hours.
   b. The length of the ——— day is set in relation to the stars. What is that length?

8. An instrument for measuring the angle from the horizon to the sun or a star is called a ———.

9. What position on the earth is indicated if Polaris, the North Star, is 10 degrees above the horizon?

10. What position on the earth is indicated if a ship's solar time is 8:00 P.M. and Greenwich time is 3:00 P.M.?

11. Challenge question: In which ocean would a ship be if the answers to exercises 9 and 10 indicated its position?

Group E Activities

1. *Making a sundial.* A sundial with a vertical or horizontal gnomon will not show the right time throughout the year. The most accurate sundial has a gnomon that points to the north celestial pole. You can have yours do this by making the angle of the gnomon equal to the latitude where you live. Then make the gnomon point directly north by establishing true

north with Polaris, the North Star. (Magnetic north is not the same as true north for most places on the earth.)

Establish 12:00 solar time by noting when the shadow of the sun is directly under the gnomon. Then with the use of a clock, mark off each hour on the sundial. If 12 o'clock solar time is 12:15 standard time, then mark the shadow an hour later (1:15) as 1 o'clock. Mark 2:15 as 2 o'clock, and so on. The times before 12:00 can be established by measuring the spaces between hours and marking off equal spaces on the other side of the gnomon.

2. *Finding solar time.* Establish the solar time for your school. This is done by devising some method for determining when the sun is highest in the sky. One way is by suspending a horizontal rod pointing north and watching the shadow that it casts. At the moment when the sun reaches its highest point, the shadow will be directly beneath the rod, and solar time will be 12:00 noon. You may want to have two clocks in your schoolroom, one labeled Standard Time and the other labeled Solar Time. At what location is solar time the same as standard time? Which time would you rather live by?

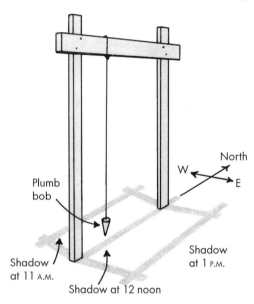

The year is set by the revolution of the earth. The earth moves in a huge, nearly circular orbit around the sun every year, traveling at about 1,140 miles per hour (1,835 km/h). Each orbit takes 365 days, 6 hours, 9 minutes, and 10 seconds, or about 365¼ days. This motion makes the sun appear to trace a path through the stars once each year. The sun's path is called the ecliptic, as you saw in the section about star maps (pages 132–133).

The ecliptic passes through twelve constellations that every stargazer should learn to identify. The twelve constellations together are called the *zodiac.* This name comes from the same word as *zoo,* which refers to living creatures. The constellations of the zodiac have the names of animals or people. To observers

on the earth, the sun appears to pass through one constellation of the zodiac each month.

As the earth orbits the sun, the sun appears to move through the zodiac so that on July it is in Gemini and on December it is in Scorpius. The diagram on the next page will help you understand the relationship between the earth, sun, and zodiac.

We cannot actually see which constellation the sun is in, because that constellation rises and sets with the sun. The position of the sun must be determined indirectly, such as by noting which constellation is just above the western horizon after the sun sets. In the diagram, Scorpius is just above the western horizon at sunset. So the observer knows that the sun is in Libra.

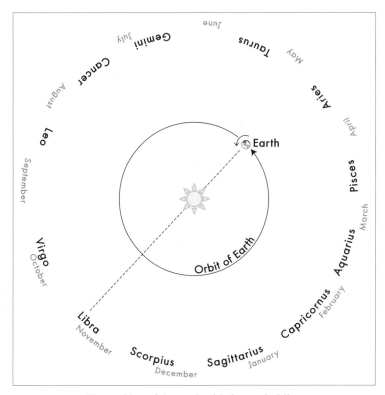

The position of the earth with the sun in Libra

Not only the sun but also the moon and planets move through the constellations of the zodiac. These heavenly bodies follow paths that are not far from the ecliptic. So if you want to look for the planets, you will find them somewhere along the band of the zodiac. If you become very familiar with the constellations of the zodiac, you will know that a "new" star-like object in this region is a planet.

You will not be able to see all twelve constellations in one evening; but if you watch the stars throughout the year, you will see an orderly progression of new constellations appearing above the eastern horizon. Every night a given star rises about 4 minutes earlier than it did the night before. That is, if a star rises tonight at 9:00 P.M., it will rise tomorrow night at 8:56 P.M. This fact is easy to prove by making some careful observations on several successive nights.

The zodiac star calendar. The sun's passage from one constellation to another once provided a very important calendar for farmers. They learned that their crops would grow better if they planted when the sun was in a certain position. For example, they might have recognized that the danger of frost was usually past after the sun entered Taurus. Would you know when that is? (See the diagram above.) Today most farmers plant their crops according to the

By giving the earth two kinds of motion at the same time, God produced both the day and the year. These units of time are exactly the right length for us. Sometimes you wish the day were a little longer so that you could get more done, but sometimes you wish it were shorter so that you would not get so tired. God in His wisdom made the day long enough to give us many working hours but not too long for our tired bodies. In His goodness, He provided the night for us to stop our work and rest awhile.

Like the day, the year is neither too long nor too short. If the year were much shorter, countries far away from the equator would not have a long enough growing season to raise crops. If the year were longer, summers would become extremely warm and winters extremely cold.

The Zodiac
The Path of the Sun, Moon, and Planets

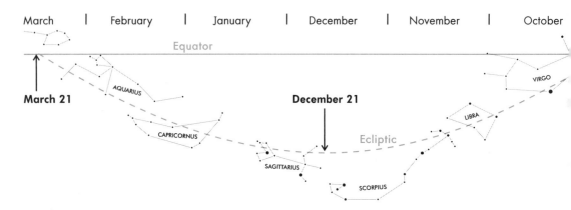

| March | February | January | December | November | October |

days on a standard calendar. Seed companies often recommend certain calendar dates as the best time for planting given crops.

Wrong use of the zodiac. It is perfectly all right for people to use the zodiac to determine when the danger of frost is past. But some people use the position of the sun, moon, and planets in the zodiac to determine the best times to do certain things. They study the stars, not to learn about God's orderly world, but to foretell the future. *Astrology* is the ancient heathen practice of fortunetelling by studying the stars.

The Bible clearly teaches against the use of astrology. "Learn not the way of the heathen, and be not dismayed at the signs of heaven; for the heathen are dismayed at them" (Jeremiah 10:2). Daniel and his friends proved that by being faithful to God, they had a much better

understanding than that of magicians and astrologers in their day (Daniel 1:20). Even the king confessed the limitations of his wise men (Daniel 4:7). Isaiah challenged the astrologers of his time to help Israel if they could, and then he predicted their total destruction (Isaiah 47:13, 14).

Some people refer to their "lucky stars." The heathen custom of choosing a star as a good luck charm was practiced already in New Testament times. When Paul was on his way to Rome, he departed from the island of Melita on a ship of Alexandria. Acts 28:11 tells us that the sign (good luck stars) of that ship was Castor and Pollux, the two heads of the Gemini Twins. We should trust in God to keep us safe, not in the stars. Neither should we swear by heaven or by the stars, for heaven is God's throne (Matthew 5:34).

─────── Study Exercises: Group F ───────

1. What movement of the earth governs the length of the year?
2. The path of the sun through the stars is called the ──────, and the constellations through which the sun passes are called the ──────.

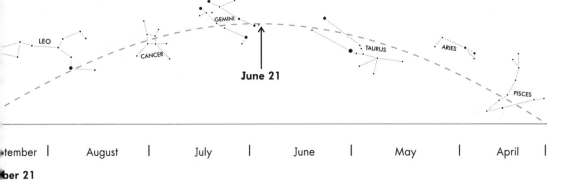

LEO   CANCER   GEMINI   June 21   TAURUS   ARIES   PISCES

| •tember | | August | | July | | June | | May | | April | |

•ber 21

3. How many constellations are in the zodiac?

4. In which constellation is the sun today?

5. Why would it be hard to find Gemini at the beginning of July?

6. Besides the sun, what other heavenly bodies are found in the zodiac?

7. If a star rose above the horizon at 10:00 P.M. on October 1, when would it rise on October 16?

8. How did farmers know when to plant their crops before the development of accurate calendars?

9. Which three statements give reasons that a Christian should not practice astrology?
   a. The Bible forbids the use of astrology.
   b. An astrologer's predictions never come to pass.
   c. Astrology leads to fear and worry about the future.
   d. Astrology leads to putting confidence in something other than God.
   e. Astrology is not as popular today as it was in Bible times.

10. According to 2 Kings 23:5, what sin relating to heavenly bodies did some people commit in Josiah's time?

Group F Activity

Demonstrating the sun's passage through the zodiac. To help you understand the sun's apparent motion through the heavens, organize the following demonstration. Choose twelve students in your class to represent the twelve constellations of the zodiac. Have them stand evenly spaced in a large circle, holding name cards for the constellations. (You could also tape the names on the wall.) Be sure the constellations are in the right order.

Have one person stand in the center of the circle to represent the sun. Perhaps he could have a band of bright yellow paper around his head, with *Sun* written on it. Then have a student walk counterclockwise around the sun, holding a globe to serve as the earth. It will become clear that as the "earth" makes its orbit, the "sun" appears to move through the zodiac. This demonstration also shows why certain constellations cannot be seen at certain times of the year.

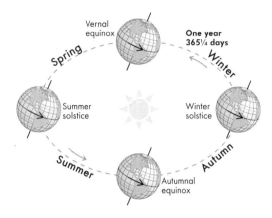

The seasons are caused by the tilt of the earth. God created the earth to have seasons: "seedtime and harvest, and cold and heat, and summer and winter." The complete cycle of the seasons is equal to one year, and the seasons change because the earth is tilted 23½ degrees on its axis. Since the direction of the earth's tilt is always the same, the Northern Hemisphere tilts toward the sun during one part of the year and away from the sun during another part of the year. The diagram above shows how this works.

The labels (such as *summer solstice*) on the diagram above are only for the Northern Hemisphere because when the Northern Hemisphere is having summer, the Southern Hemisphere is having winter. When the Northern Hemisphere is having the spring season, the Southern Hemisphere is having autumn.

Summer is the warmest season because it is the time of the most direct sunlight and the longest days. Plants grow best with bright sunlight and long daylight hours. This long growing season is God's plan to provide food. With the tilt of the earth, first one hemisphere and then the other has a growing season. This makes it possible to raise good crops even in lands far to the north and south. Without the tilt, less of the earth could grow an abundance of food.

The tilt of the earth determines some important lines of latitude. Two of these are the Tropic of Cancer and the Tropic of Capricorn, which are each 23½ degrees from the equator. These are the northern and southern edges of the region where the sun passes directly overhead. Two other important lines are the Arctic Circle and the Antarctic Circle, which are 23½ degrees from the North Pole and the South Pole. The regions within these circles have darkness all day during one part of the year and sunshine at midnight half a year later.

Four important dates are established by the tilt of the earth in relation to its orbit.

Vernal equinox. About March 20, the sun is directly above the equator and shines equally on both the Northern and Southern Hemispheres. Also on this date, the day and the night are of equal length. This is called the **vernal equinox** (ē′·kwə·noks′); *vernal* means "of spring," and *equinox* means "equal night." Calendars mark this day as the first day of spring. The actual date of the vernal equinox varies from March 19 to March 21.

Earth at Winter Solstice

Summer solstice. On June 20, 21, or 22, the sun reaches its northernmost point in the sky and passes directly overhead at the Tropic of Cancer. This is called the **summer solstice** (sol'·stis) or the first day of summer. (*Sol* means "sun" and *stice* means "standing still," which refers to the stopping of the sun in its northward advance.) From this day until winter begins, the sun will gradually be lower in the sky at noon each day. The summer solstice has the longest day and the shortest night of the year.

Autumnal equinox. On September 22 or 23, the sun is again directly above the equator, and the day and the night are of equal length. This is the **autumnal equinox,** the first day of autumn. The difference between the two equinoxes is that in March the days are becoming longer and in September the days are becoming shorter.

Winter solstice. On December 21 or 22, the sun reaches its southernmost point in the sky and passes directly overhead at the Tropic of Capricorn. This is called the **winter solstice** or the first day of winter. The winter solstice has the shortest day and the longest night of the year.

The exact date for each of these occurrences is determined by the day the sun reaches certain points on the ecliptic. Because of leap years, this does not happen on the same calendar day each year. So the dates vary somewhat as indicated above.

Remember that the seasons in the Southern Hemisphere are directly opposite from those in the Northern Hemisphere. There the sun rises higher in the northern sky each day until it reaches its highest point on the summer solstice (December 21 or 22). This is the longest day of the year, the sun is directly over the Tropic of Capricorn, and summer begins. Afterward the days become gradually shorter until winter begins on the winter solstice (June 20, 21, or 22).

The tilt of the earth explains why the ecliptic, or path of the sun, rises above and then falls below the equator on a star map. Notice that the ecliptic crosses the equator at the 12th and 24th hours of right ascension. These lines are very near the autumnal and vernal equinoxes. The position of the sun at the two solstices can also be located by their dates on the star map.

We enjoy the variety of weather produced by the seasons. The farmer works with the changing seasons in planting and harvesting his crops. Even the freezing and thawing of the ground in winter helps to make the soil loose and porous, ready to soak up the spring rains.

Apple trees need several weeks of very cold winter weather to be fruitful. We can rejoice with Noah when we see a rainbow. "While the earth remaineth, seedtime and harvest, and cold and heat, and summer and winter, and day and night shall not cease" (Genesis 8:22).

——————— Study Exercises: Group G ———————

1. Which two of the following are involved in producing the seasons?
   a. the earth's rotation on its axis
   b. the earth's revolution around the sun
   c. the earth's tilt on its axis
   d. the moon's revolution around the earth

2. a. The earth is tilted —— degrees on its axis.
   b. How does this tilt show God's wisdom in making the earth suitable for living things?

3. For which two of the following reasons is summer warmer than winter?
   a. The sun is hotter in summer.
   b. The sun is more nearly overhead in summer.
   c. The earth is closer to the sun in summer.
   d. There are more hours of sunlight each day in summer.

4. What is special about December 21 or 22 at each of the following places on the earth?
   a. north of the Arctic Circle
   b. at the Tropic of Capricorn
   c. south of the Antarctic Circle

5. Summarize the four important seasonal dates of the year by copying the table below and writing the italicized words or phrases where they belong. You will use one phrase twice.

   spring   *winter*   *vernal equinox*   *shortest day of year*
   summer   *summer solstice*   *autumnal equinox*   *day and night same length*
   autumn   *winter solstice*   *longest day of year*

Date	Name	Season That Begins	Kind of Day
March 19, 20, or 21			
June 20, 21, or 22			
September 22 or 23			
December 21 or 22			

6. Use the star map to find which constellation of the zodiac the sun is entering at the summer solstice.

7. What is one advantage of the cold winters that some parts of the earth have between their summer growing seasons?

"The Moon . . . to Rule by Night"

The moon is an amazing light in the night sky. With no water, atmosphere, or oxygen, the moon is not at all suitable for life. But as a giant reflector of light, the moon does very well at serving God's purpose for its creation.

Since the moon is the closest heavenly body, a study of it with a telescope is very rewarding. Even with the unaided eye, one can easily see the dark and light features on the moon. A telescope shows that the dark areas are smooth plains. Around these areas are some mountain ranges. Much of the face of the moon is pocked with circular craters, some with a diameter of 30 miles (48 km) or more. A number of the craters have streaks extending outward like

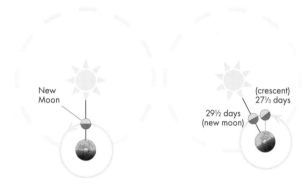

When the moon orbits the earth, it completes a 360-degree circle in 27⅓ days. But the earth is also moving in its yearly orbit around the sun. So two more days are needed to bring the moon back between the earth and the sun. The interval from one new moon to the next is 29½ days.

rays from the sun. During full moon, some prominent rays of the crater Tycho (tē′·kō) can be seen.

A sample of the lunar topography

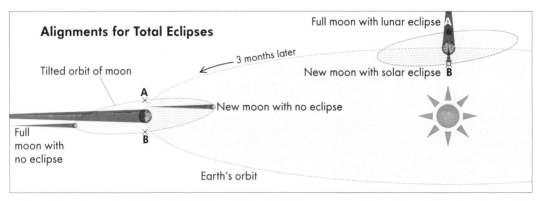

Alignments for Total Eclipses

Tilted orbit of moon

3 months later

Full moon with lunar eclipse A

New moon with solar eclipse B

New moon with no eclipse

Full moon with no eclipse

Earth's orbit

A and *B* are the only points level with the earth's orbit. Twice a year, as the earth revolves, these two points are in a straight line with the sun. If a new moon occurs during this alignment, there is a solar eclipse. If a full moon occurs during the alignment, there is a lunar eclipse. All other orbit points are too high or too low for a total eclipse to occur.

The same side of the moon always faces the earth. This is because it makes one rotation on its axis every 27⅓ days and also one revolution around the earth every 27⅓ days. Only after spacecraft were put in orbit around the moon did scientists learn what was on its opposite side. The features on that side are less interesting than on the side facing the earth. The back side of the moon has many more craters but few plains and mountains.

The moon goes through phases. The moon does not always look the same. Sometimes it looks like a small arc. Later it looks like a half circle. Still later it looks like a full circle, called the full moon. These different shapes of the moon are called *phases.* It takes the moon about 29½ days to pass through one cycle of its phases. That is very nearly one-twelfth of a year, or one month. The word *month* comes from *moon.*

The phases of the moon result from the moon's revolution around the earth. As the moon changes position in relation to

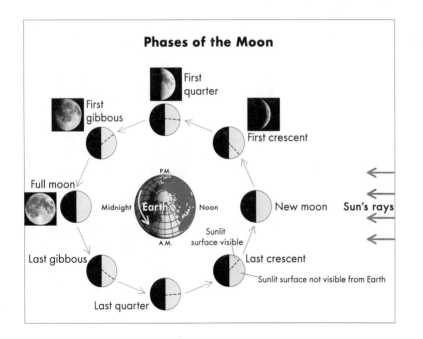

Phases of the Moon

First quarter

First gibbous

First crescent

Full moon

Midnight   Earth   Noon

P.M.

A.M.

Sunlit surface visible

New moon   Sun's rays

Last gibbous

Last crescent

Sunlit surface not visible from Earth

Last quarter

the earth and the sun, it appears to change shape. Remember, the moon simply reflects sunlight. The diagram on page 154 shows how the different phases of the moon are produced. The dotted lines help to show the part of the moon that is visible from the earth in each phase.

The night after a new moon, a very thin crescent can be seen in the west and soon sets. Each night the crescent gets wider and is higher in the western sky. Since the first quarter phase is at a right angle to the sun, it appears high in the sky just as the sun sets. When the moon is opposite the sun in the full phase, it rises in the east as the sun sets. Then just as the sun rises the next morning, the moon is setting in the west. Thus, the full moon is normally visible all night.

The moon causes eclipses. The plane of the moon's orbit is slightly tilted in relation to the plane of the earth's orbit. For this reason, the moon in the new moon phase usually passes a little above or below a line from the sun to the earth. But occasionally the moon passes directly in front of the sun. This cuts off the sun's light and causes a *solar eclipse.* The orbits of the earth and moon are so precise that the times of these solar eclipses can be calculated and predicted many years in advance.

Not all solar eclipses are alike. The size of the moon in the sky is almost the same as the size of the sun. Sometimes when there is a solar eclipse, the moon is closer to the earth in its elliptical orbit and appears a little bigger than the sun. Then there is a total solar eclipse, with the sun completely covered. During a total solar eclipse, the sky becomes so dark that

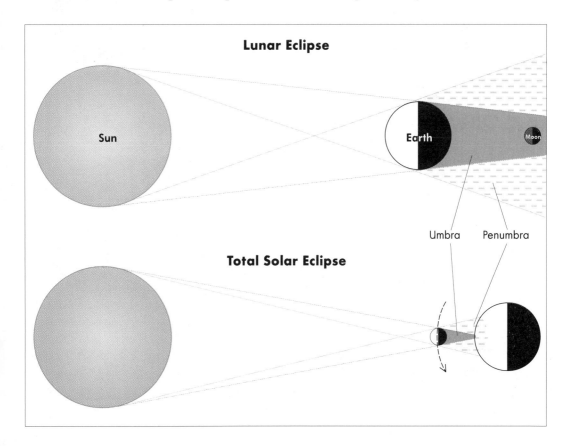

Check a science news source beforehand for the exact time and ideal viewing locations. Asterisks indicate the points of longest duration. A few paths are total for part and annular for the rest.

Two total solar eclipses will cross North America within a 7-year period. Many people will travel hundreds of miles to see these rare events. If the skies are clear, residents in southern Illinois will see a total solar eclipse twice without needing to travel.

Prior to those two events, a total solar eclipse last crossed North America in 1979. The next occurrence comes in 2045.

Total and Annular Solar Eclipse Paths: 2012–2040

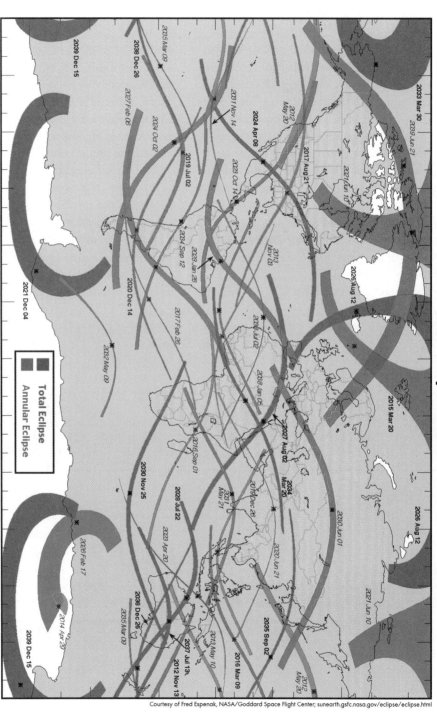

Legend:
- Total Eclipse
- Annular Eclipse

Courtesy of Fred Espenak, NASA/Goddard Space Flight Center; sunearth.gsfc.nasa.gov/eclipse/eclipse.html

To see a total solar eclipse is an awesome experience. The view is more dramatic along the center of the path because the sky is blacker there.

Scientists have compiled the times and places of eclipses thousands of years into the past as well as into the future. Chronologists find such data very useful to establish historical events. That eclipses can be predicted with such accuracy shows what an orderly universe God has created!

A lunar eclipse over Indianapolis on Feb. 20, 2008. The moon photos in this composite picture were taken over 15 minutes apart, and not everything is in proportion.

birds quiet down as if night were approaching, and some stars can be seen.

When the moon is farther away from the earth, it appears slightly smaller than the sun. Then there is an annular eclipse, with a thin ring of the sun still visible around the moon. *Annular* means "ring-shaped."

Total and annular eclipses can be seen only by persons in the strip of earth swept by the full shadow (umbra) of the moon. For people living on either side of this strip of totality, only a part of the sun is covered. The people in this area of partial shadow (penumbra) see a partial eclipse. You can expect to see several partial eclipses in your lifetime, but to live in the area of a total eclipse is a rare and special celestial event.

Sometimes during full moon, the shadow of the earth crosses the moon and causes a *lunar eclipse.* Observers can see the curved edge of the earth's shadow as it advances across the face of the moon. This curve gives visual evidence that the earth is a huge sphere.

A lunar eclipse is visible to many more people than a solar eclipse is. Everyone on the side of the earth facing the moon can see the eclipse. However, many people do not see a lunar eclipse, because it often requires missing some sleep. But to know the date and exact time when an eclipse will occur, and to see it begin exactly when it was predicted, is worth losing sleep to experience. It is a demonstration of the handiwork of God.

The moon causes ocean tides. Tides are the rise and fall of the oceans twice every day. The moon's gravity causes the water to heap up slightly on the side of the earth toward the moon. Centrifugal force also causes the water to heap up on the side opposite the moon. (See diagram on page 335.) These heaped-up areas are called high tides. Between two high tides is a low tide. As the earth rotates on its axis, the moon causes two high tides and two low tides each day.

The difference between water levels at high tide and low tide may be 10 feet (3 m) or more in some localities. At new moon and full moon, the gravity of the sun and the moon work together to make especially high tides. These were probably called spring tides because the water "springs" higher than usual.

At the quarter phases, the high tides are somewhat lower because the sun and the moon are pulling at right angles to each other. These are called neap tides.

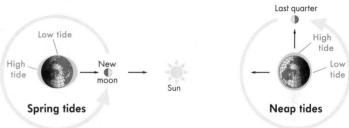

——————— Study Exercises: Group H ———————

1. Why is the moon not a suitable place to live?

2. Give three interesting features that can be seen on the surface of the moon.

3. The moon goes through one cycle of its phases as it revolves around the ——— once every ——— days.

4. Which statement best explains why the moon has different phases during the month?
   a. Differing amounts of lighted area are visible as the moon orbits the earth.
   b. The earth and moon travel around the sun together.
   c. Different sides of the moon face the earth during the month.
   d. More of the moon receives sunshine at some times than at others.

5. Tell whether each phase is best seen in the *evening,* in the *morning,* or *all night.*
   a. first crescent   b. full moon   c. last quarter

6. During which phase of the moon can a solar eclipse occur?

7. During a total solar eclipse, the moon is (closer to, farther from) the earth than during an annular eclipse.

8. During which phase of the moon can a lunar eclipse occur?

9. During a lunar eclipse, what visible evidence shows that the earth is round rather than flat?

10. Eclipses can be predicted many years in advance. What does this show about the way God made the orbit of the moon?

11. If the full moon is directly overhead when you are along an ocean, you can expect a (high, low) tide in the area.

12. Suppose there is a low tide at 12:00 noon.
    a. When will the next low tide be?
    b. When will the next high tide be?
    c. Challenge question: What is the phase of the moon?

13. Why are the tides highest during the new and full phases of the moon?

Group H Activities

1. *Viewing a solar eclipse safely.* You should never look directly at the sun. Its bright rays can cause blindness. It would be good to learn and practice ahead of time some ways to view the sun safely so that if a solar eclipse comes your way, you will be prepared to view it.

   One good way to view an eclipse of the sun is described in the Group D activity on finding the diameter of the sun (page 142). A much smaller hole can be used in the covering of the mirror, and the mirror can be placed much closer to the wall. A similar method is to make a pinhole in cardboard and hold it between the sun and a white sheet of paper. The hole will project the sun on the paper so that the eclipse can be viewed easily and safely by many observers at once.

   Another good way to observe a solar eclipse is by projecting the sun with a small telescope, as described in the Group D activity on viewing sunspots (page 141).

   Or you may wear eclipse shades or use a solar filter to view the sun. Check with an astronomy source. It is good to have a supply of such filters on hand for that special time when you and others can use them to view a solar eclipse safely.

   The use of welding helmets is also safe. The dark lens in these helmets is designed to filter out the extremely bright light from arc welding and will effectively do the same for the rays of the sun.

2. *Predicting tides by the phase of the moon.* Check a calendar to find the phase of the moon today. From that phase, calculate the time of the high and low tides for the day. For example, if it is new moon, high tides will be when the moon is overhead at 12:00 noon and twelve hours later, at midnight. Low tides will be halfway in between.

   It takes the moon 29½ days to revolve around the earth. And there are 1440 minutes in a full rotation of the earth (24 × 60). Thus, the tides shift ahead about 49 minutes every day (1440 ÷ 29½). So if today's date is three days after a quarter phase, the high tides will be about 147 minutes (3 × 49) after each 6 o'clock today, or about 8:27 A.M. and P.M. The low tides will be 6 hours after those times, around 2:27 A.M. and P.M.

"Wonders Without Number"

In exalting God, Job said, "Which alone spreadeth out the heavens. . . . Which maketh Arcturus, Orion, and Pleiades, and the chambers of the south. Which doeth great things past finding out; yea, and wonders without number" (Job 9:8–10). Since the invention of the telescope, we know even more of these wonders than Job did. Yet there were many impressive things that even Job could see with his unaided eyes.

Job had some knowledge of the stars and made specific mention of Arcturus, the brightest star in the constellation Boötes. He may have known about the variable stars that have regular changes in brightness. For example, Mira, in the neck of Cetus, varies from magnitude 3.4 to 9.3 in a 331-day cycle. That means Mira is easily visible during one part of its cycle but invisible during the other part. There are many other variable stars.

Above: The Pleiades star cluster as seen with the unaided eye. Can you find Taurus on the left side and Aries on the right side? *Right:* Pleiades magnified 4 times. Notice the variety of star colors in this region.

Occasionally a dim star explodes into a star thousands of times brighter. Such a star is called a *nova* (Latin for "new star"). For those who know the stars and constellations well, a nova is a special wonder. Novas become dim again after some time.

Binary stars. These double stars require a telescope to "split" them. Mizar (mī′·zär′), at the bend in the handle of the Big Dipper, is a binary star that shows up nicely as two stars in a small telescope. Most binary stars are two stars that are orbiting each other. The binary star at the foot of the Northern Cross (part of Cygnus) is an especially beautiful double, since one star is blue and the other is golden.

Yes, many stars do have color. The sun is slightly yellow. Antares (an·târ′·ēz), the brightest star in Scorpius, is reddish. Aldebaran (al·deb′·ər·ən) and Betelgeuse (bēt′·əl·jōōz′) are also slightly red. Bluish stars include Vega (vē′·gə) in Lyra and Rigel (rī′·jəl) in Orion.

Star clusters. Job mentioned the star cluster Pleiades (plē′·ə·dēz′). This large cluster in the constellation area of Taurus is of the open cluster type. Pleiades is sometimes called the

Seven Sisters or seven stars, as in Amos 5:8. A telescope reveals many more stars than seven; there are over 500 stars in the Pleiades cluster.

Also in Taurus is the open cluster Hyades (hī′·ə·dēz′). These star clusters are families of stars within the Milky Way. The Beehive cluster in the body of Cancer is another interesting open cluster. Between Cassiopeia and Perseus is a double cluster, so close together that they can both be seen in one view through a telescope.

A more remarkable kind of cluster is the globular cluster. In an open cluster, the stars are spaced widely enough to be seen individually with a small telescope. But in a globular cluster, thousands of stars are crowded so close together that they form a fuzzy "globe." The most impressive globular cluster of the northern sky is in Hercules. An even greater wonder is in Tucana (tōō·kā′·nə) near the south celestial pole. There are more than 100 globular clusters in the sky.

Nebulae. In the sword that hangs from the belt of Orion can be seen a misty patch of light. A telescope shows that it is a large body

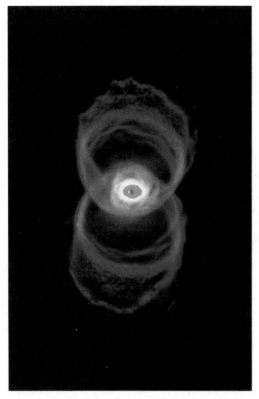

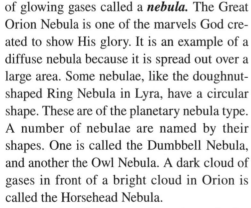

Hourglass Nebula in Musca

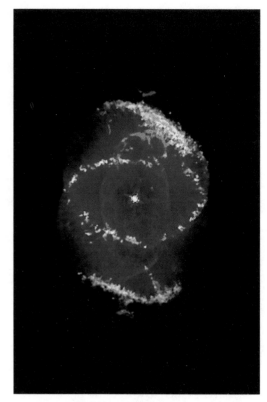

Cat's-eye Nebula in Draco

of glowing gases called a *nebula.* The Great Orion Nebula is one of the marvels God created to show His glory. It is an example of a diffuse nebula because it is spread out over a large area. Some nebulae, like the doughnut-shaped Ring Nebula in Lyra, have a circular shape. These are of the planetary nebula type. A number of nebulae are named by their shapes. One is called the Dumbbell Nebula, and another the Owl Nebula. A dark cloud of gases in front of a bright cloud in Orion is called the Horsehead Nebula.

Galaxies. There was a time when galaxies were thought to be nebulae. But the invention of big telescopes has revealed that galaxies are huge families of stars. The easiest galaxy to see is the Milky Way, which looks like a broad band of milkiness passing through Cygnus, Cassiopeia, Perseus, and Auriga.

The Milky Way appears milky because when we view that part of the sky, we are looking edgewise into the Milky Way galaxy. We see so many millions of stars that they look like a white mist. Our solar system and all the stars we can see are part of the Milky Way.

One other galaxy, located in the Andromeda constellation, is visible to the unaided eye. The Andromeda Galaxy looks like a fuzzy star and becomes a hazy patch in binoculars and small telescopes. But this galaxy and many others appear as huge masses of stars to those who use powerful telescopes. There are actually more galaxies than any person can count.

Top: This edgewise photo of the Milky Way was generated by infrared light. Our solar system is about two-thirds of the way out from the center.

Middle: Sombrero Galaxy (M104) is a spiral galaxy nearly edgewise in the constellation Virgo.

Bottom: Crown of Thorns (NGC 7049) is an interesting galaxy of type S0 in the Southern Hemisphere.

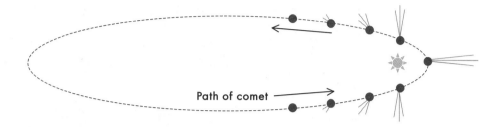

Path of comet ——→

Comets. The **comet** is an occasional visitor ranging from a tiny telescopic object to a huge body with a tail stretching halfway across the sky. A telescope will show a bright head that is the source of the vapor in the tail. A comet travels in a very elongated orbit around the sun. Periodically it returns to the vicinity of the sun, where the heat of the sun produces vapor, and the solar wind sweeps it into a long tail away from the sun.

After the comet swings around the sun, it departs tail first and the tail gradually disappears, not to form again until its next visit to the sun. The famous Halley's Comet makes a regular appearance at about 76-year intervals.

It last appeared in 1986 and will probably be visible again for several weeks in 2061.

Wonders in the atmosphere. "Falling stars" are not really stars, but pieces of rock that enter the earth's atmosphere. They move so fast that they burn up from the heat of friction with air. The glowing streak that you see is called a *meteor.* If the rock reaches the ground, it is called a *meteorite.* Very few meteorites have been found, since not many rocks survive the fiery descent through the atmosphere. No one has been known to be killed by a meteorite. God has protected us with an atmosphere that causes these rocks to burn up in their swift plunge toward the earth.

Notable Meteor Showers				
	Duration	**Peak**	**Hourly Rate**	**Notes**
Leonids	Feb. 5–Mar. 19	Feb. 26	5	Slow, bright meteors
Leonids	Mar. 21–May 13	Apr. 17	5	Slow, bright meteors
Perseids/Arietids	May 30–June 18	June 7	30	Daybreak showers
Taurids	June 24–July 6	June 29	30	Daybreak showers
Capricornids	July 15–Aug. 25	Aug. 2	5	Slow, bright meteors
Perseids	July 23–Aug. 23	Aug. 12	30	Best-known shower
Draconids	Oct. 10	Oct. 10	5	Can be spectacular
Orionids	Oct. 2–Nov. 7	Oct. 21	20	Long-lasting trails
Taurids	Sept. 15–Nov. 26	Nov. 3	20	Fireball producer
Leonids	Nov. 14–20	Nov. 17	5–30	Irregular producer
Geminids	Dec. 4–16	Dec. 13	30	Many bright meteors

Sometimes the earth passes through a region of space that has thousands of these pieces of rock. The result is a meteor shower, which can be a spectacular display with meteors every few seconds. The showers listed on the accompanying table are annual events.

Meteor showers are named for the constellations from which the meteors seem to come. The hourly rate is the number of meteors an observer would see if the radiant (point of origin) were directly overhead on a dark night. Although the figures given on page 164 are rates that can be expected, the showers can vary considerably from year to year. For example, Leonids vary on a 33-year cycle. The last major display was in 1999.

Northern lights, called *aurora borealis,* are another wonder of the nighttime sky. They appear as glowing patches, curtains, or pulses of various colors in the northern sky. Similar displays in the far southern sky are called *aurora australis.* These spectacular lights can be seen when the solar wind is caught by the magnetic field of the earth. As the charged particles stream toward the North and South Poles, they cause the upper atmosphere to glow. The aurora displays are testimony of God's provision of the earth's magnetic field to protect living things against the charged, high-speed particles from the sun.

This beautiful scene is not a sunset but rather a display of northern lights. Occasionally such light shows are seen in the middle latitudes after intense sunspot activity.

──────── Study Exercises: Group I ────────

1. Arcturus is mentioned twice in the Book of Job (9:9 and 38:32).
   a. What kind of heavenly wonder is Arcturus?
   b. In which constellation is Arcturus?

2. Many people are unaware of variable stars because
   a. they are too dim to be seen without a telescope.
   b. they change very little in brightness.
   c. they change brightness very slowly.

3. Many people are unaware of binary stars because
   a. a telescope is needed to see the individual stars.
   b. there are very few of them.
   c. they are hard to locate in the sky.

4. What difference in stars do some binary stars help us to see?

5. Pleiades is mentioned twice in the Book of Job (9:9 and 38:31).
   a. What kind of heavenly wonder is Pleiades?
   b. God's question in Job 38:31 could be asked as shown below. Fill in the missing words.

   "Can you fasten the bands of ──────, or loosen the belt of ──────?"

6. Choose the correct words.
   a. Pleiades is (an open, a globular) cluster.
   b. The cluster in Hercules is (an open, a globular) cluster.
   c. The Ring Nebula is a (diffuse, planetary) nebula.
   d. The nebula in Orion is a (diffuse, planetary) nebula.

7. Name the heavenly wonder indicated by each description.
   a. Body accompanied by a tail-shaped cloud of gases when it is near the sun.
   b. Group of stars that can be seen as individual bodies or as a fuzzy ball.
   c. Large body of glowing gases.
   d. Two stars revolving around one another in space.
   e. Streak of light formed by a rock speeding into the earth's upper atmosphere.
   f. Huge family of stars that form an island in space.
   g. Night illumination of the upper atmosphere over northern regions.
   h. Exploding star.
   i. Night illumination over southern polar regions.
   j. Rock from space that survives its plunge through the earth's atmosphere.

8. The ────── ────── is a stream of high-speed charged particles from the sun that causes northern lights and the tail of a comet.

9. The ────── ────── of the earth protects living things from high-speed solar particles.

Unit 4 Review
Review of Vocabulary

To help in the study of —1—, the sky is divided into many star pictures, or —2—. Star maps give the positions of stars by using lines of —3— to give the degrees north or south of the equator, and lines of —4— to give the hours from the vernal equinox. Men use large telescopes housed in —5— to study the heavens. They have made various —6— to give mechanical demonstrations of how the heavenly bodies move.

From ancient times, men understood that the sun follows a path called the —7— as it passes through the constellations of the —8— each year. Spring begins with the —9—, and fall begins with the —10—, both of which have days and nights of equal length. The shortest day of the year, called the —11—, marks the beginning of winter; and the longest day of the year, called the —12—, marks the beginning of summer. It is right to watch the heavenly bodies to determine when the seasons begin, but studying the stars to practice —13— is condemned in the Bible.

Occasionally there is a —14—, when the moon hides part or all of the sun. During a —15—, the shadow of the earth darkens the moon. These events happen only at the new moon and full moon —16— of the moon. Large heavenly bodies that orbit the sun are called —17—; the ones outside the earth's orbit show —18— when the earth moves past them. A body that orbits a planet is called a —19— or moon. A long tail streams out from a —20— when it makes periodic visits to the sun.

The stars are so far away that the —21— is used to state the distances. A —22— is a star whose —23—, or brightness, increases greatly in a short time. A cloud of glowing gases among the stars is called a —24—. Some fuzzy patches are not glowing gases but huge, distant families of stars called —25—.

Some sky wonders are in the earth's atmosphere. Falling rocks from space leave streaks of light called —26— as they burn up in the atmosphere, though a few reach the earth as —27—. Another atmospheric wonder is the northern lights, or —28—, caused when the magnetic field of the earth attracts high-speed solar particles. The same kind of light in the southern sky is called —29—. All these wonders show us the glory of God in His handiwork.

Multiple Choice

1. If we never studied the stars, which of the following reasons that God created the stars would have the least effect on our lives?
   a. God created the stars to mark off units of time.
   b. God created the stars to provide heat and light.
   c. God created the stars to show us His glory.

2. The study of the parallax of stars shows that
   a. the number of stars is too great to be known.
   b. the brighter the star, the smaller the magnitude number.
   c. the diameter of the earth's orbit is millions of miles.
   d. the distances to the stars are best measured in units such as light-years.

3. Which of these statements is true?
   a. Orion is a low-magnitude star in the constellation Ursa Major.
   b. Arcturus is a bright star in the constellation Cygnus.
   c. Gemini is one of the constellations of the zodiac.
   d. Mira is a binary star in the constellation Cetus.

4. In beginning to learn the locations of stars on your own, it would be best to
   a. obtain a flashlight that produces a narrow beam for pointing to stars.
   b. obtain a good-quality telescope with a sturdy mount.
   c. obtain a star-projector planetarium.
   d. obtain a star guide and learn how to use it.

5. Which of the following is *not* an important principle about sky observation?
   a. A red light hinders your night vision less than other colors do.
   b. Never look at the sun through a binocular or a telescope.
   c. Memorize many constellations on a star map before looking for them in the sky.
   d. It will take about ten minutes before you can see some of the faint stars.

6. Which line on the star map is labeled in hours?
   a. right ascension   c. declination
   b. ecliptic   d. equator

7. Which of the following aspects of a telescope is least important?
   a. the base on which the telescope is mounted
   b. the diameter of the main lens or mirror
   c. the quality of the lens or mirror
   d. the magnifying power of the lens

8. Which of these statements is false?
   a. A large planetarium in California has a 200-inch reflecting telescope.
   b. The sun and stars are still used to establish units of time.
   c. On a sundial, a shadow cast by a gnomon indicates the time of day.
   d. The moon orbits the earth about twelve times each year.

9. Which body rotates on its axis once every 24 hours?
   a. sun   b. earth   c. moon   d. Jupiter

10. People use standard time because
    a. sun time varies from one part of the year to another.
    b. sun time varies from one place on the earth to another.
    c. standard time allows clocks around the world to be set to the same time.
    d. standard time gives people more daylight hours.

11. The ecliptic is most closely related to
    a. the eclipses.   c. the zodiac.
    b. the Milky Way.   d. the tides.

12. A telescope is of little value for observing
    a. the moon.   b. a comet.   c. Jupiter.   d. a meteor.

13. Which of these facts is related to the welfare of living things on the earth?
    a. The gravity of the moon causes ocean tides.
    b. The earth's magnetic field affects the solar wind.
    c. The phases of the moon follow a regular cycle.
    d. The sun appears to pass through the twelve constellations of the zodiac.

14. To learn the schedule of the tides, it would be best to know
    a. the phase of the moon.
    b. the day of the year.
    c. the location of the sun in the zodiac.
    d. the tilt of the earth.

15. Which of these facts has the least to do with the seasons?
    a. The earth revolves around the sun once every 365¼ days.
    b. The earth's axis is tilted 23½ degrees.
    c. An equinox occurs when the sun is directly above the equator.
    d. The moon revolves around the earth every 29½ days.

16. Which one of these statements is entirely correct?
    a. The first day of summer, June 21, is the vernal equinox.
    b. The winter solstice on November 22 is the shortest day of the year.
    c. On the autumnal equinox, September 22, the day and night are equal.
    d. The first day of spring, March 21, is the longest day of the year.

17. If you were at a place south of the Antarctic Circle on December 22,
    a. the sun would pass directly overhead.
    b. the sun would not rise that day.
    c. the sun would not set that day.
    d. the sun would rise at noon.

18. Which of these descriptions is correct?
    a. The sun is a star.   c. The Milky Way is a star cluster.
    b. The moon is a planet.   d. Pleiades is a nebula.

19. Which form of energy is most closely related to the way the sun produces energy?
    a. heat energy   c. atomic energy
    b. light energy   d. chemical energy

20. Which star name is *not* mentioned in the Bible?
    a. Pleiades   c. Orion
    b. Cygnus   d. Arcturus

21. Which of these questions is the only one to which you can answer yes?
    a. Can man know how the stars came to be in the heavens?
    b. Can man cause the stars to move together or apart?
    c. Is it ever right to worship a star?
    d. Does man have a right to feel proud of his exploration of the planets?

Second Quarter Review: Units 1–4

Matching

Write the letter of the word that matches each description.

1. Disorder and randomness in energy and materials.
2. A cloudy mixture in which particles of one material are held in another material.
3. The study of energy and motion.
4. A blood vessel that carries blood away from the heart.
5. A body unit consisting of various tissues working together.
6. The outer layer of the skin.

a. artery
b. entropy
c. epidermis
d. organ
e. physics
f. solution
g. suspension
h. vein

7. A simple carbohydrate.
8. The unit for measuring the energy provided by food.
9. A condition caused by a lack of iodine.
10. A star picture in the sky.
11. One of the apparent shapes of the moon.
12. The practice of studying the stars to foretell the future.

i. astrology
j. astronomy
k. calorie
l. constellation
m. goiter
n. phase
o. starch
p. sugar

Completion

Write the correct word to complete each sentence.

13. Stored energy is called ——— energy.
14. The outward force that results when an object moves in a curved path is ——— force.
15. According to the first law of ———, energy cannot be created or destroyed.
16. At a joint, the bones are held together with strong bands called ———.
17. The body system that includes the skin is the ——— system.
18. Muscles that cause joints to unbend are called ——— muscles.
19. Some micronutrients are organic substances called ———.
20. Fingerlike ——— in the small intestine absorb food into the bloodstream.
21. An enzyme in saliva changes starch into ———.
22. A huge family of stars is called a ———.
23. A ——— is a mechanical device that shows how the heavenly bodies move.
24. A planet in ——— motion appears to move backward in its path among the stars.

Multiple Choice

Write the letter of the correct choice.

25. A material that has a definite size and shape is
    a. a solid.   b. a liquid.   c. a gas.

26. The smallest particle of a compound is
    a. an atom.   b. an element.   c. a molecule.

27. A suspension of one liquid in another liquid is called
    a. an alloy.   b. an emulsion.   c. a surfactant.

28. The —— system provides drainage and defense for the body.
    a. digestive   c. lymphatic
    b. circulatory   d. respiratory

29. The respiratory system does *not* include
    a. the lungs.   c. the alveoli.
    b. the villi.   d. the diaphragm.

30. Which part of the blood is most important for resisting infections?
    a. plasma   c. red corpuscles
    b. platelets   d. white corpuscles

31. The body uses all the following things for food *except*
    a. cellulose.   c. fats.
    b. carbohydrates.   d. protein.

32. An important mineral for helping the blood to carry oxygen is
    a. calcium.   c. iron.
    b. copper.   d. potassium.

33. A lack of vitamin C causes
    a. beriberi.   c. rickets.
    b. pellagra.   d. scurvy.

34. The ecliptic is most closely related to
    a. the eclipses.   c. the zodiac.
    b. the tides.   d. the Milky Way.

35. A light of which color is best to use for checking a star map at night?
    a. blue   c. green
    b. red   d. white

36. The northern lights are called
    a. novas.   c. aurora australis.
    b. nebulae.   d. aurora borealis.

Unit 5

Heat

Old Faithful in Yellowstone National Park is a very famous geyser. A geyser forms where groundwater drains through channels deep into the earth. As hot rock heats the water, the temperature goes far above the boiling point due to the pressure of the water above. Finally some of the water turns into steam. It expands enough to push some water to the surface. This then lightens the column of water, and more water turns into steam. The cycle escalates, and suddenly the bottom part of the channel explodes into steam and forcibly shoots everything out.

Searching for Truth

"Behold, I have created the smith that bloweth the coals in the fire, and that bringeth forth an instrument for his work; and I have created the waster to destroy" (Isaiah 54:16).

Man is able to make many instruments, including weapons, by the use of heat. People are easily intimidated by these, but this verse points out that God made both the "smith" (metalworker) and the "waster" (destroyer). We know that He also established the principles that govern heat and its uses. All these things assure us that God is in control and that He is well able to care for His people.

Searching for Understanding

1. What are you adding to a material when you heat it to make it hot?
2. Which contains more heat, an iceberg or a drop of molten iron?
3. When you take a cake out of the oven, why does it burn your finger to touch the pan but not the cake?
4. If materials generally become denser as they get colder, why does ice float?
5. How can ice be produced in the middle of the summer?

Searching for Meaning

bimetallic strip	evaporation	sublimation
British thermal unit	Fahrenheit	temperature
calorie	heat	thermistor
Celsius	latent heat	thermocouple
condensation	radiation	thermometer
conduction	refrigerant	thermostat
convection	refrigeration unit	volatile
distillation	specific heat	

Heat, an Important Form of Energy

Men have invented many devices to add heat and make things hotter, such as the stove, water heater, and welder. They have also made many devices to take away heat and make things cooler, such as the refrigerator, freezer, and air conditioner. Heat can melt ice, expand materials, and cook food. At some places, the temperature of air can vary as much as 100 degrees Fahrenheit (55 degrees Celsius) from summer to winter. But when your body temperature increases more than a few degrees, you are very sick. Heat has a very important place in our lives. In this unit, you will study principles that God has established for heat energy.

Heat Is Energy

Matter has weight and takes up space. Suppose you weigh a piece of cold iron and then make it red hot. Do you think its weight would increase, decrease, or stay the same? The iron would expand slightly, but you would not be able to detect any increase in its weight. When you add heat to a material, you are adding energy, not matter. How is it that a material has more energy when it is hot? What happens when it becomes cold?

All materials are composed of molecules that are in constant motion. Sometimes the molecules are moving faster than at other times. This kinetic energy of the molecules in motion is *heat.* As heat energy is added, the molecules move faster. The sensation you get when you touch something hot is caused by high-speed molecules striking the nerve

endings in your skin. As a material becomes colder, the molecules move more slowly.

Temperature. If heat is the kinetic energy of molecules in motion, what is temperature? It is a measure of how fast the molecules in a material are moving. The speed of the molecules determines the amount of heat and therefore the energy level. Therefore, *temperature* is a measure of the internal energy level of a material. The higher the number of degrees, the higher the speed of the molecules and the higher the internal energy level.

A *thermometer* is a device for measuring temperature. The degree markings on a thermometer are called the temperature scale. There are several temperature scales in use today. A man by the name of Gabriel D. Fahrenheit (1686–1736) established the scale now called the *Fahrenheit* scale. This temperature scale is still commonly used in the United States. On this scale, water freezes at 32 degrees and boils at 212 degrees. Salt water freezes at 0 degrees, but nothing in particular happens at 100 degrees.

The small size of the degrees on the Fahrenheit scale makes it useful for measuring body and air temperature. A simpler scale, however, is the *Celsius* scale (named for Anders Celsius), on which 0 degrees is the freezing point of water and 100 degrees is the boiling point

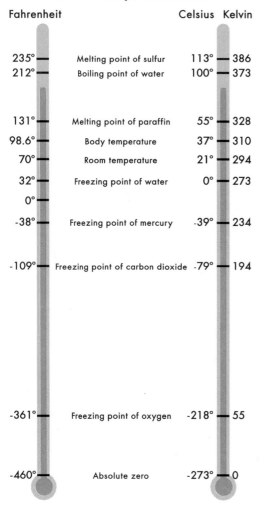

Comparison of Three Common Temperature Scales

Fahrenheit		Celsius	Kelvin
235°	Melting point of sulfur	113°	386
212°	Boiling point of water	100°	373
131°	Melting point of paraffin	55°	328
98.6°	Body temperature	37°	310
70°	Room temperature	21°	294
32°	Freezing point of water	0°	273
0°			
-38°	Freezing point of mercury	-39°	234
-109°	Freezing point of carbon dioxide	-79°	194
-361°	Freezing point of oxygen	-218°	55
-460°	Absolute zero	-273°	0

Temperature Conversion	
From Celsius to Fahrenheit	From Fahrenheit to Celsius
$F = \frac{9}{5} C + 32$	$C = \frac{5}{9} (F - 32)$
(Multiply the Celsius temperature by $\frac{9}{5}$, and add 32 to the result.)	(Subtract 32 from the Fahrenheit temperature, and multiply the result by $\frac{5}{9}$.)

of water. These two scales are compared in the diagram on page 174, which shows important points on both scales.

Sometimes a temperature is given in one scale, and you want to know the temperature on the other scale. To change from one scale to the other, use the methods on page 174.

On both the Fahrenheit and the Celsius scale, the numbers do not give the true relationship of the temperatures. For example, 50 degrees is not twice as warm as 25 degrees. Scientists prefer the Kelvin scale (named for Lord Kelvin) because it does show true temperature relationships. On this scale, 0 Kelvin is absolute zero. Nothing can be colder than absolute zero because molecules at that temperature have no heat energy at all. So an object at 200 K has twice as much heat as at 100 K. (Note that temperatures on the Kelvin scale omit the degree symbol. The temperature is given in Kelvins rather than in degrees.)

Kelvins on the Kelvin scale are the same size as degrees on the Celsius scale. So the Kelvin temperature can be found by simply adding 273 to the Celsius temperature.

—————— Study Exercises: Group A ——————

1. Give four ways that heat is used in your home.

2. Heat is —— and does not have —— or take up space.

3. In a hot object, the molecules are moving (faster, slower) than in a cold object of the same material.

4. What device is used to measure the internal energy level of a material?

5. What name is given to the units on the Fahrenheit and Celsius scales?

6. Name the temperature scale
   a. that is commonly used in the United States.
   b. that has special points at 0° and 100°.
   c. that is preferred by scientists.
   d. that has the smallest size of degrees.
   e. on which 200 is twice as hot as 100.
   f. on which absolute zero is –273°.

7. Find the equivalent temperatures indicated.
   a. 113°F on the Celsius scale
   b. 446°F on the Celsius scale
   c. 10°C on the Fahrenheit scale
   d. 20°C on the Kelvin scale
   e. Extra: –40°C on the Fahrenheit scale

8. What is absolute about absolute zero?

Group A Activity

Observing the motion of molecules. Fill one glass with very hot water and another glass with very cold water. Let both stand for about a minute until the water in them is very still. Now put several drops of food coloring in each glass. How long does it take for the color to spread uniformly through the hot water? through the cold water? What causes the difference?

Heat and the States of Matter

Temperature is directly related to the three states of matter: solid, liquid, and gas. In a cold material, the molecules are relatively close together and jostle one another within a small space. This controlled movement of the molecules gives solids their ability to hold a definite shape. When heat is added to a solid, the molecules increase in energy until they reach a temperature at which they no longer stay within a definite space. They begin sliding over each other and take the form called the liquid state. The temperature at which a solid becomes a liquid is called its melting point.

If you are cooling a liquid to make it a solid, the temperature at which it becomes a solid is called the freezing point. The melting point and the freezing point of a material are the same. For example, water freezes at 32°F and also melts at 32°F. Page 177 lists the melting and freezing points of several common materials.

A solid changes to a liquid when enough heat is added to raise its temperature to its melting point. However, some solids (like wood) enter into a chemical reaction before they can be heated to their melting points.

Some alloys (mixtures of metals) have low melting points. "Wood's metal," an alloy containing 50% bismuth and 25% lead, melts at 158°F (70°C). This metal can be used in the valves of an automatic sprinkling system. If fire breaks out in a building protected by such a system, the heat melts the bismuth alloy, the valve opens, and water sprays out. Alloys with low melting points are also used in electrical fuses, fire alarms, and safety plugs for boilers.

	The Three States of Matter		
Phase Change	**Name**	**Examples**	
Gas → liquid	Condensation	Formation of dew, liquefaction of carbon dioxide	
Gas → solid	Sublimation	Formation of frost and snow	
Liquid → gas	Evaporation	Evaporation of water or refrigerant	
Liquid → solid	Freezing	Freezing of water or a liquid metal	
Solid → gas	Sublimation	Sublimation of dry ice, freeze-drying of coffee	
Solid → liquid	Melting	Melting of snow and ice	

In a solid, the molecules vibrate very little and maintain their position in relation to each other, giving the solid its rigidity. In a liquid, the molecules move freely but maintain close contact with each other, giving a liquid its ability to flow. In a gas, the molecules fly about freely and have no contact with each other.

Melting and Freezing Points of Common Materials

Material	Temperature	
Iron	2,795°F	1,535°C
Copper	1,981°F	1,083°C
Aluminum	1,220°F	660°C
Zinc	786°F	419°C
Lead	621°F	327°C
Sulfur	235°F	113°C
Paraffin	131°F	55°C
Water	32°F	0°C
Mercury	–38°F	–39°C
Alcohol (ethyl)	–179°F	–117°C

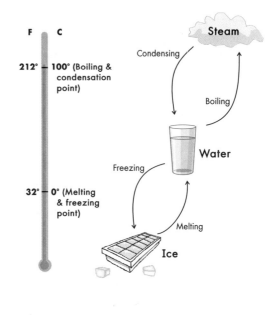

As more heat is added to a liquid, it eventually becomes so hot that the molecules no longer hold together as a liquid. Instead, they separate from each other and spread out in the surrounding space. Matter in this state is called a gas.

The temperature at which a liquid boils or changes into a gas is called its boiling point. The boiling points of several common materials are given on the right.

Years ago, alcohol was added to water in engines as an antifreeze. As the water in the engine became hot, the alcohol tended to boil off because its boiling point was lower than that of water. In time, much of the alcohol was lost and the water was no longer protected from freezing. Today ethylene glycol is used as an antifreeze. Since its boiling point is considerably higher than that of water, it is the water that boils off first; the antifreeze remains. For this reason, the antifreeze used in modern cars is called permanent antifreeze.

Boiling Points of Common Materials

Material	Temperature	
Iron	5,432°F	3,000°C
Lead	3,171°F	1744°C
Sulfur	833°F	445°C
Mercury	675°F	357°C
Ethylene glycol	388°F	198°C
Turpentine	318°F	159°C
Water	212°F	100°C
Alcohol (ethyl)	173°F	78°C

The process of a gas changing back to a liquid when it is cooled is called *condensation.* You have probably noticed the droplets of condensed water on the outside of a cold pitcher.

The diagram on page 177 shows the relationships of boiling, condensing, melting, and freezing with reference to water.

Steam, as gaseous water is called, is really colorless and is not the white cloud of condensed water vapor that is commonly called steam. However, in most bodies of steam some condensation takes place, resulting in tiny droplets of liquid water that give the steam a cloudy appearance. There is much steam in the exhaust of a jet engine. You may have noticed that the white condensation trail (contrail) begins some distance behind a high-altitude jet.

―――――― Study Exercises: Group B ――――――

1. The melting point of a material is the temperature at which it will change from a ―――to a ―――.

2. The freezing point of gold is 1,063°C. What is the melting point of gold?

3. The process of boiling is opposite to the process of ―――.

4. For each action, write whether heat is *added* or *removed.*
   a. freezing
   b. condensation
   c. changing from liquid to gas
   d. melting
   e. changing from liquid to solid
   f. boiling

5. Give two uses for a metal with a low melting point.

6. Use the tables of temperature points to give the answers.
   a. Would alcohol boil if it were placed in molten paraffin?
   b. Would paraffin melt if it were placed in boiling water?
   c. If a mixture of turpentine and alcohol were heated, which would boil first?
   d. If aluminum and zinc were heated together, which would melt first?
   e. Could you melt copper in an aluminum container?
   f. Temperatures in Antarctica, which can go below –80°C, could be measured with (a mercury, an alcohol) thermometer.

7. What property of ethylene glycol makes it better than alcohol as an antifreeze in engines?

Group B Activities

1. *Observing clear steam.* Make a steam generator with a flask or metal can and an alcohol burner. Put a stopper with a glass or metal nozzle in the opening of the generator. Heat the water in the generator very vigorously. Examine the steam that is escaping from the nozzle. How far from the end of the nozzle does the steam begin to condense?

2. *Changing the freezing point of water.* In each of six drinking cups, put 10 teaspoons of water. To each, add one of the following: 2 teaspoons of alcohol, 2 teaspoons of salt, 2 teaspoons of sugar, 2 teaspoons of antifreeze, 3 teaspoons of antifreeze, and 4 teaspoons of antifreeze. Place them all in a freezer overnight. Then set out the containers on a table, and observe their rate of melting. When there is still some solid left in the melted liquid, measure the temperature while stirring. This will give the approximate melting point. Compare the melting points of the different mixtures.

Evaporation. The passing of a liquid into a vapor or gas is called *evaporation.* Boiling is rapid evaporation throughout a liquid. But water will evaporate slowly even if it is not at its boiling point. During evaporation, some of the molecules on the surface of the water gain enough energy to fly out of the water into the air.

Two things that affect the rate of evaporation are temperature and humidity (water vapor in the air). The higher the temperature and the lower the humidity, the faster the evaporation. Clothes dry much more quickly on a dry day than on a humid day. Wind is also helpful because it brings more low-humidity air in contact with the liquid than if the air were still. Therefore, clothes on a line will dry the fastest on a warm, dry, windy day.

A third factor that affects the rate of evaporation is the bond between the surface molecules of a liquid, which is called surface tension. If the surface tension is weak, the liquid evaporates easily and is said to be *volatile* (vol′·ə·təl). Gasoline is very volatile because of its low surface tension. Diesel fuel is less volatile, and motor oil is hardly volatile at all.

If a liquid is very volatile and also flammable, its vapor is explosive. An example is gasoline; just a spark in the presence of gasoline vapors can set off a major explosion. Therefore, such liquids must be used with great care.

Sublimation. Some materials pass directly from the solid state to the gaseous state. Mothballs are an example of a solid that changes to a gas without first becoming a liquid. The process of changing from a solid directly to a gas or from a gas directly to a solid is called *sublimation.*

Carbon dioxide is another material that sublimes. Frozen carbon dioxide is very cold (–109°F or –78°C) and needs to be handled with care so it does not freeze the skin. It is called dry ice because it sublimes and does not become wet. Dry ice makes an excellent cooling agent for shipping frozen foods.

Camphor and iodine are two other materials that sublime. Even regular ice sublimes a little. Clothes hung out in freezing weather will first freeze stiff. But they will eventually become dry without the ice first melting. The water sublimes.

Distillation. A very useful application of boiling and condensation is the process of distillation. During *distillation,* a material is first changed into a gas by heating and then condensed into a liquid by cooling. Distillation requires a source of heat and a source of cooling.

Distillation is used to either purify or separate materials. For example, when hard water is boiled, only the water changes into a gas. The impurities in the water are left behind. By running this steam through a tube that is cooled with other water, a very pure form of water is

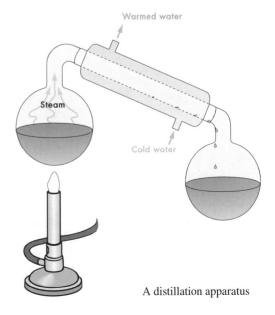

A distillation apparatus

obtained. Water purified by this process is sold in stores as distilled water. It is used in steam irons, since it does not contain minerals that will build up and clog the openings in the iron. Distilled water is also used to replace water in batteries and engine radiators.

Alcohol is purified by distillation. As a grain, such as corn or barley, ferments in water, ethyl alcohol is produced. The mixture of alcohol and water is then heated enough to boil the alcohol but not the water. The alcohol changes to a gas and is then condensed back into a liquid as it passes through water-cooled pipes. This is the source of the useful alcohol fuel and solvent that can be bought at hardware stores. Some alcohol is mixed with gasoline to make gasohol, which helps to conserve gasoline.

Ethyl alcohol is the kind that occurs in alcoholic beverages and causes drunkenness. The Bible has many warnings against this use of alcohol. If a manufacturer produces alcohol for purposes such as cleaning, he adds a poison so that it cannot be used for drinking. This is called denatured alcohol.

Fractional distillation. Liquids of different boiling points can be separated by fractional distillation. One example is the distillation of alcohol from water. Another very important application of fractional distillation is the separation of petroleum into its many parts. Crude petroleum is a mixture of different liquids, each having its own boiling point.

The petroleum is heated to change most of it into gases. These gases then enter a fractional distillation tower that has levels of decreasing temperatures. At the bottom of the tower where the temperature is highest, the tarlike asphalt condenses. Higher up, the greases condense, and still higher are lubricating oils, diesel fuel, kerosene, and heating fuel. Gasoline condenses near the top. In this way, many different useful products are obtained from petroleum.

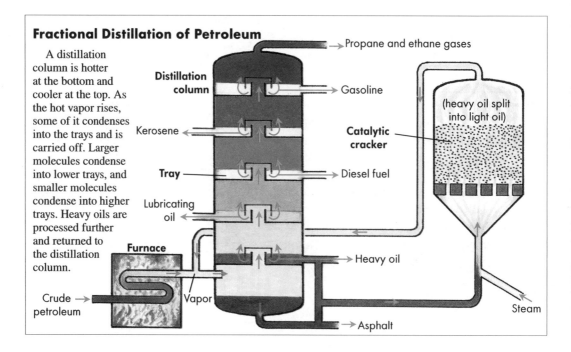

Fractional Distillation of Petroleum

A distillation column is hotter at the bottom and cooler at the top. As the hot vapor rises, some of it condenses into the trays and is carried off. Larger molecules condense into lower trays, and smaller molecules condense into higher trays. Heavy oils are processed further and returned to the distillation column.

Distillation column

Kerosene

Tray

Lubricating oil

Furnace

Crude petroleum

Vapor

Propane and ethane gases

Gasoline

Catalytic cracker

(heavy oil split into light oil)

Diesel fuel

Heavy oil

Steam

Asphalt

Destructive distillation. When soft coal is heated in the absence of oxygen, it gives off a gas that is then condensed into coal tar. This substance is used to make medicines and dyes. The solid that is left behind is coke, which is useful as a fuel in blast furnaces for removing iron from its ore. The process of driving off volatile materials with heat is called destructive distillation.

Charcoal is produced by the destructive distillation of wood. When wood is heated in the absence of oxygen, it cannot burn. Instead, the material in the wood decomposes. Most of the hydrogen, nitrogen, and oxygen escapes, and the main component left is carbon (charcoal). This is an important fuel for cooking in some countries, because charcoal burns with a small, hot flame and produces little smoke.

Charcoal is also valuable for absorbing various unwanted or harmful materials, such as in water filtering systems. It is used in gas masks to absorb poisonous gases and in certain medicines as an antidote to ingested poisons.

A modern coke factory

──────── Study Exercises: Group C ────────

1. A liquid will evaporate very slowly if the temperature is ——, the humidity is ——, and the surface tension is ——.

2. Write the correct word or phrase to name each process.
   a. Changing from a solid to a gas without becoming liquid.
   b. Changing from a liquid to a gas at any temperature.
   c. Changing from a liquid to a gas and back to a liquid.
   d. Tending to evaporate easily.
   e. Separating boiled liquids by controlling the temperature of condensation.
   f. Using heat to separate the volatile materials from a solid.

3. For each action, name the process from exercise 2 that would be used.
   a. Purifying water.
   b. Separating petroleum into various products.
   c. Drying clothes on an outdoor line.
   d. Producing coke and charcoal.
   e. Producing a gas to keep moths out of clothing.

4. Gasoline is more explosive than diesel fuel because it is more ——.

5. Why is frozen carbon dioxide called dry ice?

6. Why is the ethyl alcohol that is sold for fuel or solvent called denatured alcohol?

7. Write whether each phrase refers to *coke* or *charcoal*.
   a. made from wood
   b. made from coal
   c. used in blast furnaces
   d. used to cook food

Group C Activities

1. *Distilling water.* Set up a distillation apparatus as illustrated in this section. A flask makes a good boiler. A condensing chamber can be made with a plastic pipe having a two-hole stopper in each end. Put enough salt into the boiler water to make it taste very salty. Taste the distilled water. Why does it not taste salty? Why would it be very expensive to use distillation to make ocean water safe for irrigation?

Glass tubing

Plastic pipe

A simple condensation chamber

2. *Making charcoal.* Put several thin pieces of pine or spruce wood into a test tube. Close the test tube with a one-hole stopper that has a glass tube inserted. Heat the wood strongly with a propane torch. Ignite the volatile material that escapes from the glass tube. What is happening to the wood? How can you tell when the destructive distillation is finished?

3. *Observing the rate of evaporation.* Get 4 pans of the same shape and size. Put a cup of water in each of 3 pans, and a cup of rubbing alcohol (a volatile liquid) in the fourth pan. Place one pan of water in a warm, dry place and one pan of water in a cool, damp place, such as a basement. Keep one pan of water and the pan with rubbing alcohol in a room with normal temperature and humidity. (You could experiment with moving air by using additional containers with fans blowing over them.) After about 3 hours, measure the liquid in each container. Note how the different conditions affected the rate of evaporation.

4. *Observing the rate of sublimation.* Weigh several mothballs. Let them stand in a warm, open place. Weigh them each day to observe the amount that sublimes. Do the same for some mothballs placed in a small, tightly closed box. Why do the mothballs not sublime as fast in a closed container as in an open one? The effect is the same as that of humidity on evaporating water. As the air becomes filled with the molecules from the subliming material, fewer molecules will sublime.

How Much Heat?

The temperature of a material gives only the degree to which the material has been heated, but it does not tell the total amount of heat energy in the material. Suppose you had one piece of iron the size of a marble and another the size of a softball. If they were both 200°F, which do you think could melt more ice? It would be the larger piece, of course. Even though the two iron balls had the same temperature, the larger one would possess more heat. Even an iceberg possesses great quantities of heat. The heat of an iceberg could change more dry ice to carbon dioxide gas than a drop of molten iron could.

If two pieces of iron have the same size, the hotter object has more heat. Both the size and the temperature of an object help to determine how much heat it contains.

When comparing the amount of heat in two objects, it is helpful to have some unit for measuring heat. A unit called the *calorie* (cal.) is used to measure the amount of heat energy

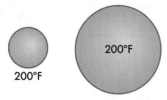

If two objects are of the same temperature and material, the larger object contains the greater amount of heat energy.

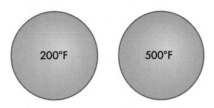

If two objects are of the same size and material, the hotter object contains the greater amount of heat energy.

that a material has and is able to give off. One calorie is the amount of heat needed to raise 1 gram of water 1 degree Celsius. This is not very much heat, since it takes a little more than 28 grams to make 1 ounce. Large amounts of heat are measured with a unit called the kilocalorie (kcal.), which equals 1,000 calories.

Since 1 calorie raises 1 gram of water 1°C, it takes 100 calories to raise 1 gram of water

from the freezing point to the boiling point. But if you have 1 gram of ice at the freezing point and start adding heat to melt it, you will find that it takes a total of 80 calories to change that ice to liquid water. What will the temperature of the water be after the ice is all melted? It will still be at the freezing point! The 80 calories you added to the ice did not raise the temperature one bit; it only melted the ice.

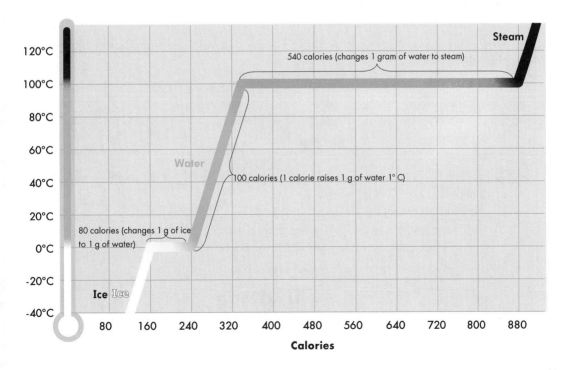

This shows why ice is a very good material to use for cooling. Every gram of ice will absorb 80 calories of heat as it melts, without getting any warmer. This also explains why it takes so long for water to turn into ice. In making ice, 80 calories of heat must be removed from every gram of water. Whether ice is being made or melted, there must be a transfer of 80 calories per gram of water.

Something similar happens when water reaches the boiling point. It takes about 540 calories of heat to change each gram of water into steam—but the temperature is still the same!

The calorie is a metric unit of measure. In the English system, the unit for measuring heat is called the **British thermal unit** (Btu). One Btu is the amount of heat needed to raise 1 pound of water 1 degree Fahrenheit. One Btu equals about 252 calories. Heaters and air conditioners are sometimes compared on the basis of their capacity in Btu's per hour.

—————— Study Exercises: Group D ——————

1. The amount of heat in an object depends on both its —— and its ——.

2. Write the number of the object in each pair that contains more heat.
   a. (1) a 2-pound piece of iron at 30°F *or* (2) a 2-pound piece of iron at 300°F
   b. (1) a 3-pound piece of copper at 250°F *or* (2) a 1-pound piece of copper at 250°F
   c. (1) a 500,000-ton iceberg *or* (2) 1 pound of molten aluminum

3. How much heat is 1 calorie?

4. How much heat is 1 kilocalorie?

5. How many calories are needed to change 1 gram of ice to liquid water?

6. How many calories are needed to bring 1 gram of liquid water from the freezing point to the boiling point?

7. How many calories are needed to change 1 gram of boiling water to steam?

8. Give the number of calories needed for each of these changes.
   a. to raise 60 grams of water from 20°C to 50°C
   b. to change 20 grams of ice to liquid water
   c. to change 5 grams of water at 100°C to steam

9. A ten-gram ice cube has more cooling ability than ten grams of ice water because
   a. the ice is colder than ice water.
   b. the ice cools over a longer period of time since it takes time to melt.
   c. the floating ice keeps the cold water near the surface.
   d. the ice water contains more heat than the ice cube.

10. What name is used for the unit of heat needed to raise 1 pound of water 1°F?

Group D Activities

1. *Computing the calories in a gram of steam.* With simple equipment as shown in the picture, you can demonstrate that it takes 540 calories per gram to change water into steam. This is done by passing steam through cold water to see how much heat is absorbed in changing steam back to water. You will need to do all the measurements in the metric

system. If you do not have a Celsius thermometer, you can convert a Fahrenheit reading to Celsius with the formula $C = \frac{5}{9}(F - 32)$.

For this experiment, prepare a data sheet like the one below. Do the measurements very carefully and the calculations accurately. If you use at least ¾ cup of cold water in an insulated cup at the beginning of the experiment, you will get reasonably close to 540 calories. Do not expect to get exactly this number. Discuss in class the possible causes of error in this experiment. Suggest ways to make the results more accurate.

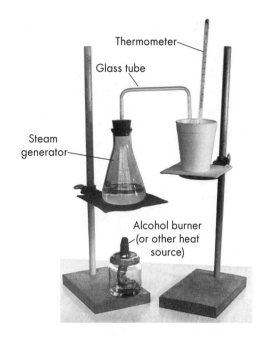

Thermometer

Glass tube

Steam generator

Alcohol burner (or other heat source)

Data Sheet	
A. Weight of water in insulated cup.	_____ g
B. Temperature of water in insulated cup.	_____ °C
C. Weight of water in insulated cup after steam has bubbled in it about 5 minutes.	_____ g
D. Temperature of water in item C.	_____ °C
E. Weight of steam that condensed into water (item C minus item A).	_____ g
F. Change in temperature of water (item D minus item B).	_____ °C
G. Number of calories absorbed by water in insulated cup (item A times item F). (Each gram of water raised 1°C equals 1 calorie.)	_____ cal.
H. Number of calories removed from each gram of steam (item G divided by item E).	_____ cal.

2. *Finding the calories needed to melt a gram of ice.* Design an experiment similar to the activity described above. Use an ice cube instead of the steam generator. You will need to make the following changes on the data sheet: (C) Weight of water in insulated cup after ice cube is added; (D) Temperature of water after ice cube is melted; (E) Weight of ice cube; (F) Change in temperature of water (item B minus item D); (G) Number of calories given up by water in insulated cup. (Each gram of water lowered 1°C equals 1 calorie.); and (H) Number of calories needed to melt each gram of ice.

Latent heat. When a gram of ice turns to water, it absorbs 80 calories of heat without becoming any warmer. The added heat is used to change the water from one state to another. Something similar happens when boiling water changes to steam: heat is added without increasing the temperature. In both cases, the extra or "hidden" energy breaks the bonds holding the molecules together and allows them to move more freely. This hidden energy is called *latent heat* (lāt′·ənt).

The latent heat in liquid water is very useful to keep plants from freezing. If an orchardist or a gardener sprinkles water on his plants on a cold morning, every gram of water that freezes will give off 80 calories of heat energy. That is enough heat to keep the blossoms and leaves from freezing, even though they may be covered with ice. The freezing point of the juices in the cells is a little lower than the freezing point of plain water.

You may have used the latent heat of water to make ice cream. Crushed ice is put around the tank of an ice cream freezer. Then salt is sprinkled on the ice, which forces the ice to melt. Every gram of ice that melts must get 80 calories of heat from its surroundings to become liquid water. This lowers the temperature of the tank and freezes the ice cream.

Road salt (calcium chloride) is often put on roads in winter to force snow and ice to melt. The melting ice makes the road colder; but the road is safer, since it is covered with liquid water instead of snow or ice.

The latent heat of water vapor explains why evaporation is a cooling process. When water is exposed to air, high-energy molecules jump into the air and become a gas. For every gram of water that evaporates, 540 calories of heat are taken from the liquid water. This makes the liquid cooler. God put this principle to good use in the human body. When the body becomes overheated, the skin produces sweat to cool it by evaporation. That is why air blowing over your sweating body makes you feel cooler.

In very dry climates, as in the southwestern United States, the air in houses can be cooled by evaporation. A fan draws air through a wet screen and blows it into the house. The dry moving air evaporates the water so quickly that many calories are removed from the air. This makes the air cooler and also raises its

Specific Heat of Selected Materials	
Specific heat = Calories needed to raise 1 gram of given material 1 degree Celsius	
Material	**Calories**
hydrogen	2.418
ammonia	1.125
water (liquid)	1.00
glycerin	0.576
ethyl alcohol	0.54
ice	0.50
steam	0.48
air	0.24
aluminum	0.22
glass	0.16
iron	0.12
copper	0.092
mercury	0.033
lead	0.031

humidity. Both make the air in the house more comfortable.

Specific heat. Some materials absorb more calories of heat than others as they become warmer. One of the very best absorbers of heat is water. The number of calories needed to raise 1 gram of a material 1 degree Celsius is called its *specific heat.* The specific heat of water is 1, since every calorie added causes a temperature increase of 1 degree Celsius. Iron, with a specific heat of 0.12, absorbs less heat than water; 0.12 calorie will raise 1 gram of iron 1 degree Celsius. The table on page 186 gives the specific heat of various materials.

This table shows why water is such a good material to circulate in heating and cooling systems. Many solar collectors use water to store heat from the sun. The oceans and lakes are excellent natural solar collectors. They can absorb great quantities of heat and then release that heat at night and in winter. God wisely gave water a high specific heat.

——————— Study Exercises: Group E ———————

1. Adding heat can make a liquid hotter and thus raise its ———, or it can make a liquid boil and thus change its ———.

2. Which two things involve the adding of latent heat?
   a. Bringing water from the boiling point to the freezing point.
   b. Changing liquid water to ice.
   c. Changing steam or water vapor to liquid water.
   d. Changing ice to liquid water.
   e. Changing liquid water to steam.
   f. Bringing water from the freezing point to the boiling point.

3. As a bucket of water freezes, it (absorbs, releases) heat.

4. Putting salt on ice makes it colder because
   a. the salt absorbs heat from the ice.
   b. the salt forces the ice to melt.
   c. the salt makes the ice freeze more completely.
   d. the salt raises the freezing point of the ice.

5. What does sweat do to make you cooler?

6. Why does blowing on hot soup cool the soup?

7. Why would an aluminum spoon take more heat out of a cup of hot chocolate than an iron spoon would?

8. Which item in each pair contains more heat if both are at the same temperature?
   a. a pound of air or a pound of steam
   b. a pound of water or a pound of alcohol
   c. a pound of copper or a pound of glass

9. The specific heat of liquid ammonia is 1.125. This means that 1 gram will gain ——— calories with a temperature increase of ——— degree(s) Celsius.

Group E Activities

1. *Making ice colder.* Make some crushed ice by pounding ice cubes in a cloth with a hammer. Sprinkle salt liberally on the crushed ice. Measure the temperature of the ice and salt mixture. How low does the temperature go? The lowest temperature attainable by this method is the basis for 0 degrees on the Fahrenheit scale.

2. *Observing the cooling effect of evaporation.* Cover the bulb of a thermometer with a small piece of cotton cloth. Tie it fast just above the bulb. Let the thermometer stand until it shows the room temperature, and record this beginning temperature. Then stick the bulb of the thermometer in alcohol so that the cloth is soaked with alcohol. Let the thermometer stand for 1 minute, and record the temperature.

   Remove the cloth, and let the thermometer return to the beginning temperature. Again tie the cloth to the thermometer, and soak it with alcohol, but this time set it in a strong breeze from a fan for 1 minute. Now note the temperature. Why is it different from the temperature after the first test?

   Repeat the experiment with water instead of alcohol. Why are the results different?

Expansion

Solids, liquids, and gases expand as their temperature increases. This principle is the basis of the thermometer. As the liquid mercury or alcohol in a thermometer becomes warmer, it expands and travels up the slender tube. A sensitive air-bulb thermometer uses the expansion of air to measure temperature. As the air in the bulb becomes warmer and expands, the water in the tube goes down.

The force of expanding solids is enough to break glass jars when they are heated unevenly. Rocks are sometimes cracked as they expand in the hot sun during the day and contract at night. The expansion of solids must be considered when building bridges, highways, and long pipelines. Usually expansion joints are

The finger joints at the beginning of this bridge allow about a foot (30 cm) of expansion or contraction.

built into a long structure to permit the parts to slide over one another.

Different materials expand different amounts when heated. The amount of expansion is stated as the coefficient of linear expansion. Multiplying that number times the length will give the amount of expansion for each Celsius degree of increase in temperature. That expansion may not seem like very much, but it is enough that it cannot be ignored.

This principle of expansion has many practical uses in the workplace. Mechanics use heat to expand the steel around a tight bolt so they can remove it, or to expand a tight bearing or collar so they can slide it onto a shaft (or take it off). Workers in steel construction heat rivets red hot before hammering them into place. As the rivets cool, they become very tight.

Contraction and expansion can be used to separate two water glasses that are stuck together. The inside glass is filled with cold water to make it contract, and the outside glass is lowered into hot water to make it expand. Then the water glasses can be separated easily. A tight lid can be removed from a jar by running hot water over the lid. The metal lid expands more than the glass and becomes loose enough to make removal easier.

Regular glass breaks easily. If one part of a glass jar is made much hotter than another part, the uneven expansion that results can break the jar. That is why you must be careful not to pour very hot liquid into a cold jar too quickly. It is why you must not put a glass dish on a hot stove. However, if the glass contains about 13 percent boric oxide (called borosilicate glass), the coefficient of linear expansion is only about one-third that of regular glass. This kind of glass is useful for kitchen and

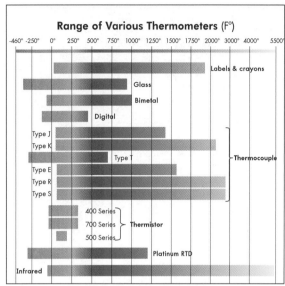

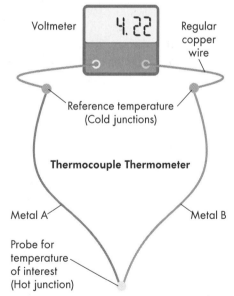

A simple circuit with two different metal wires joined at one end generates electric current when heat is applied to the junction. The hot temperature is calculated from the amount of current and the reference temperature. If such a circuit is attached to a battery instead of a voltmeter, the junction becomes warm and the reference points become cool. Special generators and certain refrigeration units employ these principles, but the current technology has too little efficiency for widespread adoption.

Coefficient of Linear Expansion for Common Materials	
aluminum	0.000023
brass	0.000019
copper	0.000017
nickel	0.000013
steel	0.000013
iron	0.000012
glass (ordinary)	0.000009
glass (borosilicate*)	0.0000032

*such as Pyrex

Problem: What will be the expansion of a piece of aluminum 20 feet long (6 m) if its temperature is increased 30°C?

Solution: The coefficient of linear expansion for aluminum is 0.000023.

$$20 \times 30 \times 0.000023 = \textbf{0.0138 foot}$$
$$0.0138 \text{ foot} = \textbf{0.1656 inch}$$

Metric: $6 \times 30 \times 0.000023 = \textbf{0.00414 m (0.414 cm)}$

laboratory utensils that need to withstand great changes in temperature.

The bimetallic strip. From the table, you can see that brass expands more than iron when the temperature is raised. If a strip is made with these two metals fastened together, the strip will bend when the temperature changes, since one side expands more than the other side. This device is called a *bimetallic strip,* since it is made of two metals.

Many thermometers contain a bimetallic strip that moves a pointer along a dial as the temperature changes. Electric circuit breakers have a bimetallic strip that opens a switch when the strip gets too hot. The common *thermostat* that regulates the temperature of a room, a toaster, an iron, or an oven contains an electric switch operated by a bimetallic strip.

An important exception. Almost all materials contract as they cool. That rule is also true for water until it reaches about 4°C. From that temperature down to freezing, the water actually expands. When the water is all frozen, it has expanded 9 percent. This outstanding exception to the contraction law has major consequences.

Freezing water produces great pressure—a pressure strong enough to burst water pipes. Water frozen in plastic jugs will cause the sides of the jug to bulge. Freezing water in a glass jar can break the jar. These are some problems caused by the expansion of freezing water.

But if it were not for this expansion, ice would not float. If ice did not float, streams and lakes would freeze solid from the bottom to the top in winter. This would kill fish, stop the rivers and streams, and then produce ruinous floods because the ice would thaw for months and months.

Common Thermocouples		
Types	**Kind of Metal Wires**	
B	+	70% platinum, 30% rhodium
	–	94% platinum, 6% rhodium
E	+	A nickel–chromium alloy
	–	A copper–nickel alloy
J	+	Iron
	–	A copper–nickel alloy
K	+	A nickel–chromium alloy
	–	A nickel–aluminum alloy
R	+	87% platinum, 13% rhodium
	–	Platinum
S	+	90% platinum, 10% rhodium
	–	Platinum
T	+	Copper
	–	A copper–nickel alloy

Here is a testimony to a wise Creator. God knew that living things depend on water. He knew that a cover of ice on the surface of water bodies would protect the living things in the water beneath the ice. He made an exception to the law of contraction and so protected living things from possible extinction. God reminded Job of this wise exception in Job 38:30: "The waters are hid as with a stone, and the face of the deep is frozen."

Other thermometers. Besides the liquid and bimetallic thermometers, there are several other kinds of thermometers. The filled-system thermometer has a curved metal tube that is filled with liquid or gas. An increase in temperature causes the liquid or gas to expand. The pressure forces the curve of the metal tube to flex outward. These changes move a pointer on a temperature dial.

A *thermocouple* thermometer has two kinds of metal wire joined to make one circuit. Varying amounts of heat cause the wires (thermocouple) to produce a varying current of electricity. A sensitive voltmeter measures the current and shows the temperature on a digital display. Thermocouples can measure a wide range of temperatures, from –300°F to 3,000°F (–185°C to

Clinical infrared thermometer

1,650°C). Different types have different kinds of wire suited for specific temperature ranges. Thermocouples are widely used in industry to measure temperatures of metals, plastics, furnaces, and many other things.

A *thermistor* thermometer has an electrical conductor that changes resistance with changes of temperature. When the temperature rises, the resistance increases. An electronic device detects these changes and indicates the temperature on a digital display. Thermistors are most sensitive and accurate in the temperature range of –40°F to 300°F (–40°C to 150°C). This makes them very useful for measuring air and body temperatures and for other biological purposes.

Infrared temperature measurement is becoming increasingly popular. Infrared thermometers measure heat energy that is absorbed by or radiated from an object. They are often called non-contact thermometers because they can measure temperature without touching the material. This makes them useful for moving objects such as rollers, conveyor belts, and bearings.

Liquid crystal thermometers are another kind. These indicate temperature by the different colors of certain materials that change from liquid to crystal form at certain temperatures.

———————— Study Exercises: Group F ————————

1. Most materials expand when they are (cooled, heated).

2. True or false? Some thermometers work on the basis of expanding solids, liquids, or gases.

3. What devices are used in long bridges to prevent damage due to changing temperatures?

4. The Golden Gate Bridge is 4,200 feet long (1,280 m) and is made of steel. The temperature at San Francisco, where this bridge is located, varies about 28°C. How many feet or meters would the bridge expand from the coldest to the warmest temperature?

5. A metal roof makes creaking noises when the heat of the sun strikes it. Which kind would tend to make more noises because of expansion: an aluminum or a steel roof?

6. If a glass jar is to be heated, why would it be better to heat it in the oven rather than on a burner on top of the stove?

7. Give two examples to show that the expansion of solids due to heat can be useful.

8. What property allows borosilicate glass to be used over the heating element of a stove?

9. Why does a bimetallic strip bend when it is heated?

10. Name two devices that contain bimetallic strips.

11. a. How is water an exception to the law of contraction?
    b. How does this exception demonstrate the wisdom of God?

12. A ——— is a device that works by electricity generated when heat is applied to two different kinds of wire joined together.

13. A ——— is a device with a conductor that changes in electrical resistance with changes in temperature.

Group F Activities

1. *Observing the contraction and expansion of air.* Inflate a round balloon, and fasten the opening. Measure the circumference of the balloon with a string held snugly around it. Take the balloon to the cold out-of-doors (if it is winter), or put it in a freezer for a while. Measure its circumference as before. What is the reason for the difference? Hold the balloon over a warm radiator or in front of an electric heater. Does this make it bigger or smaller?

2. *Making a liquid thermometer.* Fill a small glass bottle or bulb with colored water. Stop it with a one-hole stopper that has a small-diameter glass tube about one foot long. Be sure there is no air bubble left in the bottle. Tape a strip of heavy white paper behind the glass tube. Calibrate the thermometer by letting it stand beside a regular thermometer and marking various points along the scale. The water thermometer will take some time to adjust to new temperatures because of the quantity of water that needs to be heated or cooled. You can observe more rapid change in the level of the water by holding the bottle in your hand to warm it.

3. *Making an air-bulb thermometer.* Fasten a small glass bottle or bulb to one end of a foot-long glass tube with a one-hole stopper. Put the other end of the glass tube in a container of colored water. Hold your hands around the bulb until some air bubbles push out the bottom of the glass tube. Then as the bulb returns to room temperature, the colored water will come up the tube. The calibration of an air-bulb thermometer is upside-down from that of liquid thermometers. Air-bulb thermometers are affected by changes in air pressure and so are not accurate from day to day.

4. *Observing the expansion of metals.* Obtain a piece of steel, copper, or aluminum pipe with a diameter of ½ inch (1 cm) and a length of 2 feet (60 cm) or more. Clamp one end of the pipe to a board, as shown in the photograph. Put a piece of sheet glass under the other end. Push a needle or a long pin through the center of a straw, and place the needle between the pipe and the glass. Blow steam or the flame of a propane torch through the pipe. As the metal expands, the needle will turn. This motion will be amplified by the straw so that a very slight expansion can be detected. The straw will return to its original position as the pipe cools. Note: If the flame of a propane torch is used to heat the pipe, put some insulating material between the pipe and the board so the surface of the board does not become scorched.

Transfer of Heat

How does heat energy get from hot objects to cold objects? If you hold an iron rod with one end in a fire, the entire rod will eventually get hot. You know that a hot stove in one corner of a room will heat the whole room. You can feel the heat that travels from the sun on a clear day. These examples illustrate the three methods of transferring heat: conduction, convection, and radiation.

Conduction. The speeding molecules inside a material bump into each other. When a faster moving molecule bumps into a slower moving molecule, the second molecule receives more energy and begins moving faster also. In this way, heat travels through a material by the transfer of energy from one molecule to another. The transfer of heat in this way is called *conduction.* It is conduction that carries heat energy from a hot pan to the end of the handle.

Some materials conduct heat faster than others. Most solids conduct heat better than liquids or gases because the molecular structure is dense. Most metals are good conductors of heat energy. Silver, copper, and aluminum are the best conductors of heat. Aluminum makes very good cooking pans. Many pans are made of steel, which is not as good a conductor as aluminum. Sometimes a copper bottom is put on steel pans. The copper spreads the heat rapidly over the bottom of the pan. Many electronic devices have aluminum heat sinks with fins to conduct the heat away from parts that would be damaged by too much heat.

Materials such as wood and cloth are not good heat conductors. They are called heat

Heat Conductivity of Selected Materials			
Conductors	Calorie Rate	Insulators	Calorie Rate
Silver	100	Ice	0.40
Copper	92	Concrete	0.22
Aluminum	48	Glass	0.20
Brass	26	Water	0.14
Iron	16	Wood	0.02
Steel	11	Air	0.0057
		Vacuum	0.0

The rate is determined by the number of calories moving per second through a given length and diameter of a material at a certain temperature.

Cotton fibers magnified 30 times. The loose fibers create many air pockets and produce high R-value. The R-value is a measure of how well a material resists heat transfer. It is found by placing the material between two plates in a laboratory apparatus and measuring the heat flow through it.

insulators. Since they will not conduct heat very well, they hinder the transfer of heat by conduction. The common hot pad is designed to slow the conduction of heat to your hand or the table. The reason you can touch a hot cake but not the pan is because the cake is a good insulator but the pan is a good conductor.

The table above shows the relative heat conduction of various materials, with silver rated at 100. Nothing conducts heat better than silver.

Many houses have fiberglass insulation in the ceiling and outside walls to prevent heat from escaping in the winter. This fiberglass contains many air spaces. House insulation is rated by its resistance to heat flow (R-value), which gives a basis for comparing the insulating qualities of different materials and for knowing when a house has enough insulation. For example, fiberglass insulation 3½ inches thick is rated either R-11 or R-13. A rating of R-19 is recommended for house walls in areas with especially cold winters.

One of the best insulators is air. Thermal

windows have two sheets of glass separated by an insulating air space to slow the conduction of heat to the outside. Our clothes use air for insulation. The woven fibers of cloth contain many insulating air spaces. The air between layers of clothes makes additional insulation for the body.

The very best insulator is a vacuum because it contains no molecules to conduct heat. This is the principle of the vacuum bottle, which keeps drinks hot or cold by means of a partial vacuum between its double walls.

Convection. A method called convection is another means of transferring heat. *Convection* is the circulation of a liquid or gas due to uneven temperature. As a liquid or gas becomes warmer, it expands. This makes the warm fluid less dense than the surrounding liquid or gas. The less dense fluid is then pushed up by buoyant force just as a block of wood is pushed up by the denser water around it. This rising of warm fluid and sinking of cool fluid sets up a cycle of moving fluid called a convectional current.

Convectional currents cause the water to move about in a pan that is being heated on a stove. Convectional currents cause the air to rise above a heater or radiator, travel across the ceiling, and come down in the far corners of a room.

God uses giant convectional currents in the atmosphere to bring rain and other kinds of weather. You will read about the forming of convectional currents in the atmosphere when you study the unit on weather. Convection is a very useful way to move heat from one place to another and is a much faster method of heat transfer than conduction. Convection can happen only in fluids (liquids or gases).

Radiation. God has provided yet another way to transfer heat energy. The heat of the sun could not get to the earth by either conduction or convection, because the space between the sun and the earth is an almost perfect vacuum. God established *radiation* to transfer heat by the rays of the sun. Just how sunlight carries heat millions of miles through empty space is not fully understood. Light and similar heat-producing rays move out from hot objects and change back into heat when they are absorbed by other objects. In fact, all objects warmer than absolute zero emit some heat-producing rays. Radiation, the fastest method of transferring heat, operates at the speed of light. The heat from the sun gets to the earth in a little over 8 minutes.

Some materials absorb radiant heat energy better than other materials do. Black or dark objects absorb radiant heat better than white or light-colored objects do. A black asphalt road is hotter on a summer day than a white concrete road. Light-colored clothes are cooler on a sunny day than dark-colored clothes. While a shiny white metal may conduct heat well, it will absorb very little radiant heat.

All warm objects give off heat by radiation. Consider an open fire. Someone standing near the fire is warmed by radiation, not conduction. (Air is an insulator, a poor conductor.) Neither is the person warmed by convection, for the warm air currents are rising, not moving outward.

A surface that is a good absorber of heat is also a good radiator of heat energy. Thus, the roof of a black car radiates heat faster during the night than the roof of a white car. That is why frost forms sooner on the black roof than on the white roof. If an object is hot enough, it will also radiate light. That is why a hot black iron can glow red.

An infrared lamp radiates heat.

Some buildings are warmed by radiant electric or gas heaters. These serve especially well in mechanics' shops and other buildings where large doors are frequently opened. The radiant heaters are mounted near the ceiling, and they warm objects beneath them (the floor, equipment, and people) more than the air in the building. When the doors are opened, not as much heat is lost because it is stored in the objects more than in the inside air. Smaller radiant heaters are used for brooding poultry and piglets.

God has provided many sources of heat. The sources are usually concentrated, as in the sun or a fire. God has wisely provided three ways to transfer that heat to where it is needed.

Study Exercises: Group G

1. Write whether each description refers to *conduction, convection,* or *radiation.*
   a. Is the fastest of the three methods of heat transfer.
   b. Can take place only in a liquid or gas.
   c. Transfers heat through a vacuum or an empty space.
   d. Involves heat passing from one molecule to another in a material.
   e. Explains how a stove in one corner of a room can heat all the air in the room.
   f. Explains how the handle of a heated frying pan becomes hot.
   g. Explains how heat energy is transferred from the sun to the earth.
   h. Works on the principle that fluids expand and rise when they are heated.
   i. Takes place more rapidly in aluminum than in wood.
   j. Is reduced by a high R-value.

2. Why are many cooking and baking pans made of aluminum?

3. Name two good heat insulators.

4. Give two practical examples to illustrate a useful heat insulator.

5. Why is a vacuum a good insulator?

6. Fluids circulate by convection because (2 answers)
   a. warm fluids always move upward.
   b. cold fluids are more dense than warm fluids.
   c. buoyant force pulls cold fluids down.
   d. warm fluids do not mix with cold fluids.

7. When ventilating a warm room, why is it good to open a window at both the top and the bottom?

8. In a room heated by convection, the (ceiling, floor) will become warm first.

9. True or false? An object that is a good absorber of radiant energy is a poor radiator.

Group G Activities

1. *Observing the rate of conduction.* Fasten thumbtacks to a metal rod by dipping the heads into molten paraffin or candle wax and sticking them on the rod. Space the tacks every inch along the rod. Heat one end of the rod. When enough heat reaches each tack, the wax will melt and the tack will fall. If you can find equal diameter rods of aluminum, copper,

brass, and steel, fasten tacks to each of them and apply equal heat to the ends of all four. In this way you can observe different rates of conduction.

2. *Using a convection box.* Make a box 10 inches wide, 4 inches deep, and 16 inches high (25 cm by 10 cm by 40 cm), with clear plastic or glass over one of the 10-by-16-inch sides. Make sliding doors at the top and bottom of the two 4-by-16-inch sides (four doors in all). Make one of the lower doors big enough so that you can put your hand through. Put a short candle in the center of the bottom. Light the candle, and experiment with opening and closing the doors to answer the following questions.

    How long will the candle burn if all the doors are closed? Will the candle burn well with only one bottom door open? with both bottom doors open? with only one top door open? with both top doors open? What happens to the flame if only one top and one bottom door are open?

3. *Testing radiant heat absorption.* Get two identical bottles or bulbs that you can fit with one-hole stoppers. Paint one bottle flat black, or smoke it with a kerosene flame. Paint the other bottle white, or leave it unpainted. Connect the two bottles with a tube of clear glass or plastic. Place a small amount of colored water in the tube between the bottles. Now place a bright light bulb exactly halfway between the two bottles. Which way does the colored water in the tube move? What does this show? Place the two bottles in bright sunlight, and observe as before.

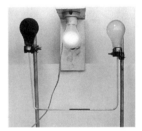

Heat and Pressure

As with other materials, gases expand when they are heated. This is what causes warm air to rise. But if a gas is heated in a closed container that does not allow it to expand, the pressure of the gas increases. Your mother may use heat to produce pressure in a pressure cooker. The pressure inside the container increases because the steam cannot escape.

When water is boiling and the steam is allowed to expand, the water never gets hotter than 212°F. Any water that does get hotter turns into steam and leaves the pan. But under pressure, as in a pressure cooker, the boiling point of water is higher. The water then gets hotter than 212°F and cooks the potatoes or meat faster. That increased temperature is the advantage of cooking with a pressure cooker.

An increase in pressure raises the boiling point of a liquid.

On the other hand, a decrease in pressure lowers the boiling point of a liquid. It takes longer to cook potatoes on the top of a high mountain than it does in the valley. This is because the air pressure on the mountaintop is lower and water boils at a lower temperature there. Of course, a pressure cooker would solve the problem of low pressure on a mountaintop.

The table on the next page shows the relationship of pressure to the boiling point of water. The first four lines show the increase in boiling point if a pressure cooker is used at sea level. At that altitude, the natural air pressure is 14.7 pounds per square inch, and water boils at 212°F if no further pressure is added.

Altitude in Feet	Pressure (Pounds Per Square Inch)	Boiling Point	
		F	C
	14.7 + 20	259°	126°
	14.7 + 15	250°	121°
0 (sea level)	14.7 + 10	239°	115°
	14.7 + 5	226°	108°
	14.7	212°	100°
2,000	13.7	208°	98°
4,000	12.7	205°	96°
6,000	11.8	201°	94°
10,000	10.2	194°	90°

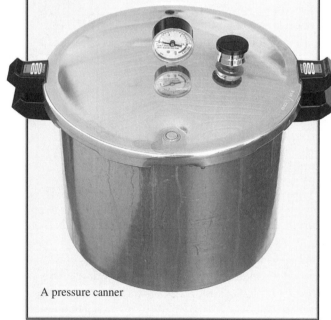

A pressure canner

rubber safety valves. If the pressure rises too high, the rubber valve will blow out and relieve the pressure. Have you ever noticed a label on an aerosol can, such as spray paint, warning you not to throw it into a fire? Heat can produce dangerous pressures.

Heat can also produce useful pressures. One of the most important uses of heat is for producing hot expanding gases to drive steam, gasoline, jet, and rocket engines. In piston engines, the pressure of the hot steam or burning gasoline is put to work in forcing the piston downward. Most transportation today depends on high-pressure gases from the heat of burning fuel.

Pressure and temperature. When a gas is compressed, the energy of the moving molecules is confined to a smaller space. This raises the temperature of the gas. For example, in a typical diesel engine, the pistons compress the air in the cylinders to $\frac{1}{22}$ (4.5%) of its original volume. This raises the temperature to over 1,500°F (815°C), which is hot enough to ignite the diesel fuel that the fuel injector sprays into the cylinders. Air compressors also get hot from the heat of compressing air. That is why air compressors have cooling fins around the pump.

When a gas expands, the energy is spread over a greater volume and the temperature drops. This cooling is used to make dry ice when compressed carbon dioxide gas is allowed to expand rapidly. The snowlike flakes can then be compressed to form blocks of dry ice. The cooling from expansion also has much to do with producing rain, as you will see in the unit about weather.

If the pressure cooker is operated at 5 pounds of pressure, the boiling point rises to 226°; at 10 pounds, it rises to 239°; and so on.

Heating a closed container can be dangerous. If the pressure becomes too great for the walls of the container to withstand, the container will explode. That is why pressure cookers have

—————————— Study Exercises: Group H ——————————

1. What happens to a gas if it is heated in a container that does not allow it to expand?

2. How does an increase of pressure affect the boiling point of a liquid?

3. True or false? The temperature of boiling water can be increased by adding more heat.

4. What is the advantage of cooking food in a pressure cooker?

5. Answer the following questions by using the table that shows the relationship of pressure to the boiling point of water.
   a. When the temperature inside a pressure cooker is 250°F, how many pounds of pressure does the gauge show?
   b. At what Celsius temperature does water boil in Denver, Colorado, which is one mile above sea level?

6. What is the danger of using a pressure cooker with a safety valve that does not work?

7. What produces the heat that ignites the fuel in a diesel engine?

8. Summarize the main points of this section by writing *increases* or *decreases* for each blank.
   a. As a gas is heated in a closed container, the pressure ———.
   b. As the pressure of a liquid increases, its boiling point ———.
   c. As the altitude increases, the boiling point of a liquid ———.
   d. As a gas is compressed, its temperature ———.
   e. As a gas expands, its temperature ———.

Group H Activities

1. *Making water boil below 212°F.* Fill a quart jar about ⅞ full of boiling water (slowly so the jar does not break). Put a lid on the jar which will seal it. Turn the jar upside down, and pour cool water over it. Why does the water start to boil?

2. *Observing the heat caused by compression.* Pump vigorously with a tire or ball pump for about 1 minute. Then feel the side of the cylinder near the bottom, and notice how warm it is.

3. *Observing the cooling caused by expansion.* Let air out of an inflated tire or pressure tank. Blow the air over the bulb of a thermometer. How low does the temperature go?

How Can You Make Cold?

God gave us six important sources of heat: the sun, chemical reactions (especially fire), electricity, the center of the earth (geothermal heat), friction, and atomic energy. Men have learned to control and use the heat from these sources.

But sometimes we need cold temperatures.

How can we make something cold? Years ago, people cut blocks of ice from ponds in the winter and insulated them in icehouses to keep food cool in the summer. However, that is not making cold; it is simply using cold objects that already exist.

Remember that heat is a form of energy. It

is fairly simple to produce heat by burning a fuel to change its potential energy into kinetic energy. But there is no simple way to produce cold by changing kinetic energy into potential energy. Heat energy cannot be destroyed. So the only way to produce cold is to move the heat to a different place.

Electrical refrigeration. An electrical *refrigeration unit* produces cold by the two following principles: A gas becomes hot when it is compressed, and a volatile liquid becomes cold when it is allowed to evaporate. Every refrigeration unit needs a *refrigerant,* a fluid used to produce cold. A good refrigerant is a fluid that vaporizes or condenses at a low temperature, and that is nonpoisonous and non-corrosive. The refrigerant produces cold by moving the heat. It absorbs heat in one place (from the food) and releases the heat in another place (into the room).

The three main parts of an electrical refrigeration unit are the compressor, the condenser, and the evaporator. The compressor is a motor-driven pump that brings about the refrigeration cycle. It compresses the gaseous refrigerant, making it very hot. And for each degree the temperature goes up under pressure (energy added), so the temperature will drop that many degrees (energy released) when the pressure is later removed.

Leaving the compressor, the hot gas then goes through the coils in the condenser, where heat is removed by blowing air over the coils. As the refrigerant cools to room temperature, it condenses into a liquid, though still under pressure.

After leaving the condenser, the refrigerant goes through a valve that allows it to expand rapidly. As the pressure is reduced, the refrigerant rapidly turns into a gas (or

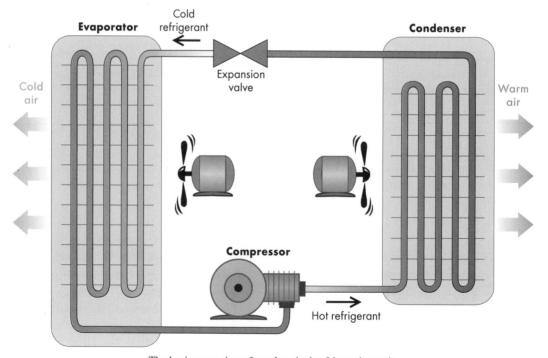

The basic operation of an electrical refrigeration unit

evaporates) and becomes very cold. The cold refrigerant passes through the coils in the evaporator, where it absorbs heat from its surroundings, thus producing cold. Then the gaseous refrigerant returns to the compressor to begin the cycle again. A thermostat (switch operated by temperature changes) turns the compressor on and off to regulate the amount of cooling.

Early electrical refrigerators used ammonia as the refrigerant. But ammonia is a poisonous, corrosive liquid. So in the 1930s, scientists developed chlorofluorocarbon compounds (containing chlorine, fluorine, and carbon) that worked well as refrigerants. These refrigerants, called Freon, were widely used not only in refrigerators but also in aerosol spray cans. But in the 1970s, concerns were raised that these refrigerants were harmful to the ozone layer of the upper atmosphere. Today the refrigerants used are hydrofluorocarbon compounds (containing hydrogen, fluorine, and carbon), such as R-134a and R-410A.

Refrigeration units are useful in many ways. Two very common applications are the refrigerator and the freezer. Dairy farmers use refrigeration units in milk coolers. Your school may have a water cooler that contains a refrigeration unit. Air conditioners use refrigeration units to remove heat from the air inside a house and carry it to the outside air. Dehumidifiers use refrigeration units to first chill the air with the evaporator to remove moisture, and then reheat the same air with the condenser.

An interesting application of the refrigeration unit is the heat pump. It has that name because it "pumps" heat from one place to another place. In summer, the heat pump works as an air conditioner by taking heat out of the house and putting it into the outside air. (The warm outside air becomes even warmer.) In winter, valves in the heat pump cause the evaporator to change into a condenser and the condenser into an evaporator. Even winter air has a little heat in it; and the heat pump takes that heat (making the outside air even colder) and brings it into the house. Heat pumps can also take geothermal heat out of well water or water circulated in underground pipes and put it into a house in winter. They take heat out of the house and put it into the water in summer.

Gas refrigeration. In a gas refrigerator, cold is produced by heating a mixture of ammonia and water

The exterior end of a window air conditioner contains the condenser. It is much like a car radiator and adds heat to the outside air.

with a gas flame, causing it to boil. This refrigerant absorbs heat in one part of the cycle and gives up heat in another part, as in electrical refrigeration, but the process is much more complex. Gas refrigerators are useful in places where electricity is not available.

Heat and its uses are gifts from God. He made the sun and other sources of heat as basic parts of our environment. He established the laws that govern heat energy, and He gave man the ability to use heat in solving many practical problems. As you study weather in the next unit, you will learn more ways that God uses heat for our benefit. "His [The sun's] going forth is from the end of the heaven, and his circuit unto the ends of it: and there is nothing hid from the heat thereof" (Psalm 19:6).

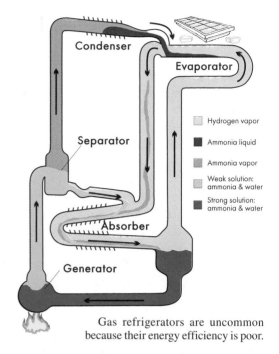

Hydrogen vapor

Ammonia liquid

Ammonia vapor

Weak solution: ammonia & water

Strong solution: ammonia & water

Gas refrigerators are uncommon because their energy efficiency is poor.

―――――――― Study Exercises: Group I ――――――――

1. Which statement does *not* help to explain why it is more difficult to produce cold than to produce heat?
   a. Heat energy cannot be destroyed.
   b. Ice is not readily available in some places.
   c. The potential energy of fuel is easily changed into kinetic energy.
   d. There is no simple way to change the kinetic energy of heat into potential energy.

2. An electrical refrigeration unit works on two basic principles.
   a. A gas becomes hot when it is ―――.
   b. A volatile liquid becomes cold when it is allowed to ―――.

3. What are the qualities of a good refrigerant?

4. Write the correct words to summarize the operation of an electrical refrigeration unit.
   a. The ――― makes the refrigerant gas (hot, cold) by compressing it.
   b. In the ―――, the refrigerant gas changes to a liquid as ――― is removed from it.
   c. In the ―――, the liquid refrigerant evaporates and produces ―――.

5. What controls the temperature level inside a refrigerator?

6. Name three devices that contain refrigeration units.

7. If a heat pump is used to heat a house, will the evaporator be inside or outside the house?

8. How is the operation of a gas refrigerator similar to that of an electrical refrigerator?

9. Why should the study of heat include a recognition of God? Give at least two reasons.

Unit 5 Review

Review of Vocabulary

The kinetic energy of the moving molecules in matter is called —1—. Absorbed energy is called —2— if it changes the state of a material without making it warmer. The internal energy level of a material is its —3—, and it is measured with an instrument called a —4—. The freezing point of water is 32 degrees on the —5— temperature scale and 0 degrees on the —6— scale. A thermometer with a —7— has wires of two different metals joined together to produce a varying current of electricity. A thermometer with a —8— detects changes in electrical resistance at different temperatures.

Heat is measured with a unit called the —9— in the metric system, and with a much larger unit called the —10— in the English system. Heat is transferred from one molecule to another by —11—. Heating a fluid causes it to expand and rise, which transfers heat by —12—. Heat is transferred through empty space by —13—. Water is an excellent substance to collect and transfer heat because it has a high —14—, which means it can absorb a large amount of heat for its volume.

A —15— liquid changes easily into a gas, which results in cooling by —16—. This principle produces cold inside a —17— by evaporating a —18—. Boiling a liquid and then using —19— to change the gas back into a liquid is called —20—. A material such as dry ice changes directly from a solid into a gas through —21—. A —22— bends when heated because the two different metals expand at different rates. One use of this device is to operate a —23— so that it regulates a furnace or refrigerator to maintain a desired temperature.

Multiple Choice

1. When heat is added to a material, the only thing you can be sure the material has gained is
   a. volume.   b. pressure.   c. weight.   d. energy.

2. A thermometer reading is to the internal energy level of a material as calories are to
   a. the temperature needed to boil water.
   b. the total amount of heat in a material.
   c. the melting point of a material.
   d. the amount of conduction of heat that takes place in a material.

3. Which of the following is true?
   a. On the Celsius temperature scale, water boils at 100°.
   b. On the Fahrenheit temperature scale, water freezes at 0°.
   c. On the Kelvin temperature scale, absolute zero is –273°.
   d. A degree on the Fahrenheit scale is about twice as big as a degree on the Celsius scale.

4. Thermometers are made to work on all the following principles *except*
   a. heated fluids tend to rise and set up convectional currents.
   b. materials expand when heated.
   c. increased heat causes an increase in electrical resistance.
   d. heated wires of different metals joined together produce variations in electricity.

5. Which statement is true?
   a. Heat is removed to cause boiling.
   b. Boiling is the opposite of evaporation.
   c. More heat is needed per gram to boil than to melt.
   d. Water boils at a lower temperature than alcohol.

6. Some bismuth alloys have low melting points. This means that
   a. these alloys contract very little as they melt.
   b. their melting points are lower than their freezing points.
   c. these alloys melt slowly.
   d. these alloys melt at low temperatures.

7. Which statement is *not* true of the British thermal unit?
   a. It is equal to about 252 calories.
   b. It is a unit of heat for comparing air conditioners.
   c. It takes 540 of them to change 1 gram of boiling water to steam.
   d. It is the amount of heat needed to raise 1 pound of water 1°F.

8. Which of the following would give up the most heat if cooled to 0°C?
   a. 1,000 grams of water at 40°C
   b. 500 grams of water at 60°C
   c. 500 grams of water at 40°C
   d. 100 grams of water at 100°C

9. How much heat would it take to change 1 gram of ice to steam?
   a. 520 calories   c. 640 calories
   b. 100 calories   d. 720 calories

10. All the following take place inside a refrigeration unit *except*
    a. condensation.   c. compression.
    b. evaporation.   d. freezing.

11. Charcoal is a product of
    a. partial condensation.   c. destructive distillation.
    b. fractional distillation.   d. volatile evaporation.

12. Which of these is an important exception to the general rules that God established for heat?
    a. Brass expands more than iron when heated.
    b. Silver has the highest conductivity.
    c. Water expands slightly as it freezes.
    d. Freon vaporizes at a low temperature.

13. Distillation would be a good method to
    a. separate a liquid from a solid.
    b. produce cold when a refrigeration unit is not available.
    c. raise the boiling point of water.
    d. test the ability of an insulator to stop heat conduction.

14. Which statement about transferring heat is true?
    a. Conduction cannot take place in a liquid.
    b. Convection requires direct contact between a cold object and a heat source.
    c. Convection is the fastest of the three methods of heat transfer.
    d. Radiation happens best between black objects.

15. Which of the following materials are correctly described?
    a. Oil is very volatile.
    c. Air is a good insulator.
    b. Salt crystals sublime.
    d. Silver is a good absorber of radiant heat.

16. Iron and nickel would not make a satisfactory bimetallic strip because
    a. the specific heats of nickel and iron are about the same.
    b. iron and nickel unite to form an alloy when they are heated.
    c. iron and nickel have almost the same coefficient of linear expansion.
    d. iron is a better conductor of heat than nickel.

17. In a refrigeration unit, the actual cooling takes place in the
    a. thermostat.
    c. compressor.
    b. evaporator.
    d. condenser.

18. Which statement gives a correct explanation of how a device operates?
    a. A thermometer indicates temperature by the difference in heat conduction.
    b. A gas refrigerator produces cold by the condensation of a gas.
    c. A pressure cooker raises the boiling point by not letting a gas expand.
    d. An ice cream freezer produces cold when salt forces water to freeze.

19. Hot air rises by convection because
    a. it is pushed upward by the denser cool air around it.
    b. hot fluids always have a tendency to go up.
    c. the heat gives the molecules energy to move upward.
    d. hot air is attracted to the heat of the sun.

20. The liquid state of a material has more heat than when the material is
    a. under pressure.
    c. evaporating.
    b. in the solid state.
    d. in the gaseous state.

Unit 6

Weather

Rain-bearing cumulus clouds like these can produce sudden downpours. A knowledge of clouds is very helpful in understanding storms and in predicting the weather. The rainbow reminds us that God is in control of the weather.

Searching for Truth

"He answered and said unto them, When it is evening, ye say, It will be fair weather: for the sky is red. And in the morning, It will be foul weather to day: for the sky is red and lowring. O ye hypocrites, ye can discern the face of the sky; but can ye not discern the signs of the times?" (Matthew 16:2, 3).

Christ was not condemning the ability of the Pharisees to predict the weather. He knew that the weather follows orderly patterns that can be discovered by men, and that certain signs can be used to predict the coming weather. Christ was condemning the Pharisees for discerning weather signs but not discerning His miracles as signs that their Messiah had come.

God has given men the ability to observe the weather and then to make generalizations based on those observations. Since people have observed many times that fair weather follows a red evening sky, they can use a red evening sky as a sign of coming clear skies. In this unit you will learn many relationships between signs and the weather that follows. The use of signs to predict the weather is possible because of the orderly way that God made the world.

Searching for Understanding

1. What makes the wind blow?
2. What makes rain?
3. How can hail form in the middle of the summer?
4. How can clouds be used to predict the weather?
5. What determines the kind of weather we will have?

Searching for Meaning

anemometer	El Niño	precipitation
barometer	front	psychrometer
cirrus cloud	humidity	relative humidity
climate	hurricane	stratus cloud
Coriolis effect	hygrometer	tornado
cumulus cloud	meteorology	typhoon
dew point	nimbus	

The Orderliness of Weather

Weather has much to do with our daily activities. A farmer would not want to mow hay if he knew it was going to rain the next day. A housekeeper considers whether it will rain before hanging out clothes to dry. Airline pilots are careful to avoid storms. State highway departments try to stay prepared for snowy or icy conditions. Weather is of such interest to everyone that it is often a subject of conversation.

The study of weather is called *meteorology.* This branch of science is a difficult one because weather is complex and changeable. The expression "changeable as the weather" has come to mean a tendency to frequent and unpredictable change. Weather does change frequently, but it is not true that it is altogether unpredictable. Weather predictions are just more difficult to make than other science predictions, such as the speed of falling objects and the boiling point of water at various altitudes.

Modern weather forecasting is based on the assumption that the weather at any particular moment is the result of many conditions that existed previously in a given area and surrounding areas. The Christian sees beyond the mere mechanics of weather to God, who is the origin of weather and its laws. The Christian knows that God has the power to miraculously interrupt these laws as He did when Christ stilled the tempest.

Weather is a testimony to the great wisdom of God in providing for the welfare of mankind. "Nevertheless he left not himself without witness, in that he did good, and gave us rain from heaven, and fruitful seasons, filling our hearts with food and gladness" (Acts 14:17). Weather is in God's control. Certain laws are in operation, but even these laws are of God. "He maketh his sun to rise on the evil and on the good, and sendeth rain on the just and on the unjust" (Matthew 5:45).

Meteorologists have discovered much order in weather formation. They have learned that cycles, patterns, and laws are at work even when weather conditions seem disorderly. "The wind goeth toward the south, and turneth about unto the north; it whirleth about continually, and the wind returneth again according to his circuits" (Ecclesiastes 1:6). The wind seems particularly changeable, yet there are many laws at work that determine which way the wind will blow.

What Makes the Wind Blow?

This is a simple question that a child might ask on a windy day, though a complete answer is beyond the comprehension of a six-year-old. Even meteorologists do not completely understand the wind patterns of some storms. But you should be able to understand the basic principles of the wind.

The wind blows because of convectional currents produced by unequal heating of the air. Heated air expands and becomes less dense. The pressure of warm, light air is less than that of cold, heavy air. So the high-pressure cold air sinks and moves over the surface to take the place of the rising low-pressure warm air. That is a simple explanation of why the wind blows.

Moving air, called wind, has much energy. Wind can push boats and lift kites. High winds can uproot trees. Where does the wind get all its energy? Since unequal heating of air is what makes the wind blow, and since the heat comes from the sun, all the energy of the wind comes from the sun. Most of the air is not heated directly by the sun but indirectly by the land or water that the air passes over.

There are three wind systems. One is the global wind system with huge cycles that cover the whole earth. Another is pressure cell wind

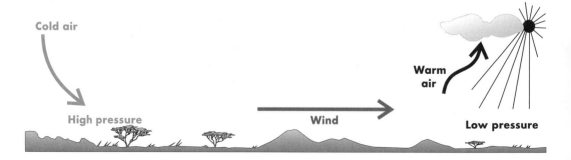

Cold air

High pressure

Wind

Warm air

Low pressure

systems that develop over large areas of land or sea. A third is local wind systems that result from local factors and may last only a few hours. The combination of these three systems determines which way the wind will blow at any given time and place.

The global wind system. Since the sun shines most directly on the equator, a belt of air around the equator becomes especially warm. This warm, moist air rises to produce the abundant rain of the tropical rain forests of South America and Africa. The equatorial belt of warm, humid, still air is called the doldrums. The air rises into the upper atmosphere and moves toward the poles until it sinks at about 30° north and south latitudes. Then it blows over the surface toward the equator again. This produces the tropical wind cycles called Hadley (had′·lē) cells.

The sinking air at the 30° latitudes is warm and dry. This causes the deserts in the southwestern United States, southern South America, northern and southern Africa, and Australia. It also causes the calm regions known as the horse latitudes, apparently so called because horses were often thrown from sailing ships to conserve water when the wind would not blow for weeks. The horse latitudes are high-pressure calm regions, and the doldrums are low-pressure calm regions.

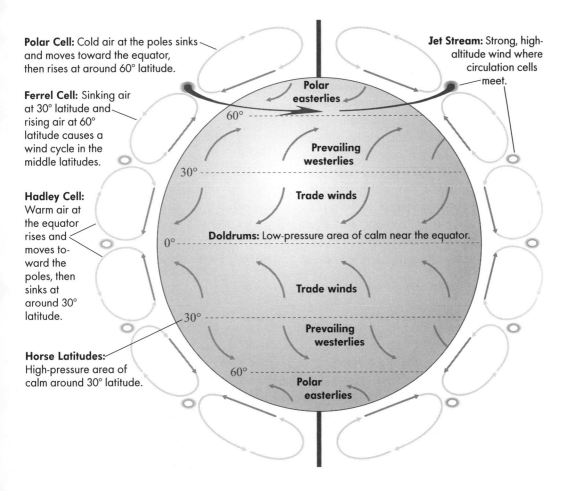

Polar Cell: Cold air at the poles sinks and moves toward the equator, then rises at around 60° latitude.

Ferrel Cell: Sinking air at 30° latitude and rising air at 60° latitude causes a wind cycle in the middle latitudes.

Hadley Cell: Warm air at the equator rises and moves toward the poles, then sinks at around 30° latitude.

Horse Latitudes: High-pressure area of calm around 30° latitude.

Jet Stream: Strong, high-altitude wind where circulation cells meet.

Doldrums: Low-pressure area of calm near the equator.

Polar easterlies

Prevailing westerlies

Trade winds

Trade winds

Prevailing westerlies

Polar easterlies

60°

30°

0°

30°

60°

Air at the North and South Poles does not receive much solar energy, so it becomes cold and dense. This creates high-pressure areas with air moving outward. About 30 degrees from the poles, the air rises. From there it returns to the polar regions to complete the polar wind cells. Between the polar cells and the Hadley cells are the Ferrel (fer′·əl) cells, which rotate in the opposite direction.

If the earth were not rotating, these giant wind cycles would blow due north and south by the basic principles of convectional currents. But the earth is rotating. This causes winds moving from the north in the Northern Hemisphere to turn west (forming east winds), and winds moving from the south to turn east (forming west winds). In the Southern Hemisphere, however, the winds turn in the opposite direction. The result is that the middle latitudes 30°–60° have prevailing westerlies. The regions just north and south of the equator have trade winds from the east, and the polar regions also have prevailing easterlies.

This shifting of wind direction because of the rotation of the earth is called the *Coriolis effect* (kôr′·ē·ō′·lis). The Coriolis effect also has important results in the next wind cycles you will study.

Included with the global wind cycles are high-speed channels of air called jet streams. The speed of these high-altitude streams can

average anywhere from 50 to 200 miles per hour (80 to 320 km/h). Some jet streams are over 100 miles wide (160 km) and a few miles thick. Others are a cylinder a mile or two in diameter with the greatest wind speed at the core. There are polar and middle latitude jet streams blowing from west to east, and an equatorial jet stream blowing from east to west.

The positions of jet streams change throughout the year. Usually the jet stream over the United States is near the Canadian border in summer, but in winter it moves to the southern half of the country. Jet streams have a direct influence on the weather patterns at ground level. They can increase the wind and rainfall brought by storms.

Winds are deflected to the right in the Northern Hemisphere.

Winds are deflected to the left in the Southern Hemisphere.

-------- Study Exercises: Group A --------

1. What name is given to the study of weather?

2. What possible cause for some weather changes is outside the ability of scientists to discover?

3. a. The wind blows from an area of —— pressure to an area of —— pressure.
   b. Differences in air pressure are caused by differences in —— of the air.

4. a. What gives the wind its energy?
   b. Explain how this energy source causes low air pressure.

5. Name the three systems of wind cycles.

6. a. What fact about the earth makes winds that blow north and south turn aside and blow from the west or the east?
   b. What term refers to the cause of this turning aside?

7. Name each part of the global wind system.
   a. Easterly winds on both sides of the equator.
   b. Low-pressure belts over the equator that are calm and rainy.
   c. High-pressure belts around 30° latitude that are calm and dry.
   d. Winds over the United States and southern Europe.
   e. Winds in the polar regions.
   f. High-speed, high-altitude channels of air.

Group A Activities

1. *Demonstrating the Coriolis effect.* Tape a paper to a rotating surface, such as the turntable of a portable lazy Susan. While the paper turns counterclockwise at a slow, uniform rate, draw a line straight out from the center. The result will illustrate the cause of the prevailing easterlies in the polar wind cycles. Draw another line from the edge toward the center. This illustrates the cause of the prevailing westerlies of the middle latitudes.

2. *Recording the wind direction.* Make a simple weathervane by using a stick with a vertical blade fastened to one end and a weight fastened to the other end. Drill a hole vertically through the stick at the center of gravity. Mount it on the end of another stick with a nail slightly smaller than the hole. Fasten cross sticks to the mount with *N, E, S,* and *W* on the ends. Fasten the weathervane to a post, with the directions proper for your location. Record wind direction 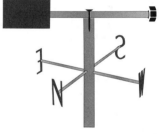 in the middle of the forenoon and in late afternoon. Do this over two or three weeks, and compare your observations with the global wind cycle for your area.

Pressure cell wind systems. Air tends to take the temperature and moisture of the surface over which it is located. For example, if a mass of air lies over a body of warm water, the air will become warm and moist. This produces a warm low-pressure cell. Air moves toward the center of the low-pressure area; and as it does so, the Coriolis effect causes the air mass to rotate. A low-pressure cell rotates counterclockwise in the Northern Hemisphere and clockwise in the Southern Hemisphere.

A mass of air over cold land will become cold and dry. This makes a high-pressure cell with the air moving outward from the center and again rotating because of the Coriolis effect. A high-pressure cell rotates clockwise in the Northern Hemisphere and counterclockwise in the Southern Hemisphere.

Since most low-pressure cells contain moist air, they tend to bring cloudy and rainy weather. In contrast, the cold, dry air of a high-pressure cell brings sunny weather. Often a

Wind-chill Equivalent

	Wind Speed (mph)							
	5	**10**	**15**	**20**	**25**	**30**	**35**	**40**
35°F	31	27	25	24	23	22	21	20
30°F	25	21	19	17	16	15	14	13
25°F	19	15	13	11	9	8	7	6
20°F	13	9	6	4	3	1	0	-1
15°F	7	3	0	-2	-4	-5	-7	-8
10°F	1	-4	-7	-9	-11	-12	-14	-15
5°F	-5	-10	-13	-15	-17	-19	-21	-22
0°F	-11	-16	-19	-22	-24	-26	-27	-29
-5°F	-16	-22	-26	-29	-31	-33	-34	-36
-10°F	-22	-28	-32	-35	-37	-39	-41	-43
-15°F	-28	-35	-39	-42	-44	-46	-48	-50
-20°F	-34	-41	-45	-48	-51	-53	-55	-57

At 10°F, a wind of 5 miles per hour chills exposed skin as much as calm air does at 1°F.

high-pressure cell with its clear weather follows a low-pressure cell.

You can tell in two ways when a low-pressure cell is approaching. Often the wind will shift from the prevailing westerly direction of the global system. In much of the United States, a shift to an east or a south wind is a sign of a coming low-pressure cell. A drop in air pressure is another sign that a low-pressure cell is coming. So it would not be a good time to mow hay if the wind direction changed to the southeast and the air pressure was falling.

These pressure cell wind systems mainly affect the middle latitudes. Low-pressure cells and high-pressure cells, each with a diameter of about 1,000 miles (1,600 km), move very slowly across the land and may take several days to pass over. They tend to move in the direction of the global wind cycles, which in the case of the middle latitudes is from west to east.

Local wind systems. Wind is also generated by local conditions. The most common local wind cycles are the sea–land cycles. During the day, sunlight quickly heats the top few inches of land, and it becomes much warmer than a nearby body of water. This causes convectional currents of rising air over the warm land, which is replaced by cool air blowing from the sea. The cool air from the sea is called a sea breeze.

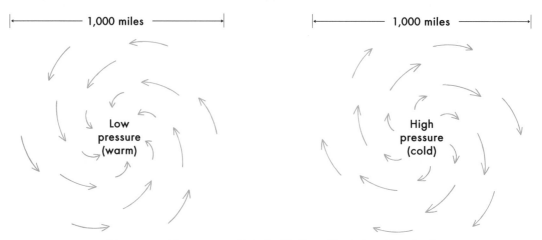

Air movement in pressure cells in the Northern Hemisphere (top view)

Sea breeze

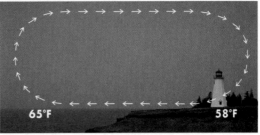

Land breeze

At night the opposite occurs. The land cools off rapidly while the water holds the heat it collected during the day. Convectional currents form over the sea, with the rising warm air replaced by cooler air coming from the land. These winds are called land breezes. Local land and sea breezes can occur on a daily basis.

The same principle works on a seasonal scale in Southeast Asia. The famous monsoon winds blow in from the Indian Ocean during the summer months when the land is hot, and they blow out toward the ocean during the winter months. That gives India a very rainy season during summer and a cooler dry season in winter.

Land features, cities, and forests (especially forest fires) affect the wind and sometimes cause local wind patterns of their own. Even a large area of snow on a mountain can create dense cold air that rushes down the mountainside. Or the sun shining on tall mountains can make the higher air warmer than the valley air. This sets up a valley wind that blows up the side of the mountain.

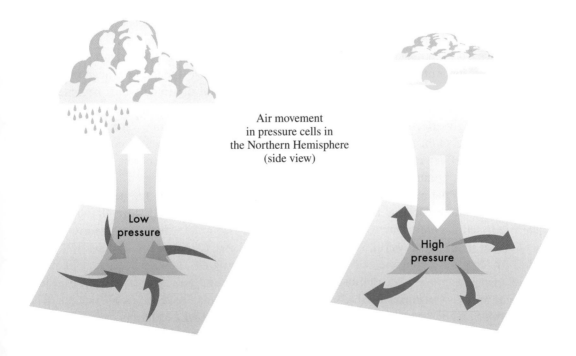

Air movement
in pressure cells in
the Northern Hemisphere
(side view)

Low
pressure

High
pressure

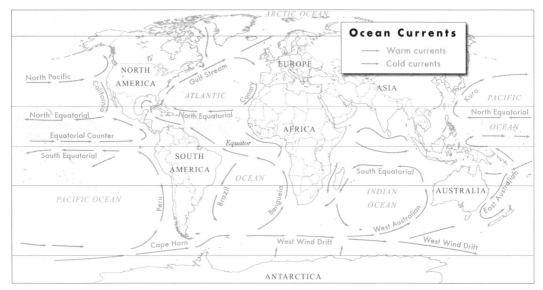

Ocean currents of the world

Therefore, the answer to the question "What makes the wind blow?" can be simply given as differences in air pressure due to differences in heating. It is more complicated to answer the question, "Which direction does the wind blow?" As you have seen, the direction of the wind is a result of three wind systems at work. The wind is the product of such a variety of conditions that it is often hard to predict just which way the wind will blow.

Effects of ocean currents. Global wind cycles are strongly affected by the temperatures and currents of the ocean, and the wind cycles in turn affect *climates* (long-term weather patterns) around the world. Warm, tropical waters in the Northern Hemisphere flow in a clockwise current, following the trade winds and the prevailing westerlies. In the Southern Hemisphere, warm water flows toward Antarctica in a counterclockwise current, following the prevailing easterlies. Land areas near warm ocean currents usually have a warm, wet climate. Areas near cold currents are usually cool and dry.

The Pacific, the largest ocean, has the greatest effect on climates. The cold Peru current gives western South America its cool desert conditions, whereas the warm current in the western Pacific makes Southeast Asia and Australia warm and wet.

Sometimes water temperatures in the central Pacific rise enough to change the ocean currents, and the trade winds become very weak. In this condition, known as *El Niño* (el nēn′·yō), the rising warm air and heavy rains of the western Pacific move to the central or eastern part of the ocean. This may bring drenching rains to barren areas of western South America while the heavily populated areas of Southeast Asia suffer drought.

El Niño can affect climates around the world. Areas that normally have wet summers and cold, snowy winters may have dry summers and mild winters during El Niño. Dry areas may have wet, cool winters. The unusual weather conditions over the Pacific

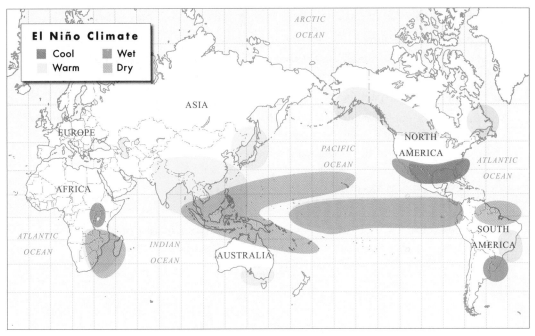

Common effects of El Niño on winter weather

may affect the jet stream in such a way that it sweeps across the central Atlantic and trims off the top of tropical storms before they develop into strong hurricanes.

El Niño appears to come in cycles of two to seven years. Its occurrence in 1997–98 was labeled "the El Niño of the century" in the United States because of the record-breaking rainfall and warm winter temperatures. The next El Niño effects were not observed until 2002-03, but then they occurred again in 2004-05 and in 2006-07. Every occurrence changes the normal weather in one way or another, and the pattern is different each time.

──────── Study Exercises: Group B ────────

1. An air mass tends to get its properties from
   a. contact with other air masses.
   b. contact with the earth's surface.
   c. the sun shining through it.
   d. the direction the wind is blowing.

2. A large cell of warm air tends to form an area of (high, low) pressure.

3. Tell whether each description refers to a low-pressure cell (*L*) or a high-pressure cell (*H*).
   a. Produced over cold land.
   b. Air moving away from center.
   c. Counterclockwise rotation in the Northern Hemisphere.
   d. Often brings cloudy and rainy weather.
   e. Cool, sinking air at the center.
   f. Marked by changing wind direction and falling air pressure.

4. Which latitudes of the earth have low-pressure cells and high-pressure cells?

5. True or false? The land heats up faster than the sea during the day.

6. The principle most directly involved in sea–land wind cycles is (conduction, convection, radiation).

7. At night you can expect the wind to blow (in from the sea, out toward the sea).

8. Which one of the following is a way that monsoons and sea–land breezes are alike?
   a. Cause (difference in temperature between water and land).
   b. Structure (part of the global wind cycles).
   c. Length (how long they last).
   d. Results (violent storms).

9. Which choice in number 8 is a way that monsoons and sea–land breezes are different?

10. a. Ocean currents affect the ———— (long-term weather patterns) of coastal areas.
    b. Changes in the currents of the Pacific Ocean produce an occurrence called ————, which can cause unusual weather around the world.

Moisture in the Air

Water in Saturated Air			
Temperature		Weight of Water Per Unit of Air	
F	C	oz. per cu. yd.	grams per m3
86°	30°	0.82	30.4
68°	20°	0.47	17.3
50°	10°	0.25	9.4
32°	0°	0.13	4.9

Dry air is a mixture composed mainly of about 20% oxygen and 80% nitrogen, but air usually contains water vapor. If you leave an open pan of water exposed to the air long enough, it will become empty. The water evaporates and mixes with the air above its surface. The moisture in air is called *humidity.* Air that is very humid contains much water vapor. When the humidity is low, the air is said to be dry. In air of low humidity, sweat evaporates quickly and you feel cool even on a warm day. But if the humidity is high, sweat evaporates slowly and you feel uncomfortably warm.

Air at a certain temperature can hold only a certain amount of water vapor. When air contains all the water it can hold, it is said to be saturated.

How is rain produced? As you can see on the table at the left, warm air can hold considerably more water vapor than cold air. Saturated air at room temperature (68°F) holds more than three times as much moisture as air at freezing temperature (32°F). When air at any temperature is saturated, it is said to have a relative humidity of 100%. *Relative humidity* is a ratio stating the amount of moisture in the air as compared with the total amount of moisture it could hold if saturated. Thus, a relative humidity of 60% means that the air is holding 60% of the moisture it could hold at that temperature.

Heat Index

	Relative Humidity (%)										
	0	**10**	**20**	**30**	**40**	**50**	**60**	**70**	**80**	**90**	**100**
100°F	91	95	99	104	110	120	132	144			
95°F	87	90	93	96	101	107	114	124	136		
90°F	83	85	87	90	93	96	100	106	113	122	
85°F	76	80	82	84	86	88	90	93	97	102	108
80°F	73	75	77	78	79	81	82	85	86	88	91
75°F	69	70	72	73	74	75	76	77	78	79	80
70°F	64	65	66	67	68	69	70	70	71	71	72

The relative humidity can make a temperature of 85°F feel as low as 76°F or as high as 108°F.

Air with a relative humidity of 30% or less is considered very dry. A warm day with a relative humidity over 80% is very humid and may be called sultry or muggy.

Raising the relative humidity can be done in two ways. One way is to add more moisture to the air by evaporation. But this is practical only to a certain point; for as air gets close to

Other Water Vapor Facts

1. At sea level, dry air at 68°F weighs 1200 g/m³.

2. At sea level, humid air at 68°F weighs 1190 g/m³.

3. Water vapor is much lighter than most other atmospheric gases. Suppose a room of dry air contains 99% nitrogen and oxygen. If 4% water vapor is added, 4% of the nitrogen and oxygen must leave. Then the overall weight of the air (95% nitrogen and oxygen and 4% water vapor) becomes somewhat lighter.

4. The atmosphere can have nearly 0% water vapor in dry polar areas and as much as 7% water vapor in very hot, humid conditions.

saturation, the rate of evaporation slows down to almost zero.

An easier way to saturate air is to cool it. Since cold air cannot hold as much moisture as warm air, the relative humidity will rise as the temperature falls. At some point in the cooling process, the relative humidity of the air reaches 100%. The temperature at which a certain body of air becomes saturated as it is cooled is called its *dew point.* At the dew point, the air is holding all the moisture possible at that temperature.

What happens if a body of air is cooled below its dew point? Moisture condenses out of the air. Such moisture is called *precipitation.* Rain and snow are forms of precipitation. The kind of precipitation depends on the temperature of the air.

If a body of air is cooled below its dew point when the dew point is above freezing, the precipitation is rain. If precipitation takes place when the dew point is below freezing, the moisture comes out of the air as ice crystals called snowflakes. If rain falls through a layer of freezing air on its way to the ground, it may freeze into ice pellets called sleet. Sometimes it stays liquid but freezes on contact when it lands. This is called freezing rain, and it forms a glaze of ice on the ground and other surfaces. If the moisture in the air condenses on cool grass or objects, it forms either dew or frost, depending on the air temperature.

Rain, snow, sleet, dew, and frost are all forms of precipitation. Hail, another form of precipitation, is discussed in a later section about storms. These forms of precipitation are part of the marvelous water cycle that God designed to water the earth. "All the rivers run into the sea; yet the sea is not full; unto the place from whence the rivers come, thither they return again" (Ecclesiastes 1:7).

The water in the oceans is continually evaporating into the air. The wind carries these moist air masses over the land, where they are cooled and give up their moisture in the form of precipitation. Some of the water evaporates from the land and vegetation.

Streams, rivers, and underground channels carry the rest of the water back to the oceans to complete the cycle. The water cycle is a wonderful provision of God to send us rain from heaven and give us fruitful seasons. "He watereth the hills from his chambers: the earth is satisfied with the fruit of thy works" (Psalm 104:13).

"The treasures of the snow" (Job 38:22) are an example of God's orderly world. As ice crystals grow, they form beautiful, six-sided arrangements; but no two snowflakes are alike.

The Water Cycle

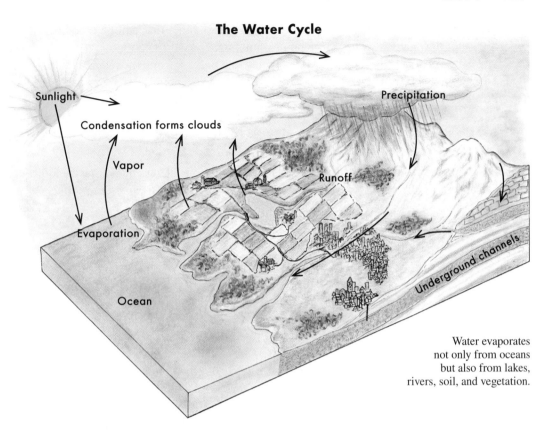

Water evaporates
not only from oceans
but also from lakes,
rivers, soil, and vegetation.

————— Study Exercises: Group C —————

1. Moisture enters the air when water ——— from vegetation, rivers, lakes, and oceans.

2. Warm air holds (more, less) moisture than cold air.

3. Air is ——— when it is holding all the moisture possible at a particular temperature.

4. What is meant by a relative humidity of 80%?

5. Which of these air conditions would be most likely to produce rain before long?
   a. 60% relative humidity and getting cooler
   b. 75% relative humidity and getting warmer
   c. 45% relative humidity and getting warmer
   d. 20% relative humidity and getting cooler

6. As the temperature of air decreases, its relative humidity (increases, stays the same, decreases).

7. What is the dew point?

8. Air that is cooled below the dew point will produce some form of ———.

9. For each set of conditions, name the form of precipitation that will be produced. Write *none* for the one that will produce no precipitation.
   a. High air is below the dew point and below the freezing point of water.
   b. Ground is freezing and temperature is below the dew point.
   c. Dew point is above freezing in higher air but below freezing in lower air.
   d. Temperature is above the dew point and below freezing.
   e. Ground is below the dew point and above freezing.
   f. High air is below the dew point and above the freezing point of water.

10. Illustrate the water cycle by drawing a diagram like the one below and putting the following five words in the correct spaces: *rivers, clouds, precipitation, wind, evaporation.*

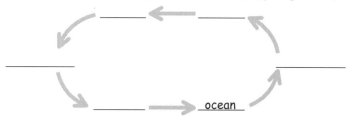

Group C Activities

1. *Finding the dew point of air in your classroom.* Obtain a metal can or a cup with smooth metallic sides. A glass container will also work, but the condensation will be easier to detect on a metal surface. Fill the cup with ice water, and let it stand until the temperature of the can is low enough to cause moisture to condense on the outside of the cup. Use a thermometer to find the temperature of the inside of the can. The temperature at which moisture condenses out of the air is its dew point.

2. *Demonstrating the water cycle.* Set up a small, shallow pan with water over an alcohol burner. (A propane torch will do.) A few inches above this pan, support a large aluminum cake pan full of water and ice. The cake pan should be slightly inclined, with the higher end above the small pan. Use aluminum foil with a number of ridges in it to make a drain from the lower end of the cake pan to the small pan. Now boil the water in the small pan. What is the source of energy to operate your water cycle? Tell what is represented by each part of your model.

Clouds. The clouds have inspired man's imagination for centuries. Poets speak of clouds as symbols both of peace and glory and of misery and gloom. A sunset adorned with clouds or a blue sky full of large, billowy clouds does inspire us with its beauty. A gray, cloudy, or foggy day can have the opposite effect on our emotions. But the clouds are one major part of God's system for providing precipitation.

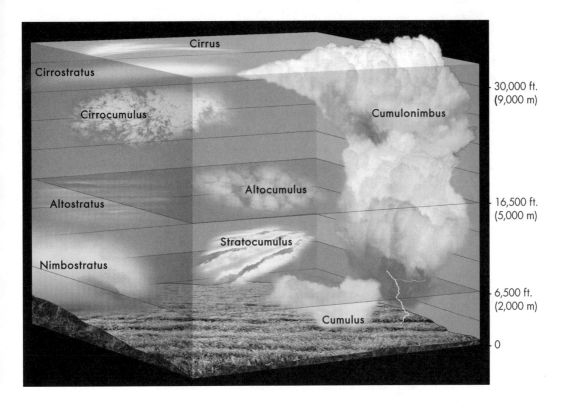

A cloud is a mass of saturated air in which some of the moisture is condensed to form a suspension of tiny water droplets or ice crystals. The size and shape of the clouds are constantly changing. No two clouds look exactly alike. Yet because of certain similarities, clouds can be placed in various categories.

Cumulus clouds (kyōōm'·yə·ləs) are the fluffy, piled-up clouds that look like mounds of soapsuds, balls of cotton, or heads of cauliflower. They are usually a sign of fair weather unless they develop into very tall and dark thunderheads. If rain does come from cumulus clouds, it is sometimes in the form of a heavy downpour.

Cirrus clouds are thin, featherlike clouds composed entirely of ice crystals. These clouds form at a very high altitude. Sometimes they have the wispy appearance of thin gauze.

Occasionally they are scattered and look like wind-blown horsetails—which gives them the common name mare's-tails. Cirrus clouds are so thin that you can see the sun through them. Although you cannot expect rain from such clouds, they are a sign that precipitation is likely within the next few days. The ice crystals within cirrus clouds can cause a ring around the sun or moon, as well as bright spots called sundogs on either side of the sun. These rings and spots may indicate coming precipitation.

Stratus clouds (strā'·təs, strat'·əs) are broad layers of clouds that cause an overcast sky. These are the lowest clouds, lying a mile or less above the ground. Stratus clouds often precede a damp, rainy spell in which a drizzle may fall for several days.

There are many variations in cloud forms. Meteorologists add certain word elements to cloud names to describe different clouds. For example, ***nimbus*** refers to rain-producing clouds in names such as *nimbostratus* and *cumulonimbus*. Cloud names with *alto* refer to certain very high clouds, such as *altostratus* and *altocumulus*. Names formed by combinations of the three basic cloud names designate still other cloud forms. *Cirrocumulus* clouds form what is called a mackerel sky and are a sign of approaching precipitation. *Stratocumulus* clouds are an almost solid layer of cumulus clouds. *Cirrostratus* clouds are a layer of clouds composed of ice crystals.

You may have wished to be up in a beautiful cloud. Any cloud looks better from the outside than from the inside. When an airplane flies into a cloud, the view out the window is like a dense fog. When you are in a fog, you are in a cloud. Fog is a stratus cloud that touches the ground.

All clouds and fog are formed in one way. When air is cooled below its dew point, tiny droplets of water or ice crystals condense to form the whitish mist that we call a cloud.

─────── Study Exercises: Group D ───────

1. What is the relative humidity inside a cloud?
2. Name a general cloud category (*cumulus, cirrus,* or *stratus*) for each description.
   a. Composed of ice crystals.
   b. Fair-weather clouds.
   c. Form broad layers.
   d. Fluffy and piled up.
   e. May become tall clouds called thunderheads.
   f. Thin and featherlike; may resemble horsetails.
   g. Cause a ring around the sun or moon.
   h. Cause an overcast sky.
   i. May produce a steady rain lasting several days.
   j. Often seen a few days before precipitation.
   k. Very high in altitude.
   l. Low, dark layers of clouds.
3. What is meant when *nimbus* is attached to the name of a cloud?
4. What does *alto* mean in cloud names, such as *altocumulus* and *altostratus*?
5. What is a fog?

How Air Is Cooled

You read in the previous section that cooling of air causes clouds to form. When air is cooled below its dew point, some moisture comes out as precipitation. One way that air becomes cold is by coming in contact with a cold surface. Fog forms when warm, moist air blows over cold water or land. During a clear night, the earth gives off radiant heat into space. This cooling of the earth's surface and the air above it is the reason for frost, dew, and morning fog. But these generally do not form under a cloudy sky, for clouds reduce the loss of heat by reflecting it back to the earth.

By far the most common way that air is cooled below its dew point is by being lifted to a higher altitude. As air rises, its pressure becomes less and the air expands. The expansion causes cooling as you learned in the unit about heat. Air cools about 5½°F for each 1,000 feet (3°C for each 300 m) of increased altitude. Therefore, anything that causes air to rise can cause clouds and possible precipitation. Air may rise by convectional, orographic, or frontal lifting.

Convectional lifting. A warm island in the ocean will cause convectional currents. Air rises above the island and forms cumulus clouds that frequently bring tropical showers. Plowed fields can also give rise to convectional currents and possible thunderheads on a calm day. This happens because a plowed field absorbs more of the sun's heat energy than nearby forests or grassy fields. Cities, large parking lots, and warm land next to the sea are other causes of convectional currents that help to produce clouds.

Orographic lifting. The word *orographic* refers to mountains. As air blows over land, hills and mountains cause it to rise. Suppose a mountain forces air to rise 6,000 feet (1,830 m). If the air had a temperature of 70°F (21°C) at sea level, it would cool to about 37°F (3°C) when it reached the top of the mountain. Some mountaintops are so cold that they are always covered with snow. Mount Kilimanjaro, in Africa, with a height of about 3.7 miles (5.9 km), always has snow on its peak, even though it is very near the equator.

Clouds often hover above or around the peaks of mountains. The top of Mount Washington in New Hampshire, with an altitude of 1.2 miles (1.9 km), is one of the windiest places on earth. The air that blows over its top is cooled so much that there is a cloud at the top of the mountain most of the time. Such clouds hovering over a mountain peak are sometimes shaped like a convex lens, thick in the middle and thin at the edges.

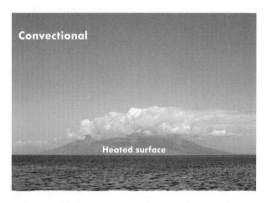

Air is lifted and cooled in three ways.

When warm, moist air rises and cools as it blows over a mountain range, the moisture precipitates out of the air on the uplifting side of the range. Then the cool, drier air moves down the opposite side and becomes warm through compression. Since much of the moisture has been removed, this air has a low relative humidity. The result is a dry area called a rain shadow on the leeward side of the mountain range. Several deserts are largely the result of this effect. The Great Basin of Nevada is very dry because air from the Pacific Ocean rises and then descends as it blows across the Sierra Nevada mountains.

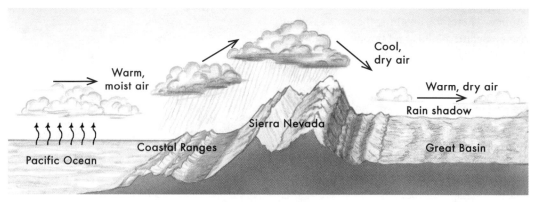

Mountains affect climate.

———— Study Exercises: Group E ————

1. At night, heat leaves the surface of the earth by (conduction, convection, radiation).

2. What happens to rising air that causes it to become cooler?

3. The Sierra Nevada mountains of California are about 10,000 feet high (3,000 m). How much would sea-level air be cooled as it rises over these mountains?

4. Clouds may form over an island in the middle of the ocean because of (conduction, convection, radiation).

5. By the process called ——— lifting, air is forced to rise as it passes over a mountain.

6. Where can you expect to find snow at the equator?

7. Why is the land drier on the east side of the Sierra Nevada mountains than on the west side?

8. Complete these statements to summarize the steps leading to precipitation.
   a. Warm, moist air is forced to ——— by a mountain, convectional current, or front.
   b. The air decreases in ——— as it increases in altitude.
   c. This causes the air to ——— in volume and ——— in temperature.
   d. The relative humidity ——— until the air reaches its dew point.
   e. Further cooling then causes ———.

Frontal lifting. The line where two air masses of different temperatures meet is called a *front*. When a moving warm air mass meets a cooler air mass, it forms a warm front; and when a moving cold air mass meets a warmer air mass, it forms a cold front. Cold fronts and warm fronts are shown on the following weather map. The triangles or semicircles are on the side of the line toward which the front is moving.

In reading about warm and cold air masses, you must understand that these terms refer to the temperatures of air masses in relation to each other. A warm air mass in winter may actually be colder than what is called a cold air mass in summer. But because the air mass is warmer than the air mass it meets, it is called a warm air mass.

As warm air collides with cold air at a warm front, the warm air is forced upward. Why does the warm air not simply push the cold air before it? The resting cold air resists moving because of its inertia. The warm air is also less dense and tends to rise, forming a thin wedge above the cold air. Cirrus clouds form along the upper edge of this wedge. For this reason, cirrus clouds are a sign of an approaching warm front.

Stratus clouds form farther down the wedge. Just ahead of where the wedge meets the ground, thicker nimbostratus clouds often form and bring a steady rain for several hours. The distance from the high cirrus clouds to the rain clouds may be 800 miles (1,300 km), and two days may pass from the time that the cirrus clouds are overhead until the nimbostratus clouds arrive. A period of warmer weather follows the passing of a warm front.

A cold front also lifts and cools warm air, but in this case the warm air is at rest. Instead of forming a thin wedge, the leading edge of a cold air mass has a steep slope that forces the warm air to rise almost vertically. If the cold front is moving slowly, nimbostratus clouds form over a wide area. But a fast-moving cold front causes large cumulus clouds and thunderstorms that produce heavy rain over small areas in a short time. A cold front is followed by a period of cool, sunny weather.

You can predict the coming of a cold front. Large, dark cumulus clouds in the west, distant thunder, and a sky becoming gradually overcast are all signs of a cold front.

There is another way to detect the approach of either a

A NOAA Weather Map*

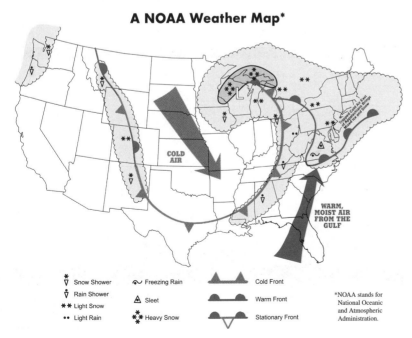

Symbol		Symbol		Front
Snow Shower		Freezing Rain		Cold Front
Rain Shower		Sleet		Warm Front
Light Snow		Heavy Snow		Stationary Front
Light Rain				

*NOAA stands for National Oceanic and Atmospheric Administration.

cold or a warm front. Because both fronts are associated with low-pressure cells, they can be predicted by a drop of air pressure. When the barometer begins to fall and continues to do so for half a day, there is reason to believe that frontal weather is on the way. Since a low-pressure cell has a counterclockwise rotation, the wind direction in the Northern Hemisphere will shift to the south and east (to the north and east in the Southern Hemisphere).

Meteorologists have identified two other kinds of fronts. One is the stationary front, which does not move. Neither the cold air mass nor the warm air mass is advancing. On a weather map, a stationary front is shown with a heavy line that alternates with triangles on one side and semicircles on the other. Rain may continue for several days along a stationary front.

Cold fronts tend to move faster than warm fronts. As a result, a cold front may overtake a warm front and merge with it to form an occluded front. The warm air is lifted off the ground; and this, of course, produces clouds and possible precipitation.

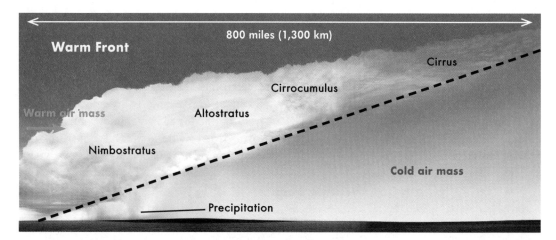

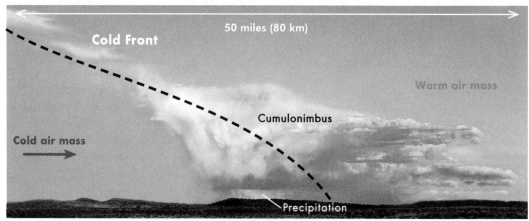

Top: The angle of the front is greatly exaggerated for the diagram. In reality, most warm fronts have only a 0.5% incline. *Bottom:* The boundary for a typical cold front is curved much like this, but the incline is not quite this steep.

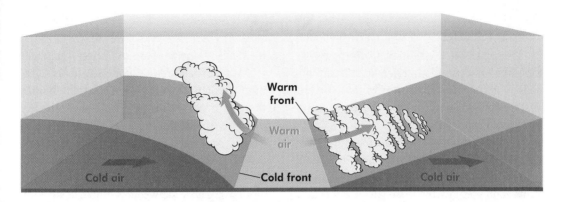

Development of an occluded front

———— **Study Exercises: Group F** ————

1. The line where two different air masses meet is called a ———.

2. If two air masses collide, the (warmer, cooler) air mass is forced upward.

3. Write *cold front* or *warm front* for each description.
   a. Usually preceded by cirrus clouds.
   b. Preceded by dark clouds forming in the west.
   c. Brings thunderstorms and heavy rain.
   d. Includes a thin wedge of warm air above cold air.
   e. Brings a steady, widespread rain.
   f. Followed by cool, fair weather.
   g. Brings rain from cumulonimbus clouds.

4. What two weather signs indicate an approaching low-pressure cell?

5. Name the front described in each sentence.
   a. A cold air mass pushes into warmer air and lifts it almost vertically.
   b. A cold front overtakes a warm front, and the warm air is pushed upward.
   c. A cold air mass and a warm air mass are side by side, without moving.
   d. A warm air mass glides up over a cold air mass.

Storms

There are three common storm patterns: the thunderstorm, the hurricane, and the tornado. All three have the ability to cause severe destruction. Yet each is very different in its form and behavior.

Thunderstorms. The thunderstorm is produced by a strong updraft of warm, moist air caused by convectional currents or a rapidly approaching cold front. This air expands and cools as it rises, with the result that its moisture condenses to form high, billowing cumulonimbus clouds that can bring pouring rain.

Sometimes water droplets freeze in the cold upper part of the thunderhead. More water freezes on the pellet to make a ball of ice called a hailstone. Additional layers of water freeze on the outside, giving the hailstone an onion-like structure. The hailstone can continue to grow as long as it is supported by the updraft, and it may even travel up and down in the cloud as it is buffeted by strong winds.

Hailstones are usually no bigger than marbles, but they can become as large as apples. Finally they grow too heavy to be held up by the updraft, or they get into a downdraft, and the hailstones fall to the ground. Hail can cause great destruction to fruit and field crops, and it can also damage roofs and break windows.

Another feature of thunderstorms is lightning, which is caused by electrical charges that build up within clouds. The electricity in a stroke of lightning may measure several million volts and several thousand amperes. A lightning stroke between a cloud and the ground may measure as much as 9 miles long (14 km). Those between clouds can be even longer.

The apparent cause of these tremendous electrical charges is the strong up-and-down movement of air and precipitation in a thunderhead, which carries electrons from one area to another. The heat of a lightning stroke is over 60,000°F (33,000°C) and causes the air to expand explosively. This produces a shock wave that we hear as thunder. Often the thunder continues rumbling for some time after the initial stroke, due to additional strokes or to echoes between clouds, hills, and forests.

Both lightning and thunder are marvelous displays of God's ability. "He directeth ... his lightning unto the ends of the earth. ... God thundereth marvellously with his voice; great things doeth he, which we cannot comprehend" (Job 37:3, 5).

A tornado is one storm that can completely destroy a house, as shown here.

According to the National Weather Service, the continental United States can expect the following storms in a typical year.

—About 10,000 severe thunderstorms
—About 1,000 tornadoes
—About 1,000 flash floods
—About 10 severe winter storms
—Threats from tropical storms and hurricanes in the North Atlantic, the Gulf of Mexico, and the Caribbean Sea

A single release of this energy in a stroke of lightning is capable of causing death, destruction, and fire. In addition, the high winds of a thunderstorm can uproot trees and damage roofs, and accompanying hail may harm crops and buildings. But thunderstorms also do much good. Lightning causes nitrogen and oxygen in the air to form a nitrogen compound that dissolves in rain and fertilizes the soil. This is part of God's nitrogen cycle by which the soil is replenished with nitrogen. Thunderstorms also bring heavy rains, which make them welcome to farmers and gardeners.

Ice crystals

Path of a hailstone

Updraft

Wall cloud

A supercell thunderhead is a very powerful thunderstorm that can spawn tornados. If a wall cloud (sometimes called a shelf cloud) underneath rotates persistently for 15 minutes, it may trigger a tornado.

—————— Study Exercises: Group G ——————

1. A thunderstorm results from the rapid transfer of heat by (conduction, convection, radiation).

2. The thunderhead is a kind of —— cloud.

3. What two things can cause the rapid updraft that produces a thunderhead?

4. Sleet is simply frozen rain, but hail is —— of ice formed by water freezing on an ice pellet in one —— upon another.

5. Hailstorms usually occur in the summer. What produces the very cold temperatures in which hailstones form?

6. What causes the loud noise from a lightning stroke?

7. Give three features of a thunderstorm that may cause destruction.

8. Give two beneficial results of a thunderstorm.

Hurricane Fran just east of Florida as viewed from space. Notice the spiraling bands of rain clouds leading inward to the eye in the center. This hurricane made landfall at North Carolina on September 5, 1996. Thirty-four people died, largely due to flash flooding.

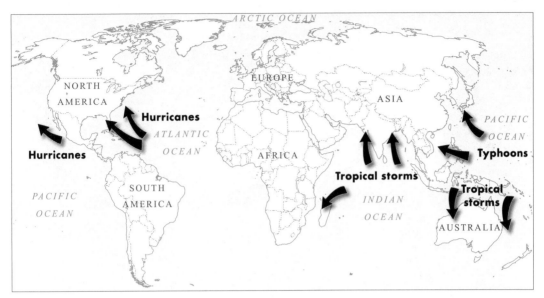

Areas of hurricanes, typhoons, and tropical storms

Hurricanes. A *hurricane* is a tropical low-pressure system that has developed into a huge storm. A hurricane always begins over warm water, which may be on either side of the equator. As the warm, moist air moves toward the center, the low-pressure air mass rotates because of the Coriolis effect. The rising warm air causes dense clouds that produce rain. Such a rotating air mass has a diameter of several hundred miles and is called a tropical storm at first. When its winds reach 75 miles per hour (119 km/h), it becomes a fully developed hurricane.

A hurricane has a clearly defined structure. In the center is the eye, consisting of a calm low-pressure area with a diameter of about 20 miles (32 km). Around this eye are concentric bands of clouds that can produce up to 2 inches (5 cm) of rain per hour. The whole storm moves from 5 to 20 miles per hour (8 to 32 km/h), and its winds may reach speeds over 150 miles per hour (240 km/h). As a hurricane moves from sea to land, a wall of seawater

The Saffir-Simpson Hurricane Scale		
Class	**Wind Speed**	**Effects**
Category 1	75–95 mph (120–153 km/h)	Light damage to trees and mobile homes.
Category 2	96–110 mph (154–177 km/h)	Great damage to trees; some roof damage.
Category 3	111–130 mph (178–210 km/h)	Trees uprooted; mobile homes destroyed.
Category 4	131–155 mph (211–250 km/h)	Windows, doors, and roofs severely damaged.
Category 5	over 155 mph (over 250 km/h)	Small buildings destroyed; others damaged heavily.

Names of Atlantic Hurricanes

2011	2012	2013	2014	2015	2016
Arlene	Alberto	Andrea	Arthur	Ana	Alex
Bret	Beryl	Barry	Bertha	Bill	Bonnie
Cindy	Chris	Chantal	Cristobal	Claudette	Colin
Don	Debby	Dorian	Dolly	Danny	Danielle
Emily	Ernesto	Erin	Edouard	Erika	Earl
Franklin	Florence	Fernand	Fay	Fred	Fiona
Gert	Gordon	Gabreille	Gonzalo	Grace	Gaston
Harvey	Helene	Humberto	Hanna	Henri	Hermine
Irene	Isaac	Ingrid	Isaias	Ida	Igor
Jose	Joyce	Jerry	Josephine	Joaquin	Julia
Katia	Kirk	Karen	Kyle	Kate	Karl
Lee	Leslie	Lorenzo	Laura	Larry	Lisa
Maria	Michael	Melissa	Marco	Mindy	Matthew
Nate	Nadine	Nestor	Nana	Nicholas	Nicole
Ophelia	Oscar	Olga	Omar	Odette	Otto
Philippe	Patty	Pablo	Paulette	Peter	Paula
Rina	Rafael	Rebekah	Rene	Rose	Richard
Sean	Sandy	Sebastien	Sally	Sam	Shary
Tammy	Tony	Tanya	Teddy	Teresa	Tomas
Vince	Valerie	Van	Vicky	Victor	Virginie
Whitney	William	Wendy	Wilfred	Wanda	Walter

called a storm surge comes rushing in. This flood and the accompanying winds can bring death and destruction to the coastal region.

If a place in the Northern Hemisphere lies directly in the path of a hurricane, the storm arrives with heavy rain and powerful winds first blowing from the north or east for several hours. This is followed by half an hour or more of calm, clear weather as the eye of the storm passes. Then the wind and rain come from the opposite direction for several more hours. People living in the predicted path of a hurricane often cover their windows with plywood and may even evacuate the area.

Hurricanes are named according to an established system. The first hurricane in each season receives a name beginning with A, the second one a name beginning with B, and so on. A six-year cycle of names is used.

If a storm with a certain name is especially severe, that name is removed from the list and replaced by another name beginning with the same letter. The hurricanes called Agnes (1972), Gilbert (1988), Andrew (1992), Mitch (1998), and Katrina (2005) are some of the ones whose names now refer only to those exceptional storms. The hurricane season normally lasts from June to September.

Hurricanes are much more common along east coasts than along west coasts because they move in a westward direction. Europe and California have almost no history of hurricanes, whereas Central America, the Philippines, and India suffer frequently from them. These storms are called *typhoons* in the western Pacific Ocean, and tropical storms in the region of Australia and the Indian Ocean. Hurricanes and tropical storms in the Southern

The Fujita-Pearson Tornado Scale		
Class	**Wind Speed**	**Effects**
F-0	40–72 mph (64–116 km/h)	Light damage to tree branches.
F-1	73–112 mph (117–180 km/h)	Windows broken; trees uprooted; shingles blown off.
F-2	113–157 mph (181–253 km/h)	Roofs removed; large trees uprooted; weak buildings destroyed.
F-3	158–206 mph (254–332 km/h)	Walls torn away; automobiles lifted; trains overturned.
F-4	207–260 mph (333–419 km/h)	Houses leveled; heavy damage from flying objects.
F-5	over 260 mph (over 419 km/h)	Houses lifted; trees debarked; steel structures damaged.

—Scale designed after Palm Sunday tornadoes of April 11, 1965

Hemisphere rotate clockwise.

Tornadoes. The *tornado* is the most violent storm of all. It occurs where the atmosphere is very unstable, usually ahead of a strong cold front. A funnel-shaped cloud of whirling wind dips down and touches the earth. With a wind velocity of up to 300 miles per hour (480 km/h), a tornado can collapse buildings and toss cars into the air as if they were toys. The lower end of the funnel is often only a few hundred feet wide, but it may reach a diameter of several thousand feet. Since it commonly touches down for only a mile or so, a tornado does not cause damage in as large an area as a hurricane. But where the tornado does strike, it can cause almost total destruction.

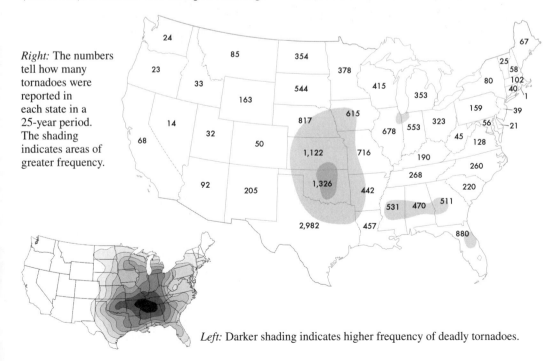

Right: The numbers tell how many tornadoes were reported in each state in a 25-year period. The shading indicates areas of greater frequency.

Left: Darker shading indicates higher frequency of deadly tornadoes.

A funnel cloud is extending downward. Surface winds often swirl before the powerful funnel reaches the ground. Sometimes the surface swirl becomes just as powerful as the tornado itself.

Some tornadoes are much more destructive than others. Mild tornadoes have winds of perhaps 50 miles per hour (80 km/h), and the most severe ones have winds over 260 miles per hour (418 km/h). Tornadoes occur in many parts of the world, but the United States has the most—from 600 to 1,000 tornadoes per year, with the majority in the central states. A tornado over water is called a waterspout.

Before a tornado strikes, the sky is often covered with clouds that have rounded pockets hanging down. Then a shape like a funnel or rope descends from the clouds. The tornado makes a loud rumbling or roaring sound like that of a freight train or a low-flying jet. Never try to run or drive away, for a number of funnels may descend from a tornado sky. You may flee from one only to run into another. The best place to be during such a storm is in a basement.

People react to storms in various ways. An ungodly person may curse because of the storm's destruction. A Christian sees the hand of God in such events and draws closer to the omnipotent Father. Some sinners yield to the Holy Spirit in a time of storm and turn to God for mercy, realizing that they deserve punishment from their Maker. We should not question God's wisdom in sending a storm, but we should be open to any lessons He may have for us in the experience.

> God moves in a mysterious way,
>> His wonders to perform;
> He plants His footsteps in the sea,
>> And rides upon the storm.
>> —*William Cowper*

———— Study Exercises: Group H ————

1. A hurricane begins as an area of very ——— pressure over a large body of warm water on either side of the ———.

2. Wind speeds of a hurricane may reach 150 miles per hour (240 km/h). How can that be if the hurricane moves only 20 miles per hour (32 km/h)?

3. Suppose a hurricane brings strong winds that blow for two hours, and then the sky becomes clear. Why should you not take this as a sign that the storm is over?

4. Hurricane Mitch was the number ——— hurricane in the 1998 season.

5. In each pair, which location is likely to have more storms of the hurricane type?
   a. west end of continent or east end of continent
   b. north of equator or south of equator (See the map.)
   c. near equator or far from equator

6. The Philippines have many tropical storms called ———.

7. a. In what way are tornadoes more violent than hurricanes?
   b. In what way does a tornado cause less destruction than a hurricane?

8. a. What country has the most tornadoes?
   b. What is a tornado called that touches down over water?

9. a. Before a tornado actually strikes, what are two visual signs that it is coming?
   b. What sound accompanies the coming of a tornado?

10. What is one good thing that can result from a destructive storm?

Group H Activities

1. *Producing a vortex: method 1.* Sprinkle several pinches of black pepper into a cup of water. Stir the water with a spoon until it is whirling rapidly. After removing the spoon, notice that the pepper is rotating faster at the center of the cup than at the edges. Also notice that the pepper at the bottom of the cup moves to the center and forms a pile, indicating that the water is flowing inward to an "updraft" created at the center. A tornado is a vortex in air.

2. *Producing a vortex: method 2.* For this demonstration, you will need a bucket or tub with a hole (measuring at least ½ inch) in the center of the bottom. Put a plug in the hole, and fill the container about half full of water. Place the bucket over another bucket that will hold all the water. Remove the plug, and watch a vortex form as the water in the center spins very rapidly, making a tornado-shaped funnel of air down to the drain.

Weather Forecasting

The practical value of meteorology, of course, is to be able to foretell the coming weather, at least to a limited extent. A knowledge of coming weather conditions is very important to farmers, construction workers, airline pilots, and road maintenance men. In fact, almost everyone receives at least some benefit from weather forecasting, even if it only helps a student to decide whether he should take boots along to school.

A good weather forecast requires more than knowing the condition of the weather at a given place and time. Meteorologists study weather conditions that have occurred over a wide area for the past several days before they make weather forecasts. But even trained meteorologists sometimes make inaccurate predictions. Only God knows all the factors that will affect the weather days and weeks in advance. Only God has power to cause certain weather at a certain place and time.

The problem of weather prediction is sometimes confused by a mixture of true and false traditions about the weather. There are several rhyming versions of the weather saying that Christ mentioned in Matthew 16:2, 3.

> Evening red, morning gray,
> Sends the traveler on his way.
>> Evening gray, morning red,
>> Brings down rain upon his head.

This weather tradition is based on actual weather patterns. However, some legends about weather are mere superstition. Many calendars label February 2 as Groundhog Day, but whether a groundhog can see his shadow that day does not determine the weather for the following weeks. Neither do the bands on autumn caterpillars indicate the kind of winter that is to follow. We should desire the truth and accept only those rules for weather prediction which are based on scientific observations and generalizations. Just because you notice one year that rain follows the blooming of lilac bushes does not mean that this is an accurate weather sign.

A number of signs about approaching weather conditions are known to be true and have already been described in this unit. But much of the ability to predict weather from local observations comes with practice. If you observe that a certain sequence of cloud and pressure conditions always precedes a certain kind of weather, you will be able to forecast the weather when those conditions are present. A single sign is not as important as a combination of signs or a sequence of signs. For example, a steady rise in air pressure, a falling temperature, a moderate wind from the west, and clearing skies with small cumulus clouds make a very good basis for concluding that a high-pressure cell has come with several days of fair weather. The more clues that are available to a weather forecaster, the better he is equipped to predict the weather in God's orderly world.

———————— Study Exercises: Group I ————————

1. Give an example of how a weather forecast would be important to a person in a certain vocation.

2. How could God foretell that Egypt would have seven years of drought, seven years before the drought started? (See Genesis 41:28–30.)

A typical workstation for a meteorologist. As weather data is processed through a formula, the meteorologist combines various possibilities to produce a comprehensive forecast. On the right screen, snow showers are shown via radar. On the middle screen, the colored areas depict temperature and precipitation variations for a 3-day forecast. The left screen represents a 1-week forecast. Each colored line reflects slightly different input data to mark the boundary of an air mass. Where the lines run together, the forecast is quite reliable. Where the lines spread out, the meteorologist takes an average but the forecast for that area will be less accurate.

3. To make a good weather forecast, which *three* of the following things would you need to know?
    a. The local weather conditions today.
    b. The local weather conditions over the past several days.
    c. The local weather conditions on this date one year ago.
    d. The weather conditions as described in a weather almanac.
    e. The weather conditions in a large area around the local area.

4. Why are the weather predictions of meteorologists fairly accurate?

5. Why are the weather predictions of meteorologists sometimes wrong?

6. What traditions about the weather should be rejected?

7. Fill in the blanks so that these statements are good weather-forecasting rules.
    a. A ——— in air pressure is a sign that a low-pressure cell is coming.
    b. A coming warm front is signaled by ——— clouds.
    c. In the Northern Hemisphere, a wind blowing from the ——— or ——— is a sign of an approaching low-pressure cell.
    d. Clouds with rounded pockets hanging from them are a sign of a possible ———.
    e. A red evening sky is a sign of ——— weather the next day.
    f. If there are many ——— clouds against a blue sky, it is a sign of fair weather.
    g. A ring around the sun or moon means that a ——— front is coming.
    h. Towering cumulonimbus clouds indicate a coming ———.

Weather Instruments

Good weather forecasting is based on accurate measurement of weather conditions over a wide area for many days. This data is gathered from weather instruments at many places. Such tools used by weather stations contain very precise electronic sensors and almost no moving parts. You may have a few simpler weather tools in your school or home. The thermometer (mechanical or electronic) is one useful weather instrument. To give the true temperature of the atmosphere, the thermometer must be outside and in a shaded spot. Some thermometers record the readings of the highest and lowest temperatures for the day.

Another important weather instrument is the *barometer* for measuring air pressure. The first kind was the mercury barometer, with a column of mercury about 30 inches (76 cm) tall. This barometer measures atmospheric pressure by how far it pushes the mercury up the tube. For this reason, atmospheric pressure is normally given in inches of mercury even if it is measured with other kinds of barometers.

The aneroid barometer has a closed flat container with a partial vacuum inside. One side of the container is fastened to a frame. The other side is fastened to a spring and lever that transfer motion to a needle in front of a dial.

Since the rising or falling of air pressure is more important than a single reading, many aneroid barometers have a second needle that can be set to the reading at a certain time. After an hour, a check of the barometer will show if the air pressure is going up or down. Falling pressure means that a low-pressure cell is probably coming with its frontal weather. Rising pressure indicates the coming of a high-pressure cell and fair weather.

A *hygrometer* measures the relative humidity of the air. Any material that will contract and expand with changes in humidity is useful for making a hygrometer. Since hair has this property, long human hairs are commonly used to make mechanical hygrometers. More accurate hygrometers measure humidity by electronic means.

The *psychrometer* has two thermometers, one with a dry bulb and the other with a small cloth tied around the bulb and made wet. Because of the evaporating water, the

Data for a weather map comes from various instruments at many different places.

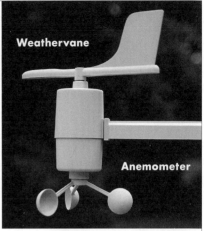

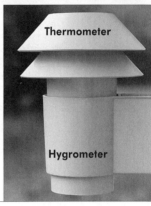

A weather station for a home may include a wireless connection to outdoor instruments. This unit also has a barometer, data recording and graphing, and atomic time-keeping features.

wet-bulb temperature is lower than the dry-bulb temperature. The lower the relative humidity, the faster the evaporation and the greater the difference in the two temperatures. To speed evaporation, some psychrometers are swung around on a sling, and some have a fan to blow air over the wet bulb. The relative humidity is found by using a table like the one on page 242.

A weathervane shows wind direction, which is stated according to the direction the wind is coming from. For example, a northwest wind is blowing from northwest to southeast. A sudden change in wind direction means there will probably be a great change in weather.

Instructions for making a simple weathervane are found on page 211.

An *anemometer* measures wind speed. One kind has a propeller fastened to the front of a weathervane; it measures wind speed by how fast the propeller turns. Another kind has three or four cup-shaped blades fastened to a vertical shaft and rotating horizontally. The cup-shaped blades catch the wind from any direction and always turn the same way. The rotating shaft of either anemometer is connected to a dial or digital display that shows the wind speed.

The amount of precipitation is measured by a rain gauge. Any simple, straight-sided

Equipment of a typical weather station. The large sphere is an antenna for Doppler radar. The three probes above the pole (see inset) are electronic sensors for wind speed and wind direction. Other equipment behind the fence: rain gauge, hygrometer, thermometer, fog sensor, precipitation sensor, cloud reader, and ice-buildup sensor.

container can be used for a rain gauge. But a rain gauge with a funnel or sloping sides multiplies the height of the water and makes it easier to get a precise reading. It is useful to remember that 10 inches of snow is roughly equal to 1 inch of rainfall. One inch of rainfall means that enough rain fell to cover a level surface of the ground with 1 inch of water if none of it soaked into the ground.

Inexpensive electronic weather units are available that simultaneously measure temperature, air pressure, wind speed, and humidity and then record the data over a period of time.

Meteorologists have yet more instruments for gathering weather data. Twice every day, many weather stations use a balloon to send a small package of instruments called a radiosonde into the upper atmosphere. Temperature, air pressure, and humidity readings are radioed to the ground. The path of the radiosonde can be tracked to find the speed and direction of winds high in the atmosphere.

Radar is used to detect falling rain or snow many miles away. Radar can even be used to learn the direction and speed of the precipitation. Weather satellites allow constant observation of moving cloud masses around the world. This is especially helpful in predicting the path of a severe thunderstorm, hurricane, typhoon, or tropical storm.

Computers have become very important as weather-forecasting tools. Mathematical formulas, based on the orderly laws of weather, are programmed into computers, which then constantly receive weather data from around the world and compute forecasts. This method does not replace the judgment of experienced meteorologists, but it does help to improve the accuracy of their predictions. Despite the order in God's world, weather is so complex that man can never hope to produce perfect forecasts.

Weather Agencies

The larger a weather organization, the more data it can assemble and the better it can make forecasts. The United States government maintains an agency called the National Weather Service to gather and interpret weather data. Every day, the National Weather Service receives thousands of weather reports from weather stations, ships, planes, and radiosondes. This information along with satellite observations is used to make nationwide forecasts for everyone interested in them. Weather maps are generated continually to summarize the mass of data that is collected. A simplified form of these maps is published in newspapers.

The weather agency of Canada is the Atmospheric Environment Service, and that of the United Nations is the World Meteorological Organization, which has weather stations around the world. Airlines and other private companies sometimes have their own weather organizations to provide forecasts needed for their businesses. The extensive effort in studying and predicting weather is evidence of the importance of weather in people's lives.

The laws that God established are His servants to bring the weather He considers best from day to day. "Thou hast established the earth. . . . They continue this day according to thine ordinances [laws]: for all are thy servants" (Psalm 119:90, 91). "The LORD hath his way in the whirlwind and in the storm, and the clouds are the dust of his feet" (Nahum 1:3). God wisely planned the variety of weather we need. Food would not grow if we had rain all the time, but neither would it grow if the sun shone all day every day. But it is helpful to know whether a certain day will be rainy or sunny. God created the weather with orderly patterns, and He gave man the ability to discover those patterns and apply them to make weather predictions. To God be the praise for weather and weather forecasting.

Study Exercises: Group J

1. Name the weather instruments described below.
   a. Measures relative humidity.
   b. Measures wind speed.
   c. Measures temperature.
   d. Measures atmospheric pressure.
   e. Measures the amount of precipitation.
   f. Measures humidity with wet-bulb and dry-bulb thermometers.
   g. Indicates wind direction.
   h. Records temperature.
   i. Records atmospheric pressure.
   j. Records humidity.

    k. Obtains data about weather at high altitudes.

    l. Detects precipitation many miles away.

    m. Allows observation of cloud patterns from outer space.

2. What United States government agency collects weather information from many sources and uses it to make weather maps?

3. Since weather patterns follow laws, in what way is it true that God sends our daily weather?

4. For what two reasons should God receive the glory for the successes of meteorology?

5. Challenge question: Why is a printed weather map always out of date?

Group J Activities

1. *Making a sling psychrometer.* Carefully remove the base from around the bulb of a thermometer. Tie a small cotton cloth around the exposed bulb. Fasten that thermometer and an identical one to either side of a ⅜-inch board that measures 1½ inches by 10 inches. Drill a ¹⁄₁₆-inch hole near the upper end of the board, and screw it to the end of a ¾-inch rod 4 inches long, which will serve as a handle. Leave the screw loose so the thermometer board can be swung in a circle.

    Wet the cloth; then swing the psychrometer vigorously in a circle for 15 seconds. Read the wet-bulb thermometer. Repeat until the temperature will not go lower. Read the dry-bulb thermometer. Subtract the two temperatures, and use the following table to find the relative humidity.

Relative Humidity Table															
Dry Bulb	**Difference in Dry-bulb and Wet-bulb Temperatures**														
	1°	*2°*	*3°*	*4°*	*5°*	*6°*	*7°*	*8°*	*9°*	*10°*	*11°*	*12°*	*13°*	*14°*	*15°*
80°F	96	91	87	83	79	76	72	68	64	61	57	54	51	47	44
76°F	96	91	87	83	78	74	70	67	63	59	55	52	48	45	42
72°F	95	91	86	82	78	73	69	65	61	57	53	49	46	42	39
68°F	95	90	85	81	76	72	67	63	59	55	51	47	43	39	35
64°F	95	90	85	79	75	70	66	61	56	52	48	43	39	35	31
60°F	94	89	83	78	73	68	63	58	53	49	44	40	35	31	24
56°F	94	88	82	77	71	66	61	55	50	45	40	35	31	26	21
52°F	94	88	81	75	69	63	58	52	46	41	36	29	25	20	15

Our Environment, p. 483

2. *Forecasting the local weather.* Observe clouds, wind direction, temperature changes, and air pressure changes throughout the day. Try to predict tomorrow's weather on the basis of your observations. Continue to do this for a week. How accurate are your predictions? How do they compare with the forecasts in a newspaper?

Unit 6 Review

Review of Vocabulary

The study of weather and the laws at work in the atmosphere is called —1—. It includes the study of prevailing winds, which turn to the west or east because of the —2—. The long-term weather pattern of a certain area is called its —3—. Unusual weather sometimes comes with an occurrence called —4—, which is due to changes in Pacific Ocean currents.

The moisture in the air, called —5—, is often stated in terms of the total amount of moisture the air can hold. The —6— of saturated air is 100%. Air can be made saturated by cooling it to its —7—. Clouds are saturated masses of air in which condensation has taken place. High-altitude clouds composed of ice crystals are called —8—. Fluffy —9— are associated with fair weather. The layers of —10— produce an overcast sky. Any clouds that produce —11—, such as rain or snow, are called —12— clouds.

Clouds and pressure changes can often be used to predict the coming of a —13—, which is the line where a cold and a warm air mass meet. The —14— over the Atlantic and the —15— over the Pacific are huge circular storms covering large areas. The smaller funnel-shaped —16— may have violent winds that cause severe destruction in a narrow path.

A good weather station is equipped with a thermometer, a —17— for measuring air pressure, a —18— for measuring humidity, and an —19— for measuring wind speed. An instrument for measuring humidity with wet-bulb and dry-bulb thermometers is called a —20—. Meteorologists use these and other instruments to study and predict the weather.

Multiple Choice

1. The main reason for winds is that air masses differ in
   a. moisture due to unequal evaporation.
   b. temperature due to changes in altitude.
   c. pressure due to unequal heating.
   d. rotation due to the Coriolis effect.

2. Prevailing westerlies are found in the middle latitudes because of
   a. the effect of the earth's rotation on air moving between the tropical and polar cycles.
   b. the inertia of warm air as it moves from a lower to a higher altitude in a jet stream.
   c. the Coriolis effect acting on reduced air pressure and causing counterclockwise rotation.
   d. the combined action of global wind systems and pressure cell wind systems at these latitudes.

3. At night you would expect air over the ocean as compared with air over nearby land to be
   a. more dense.   c. lower in relative humidity.
   b. higher in pressure.   d. higher in temperature.

4. If the atmospheric pressure is rising after a period of rainy weather, you can be fairly sure that which of the following has just passed?
   a. a low-pressure cell   c. a prevailing westerly
   b. a high-pressure cell   d. a sea breeze

5. When a certain mass of air is said to have a relative humidity of 60%, it means
   a. 60% of the air is water.
   b. there is a 60% chance of rain.
   c. 60% of the air contains moisture.
   d. the air contains 60% of the moisture it can hold.

6. You would expect to find the most moisture in
   a. cold air with high relative humidity.
   b. warm air with high relative humidity.
   c. cold air with low relative humidity.
   d. warm air with low relative humidity.

7. The most common way that air masses reach their dew point is by
   a. becoming cooler.   c. becoming warmer.
   b. gaining moisture.   d. precipitating.

8. Which is the right order for the way rain is produced from warm, moist air?
   a. low pressure, expansion, lifting, cooling, dew point, raising relative humidity
   b. expansion, low pressure, lifting, raising relative humidity, dew point, cooling
   c. lifting, low pressure, expansion, cooling, raising relative humidity, dew point
   d. cooling, lifting, raising relative humidity, expansion, low density, dew point

9. If an air mass is below its dew point and also below 32°F, it can produce precipitation in the form of
   a. rain.   b. hail.   c. sleet.   d. snow.

10. Two days before a warm front brings rain, you may see
    a. cumulus clouds.   c. stratus clouds.
    b. cirrus clouds.   d. nimbus clouds.

11. A thunderhead is
    a. a cumulonimbus cloud.   c. an altostratus cloud.
    b. a cirrostratus cloud.   d. a nimbostratus cloud.

12. Clouds are likely to form over a mountain because
    a. there is much evaporation from the vegetation on a mountain.
    b. mountains become warmer than surrounding areas.
    c. mountains become cooler than surrounding areas.
    d. air is forced to rise as it passes over a mountain.

13. Frontal weather is part of
    a. the global wind system.   c. storm wind systems.
    b. pressure cell wind systems. d. local wind systems.

14. Which of the following is typical of a warm front but not a cold front?
    a. It brings thundershowers and stormy weather.
    b. It forms a long, thin, wedge-shaped air mass.
    c. It is part of a low-pressure cell.
    d. It is preceded by wind from the south or east.

15. Which statement about the weather is *not* correct?
    a. As air rises, it expands and cools.
    b. Wind blows from an area of high pressure to an area of low pressure.
    c. As air moves north in the Northern Hemisphere, the Coriolis effect causes it to blow from the west.
    d. As the temperature of air increases, the relative humidity goes up.

16. Which of the following describes a hurricane?
    a. High-pressure cell with strong winds rotating over a small area.
    b. Winds rotating swiftly because a cold front rapidly lifts unstable air.
    c. Strong winds rotating counterclockwise in a storm several hundred miles across.
    d. Winds rotating swiftly because of different pressures at different latitudes.

17. When using a psychrometer, the lower the humidity,
    a. the higher both the wet-bulb and dry-bulb temperatures.
    b. the smaller the difference between the wet-bulb and dry-bulb temperatures.
    c. the higher the dry-bulb temperature and the lower the wet-bulb temperature.
    d. the greater the difference between the wet-bulb and dry-bulb temperatures.

18. When a meteorologist predicts that there will be rain tomorrow, you can assume
    a. that he has proof that it will rain tomorrow.
    b. that he has simply guessed there will be rain tomorrow.
    c. that the weather conditions today are similar to other days just preceding rain.
    d. that he has information on all the conditions that will influence tomorrow's weather.

19. Which sentence correctly states a weather sign?
    a. A drop in pressure is a sign of a high-pressure cell.
    b. Cirrus clouds are a sign of a warm front.
    c. A red evening sky is a sign of rain the next day.
    d. A ring around the moon is a sign of fair weather.

20. Which statement best expresses God's relationship to the daily weather?
    a. Weather operates according to laws that God established at the Creation; therefore, it has little to do with God's will today.
    b. God sends different weather from one day to the next so that it is difficult to make weather forecasts.
    c. Weather changes from day to day according to laws and patterns that God has established to fulfill His will.
    d. Most of the daily weather goes on without God's control, but sometimes He causes special weather to teach men spiritual lessons.

21. Which of these statements about weather and the Bible is false?
    a. Miracles relating to weather are recorded in both the Old and the New Testaments.
    b. The Bible mentions wonders of weather that meteorologists know about today.
    c. Some Bible statements about weather refer to more than natural weather.
    d. The Bible includes an explanation of the laws that control weather.

Third Quarter Review: Units 1–6

Matching

Write the letter of the word that matches each description.

1. An agent that allows oil particles to be suspended in water.
2. A mixture of metals.
3. The red substance in blood that helps to carry oxygen.
4. A tough tissue that connects a muscle to a bone.
5. A complex carbohydrate.
6. The action by which food moves through the alimentary canal.

    a. alloy
    b. emulsion
    c. hemoglobin
    d. ligament
    e. peristalsis
    f. starch
    g. surfactant
    h. tendon

7. The darkening of a heavenly body.
8. A large cloud of glowing gases in outer space.
9. The internal energy level of a material.
10. The amount of heat needed to raise 1 gram of water 1 degree Celsius.
11. The scientific study of weather.
12. A kind of cloud that produces rain.

    i. astronomy
    j. calorie
    k. cirrus
    l. eclipse
    m. meteorology
    n. nebula
    o. nimbus
    p. temperature

Completion

Write the correct word to complete each sentence.

13. Energy in action is called ——— energy.
14. Disorder and randomness in energy and materials is called ———.
15. Muscles that cause joints to bend are called ——— muscles.
16. Blood returns to the heart through the ———.
17. Some micronutrients are natural, inorganic elements called ———.
18. Lack of a micronutrient may cause a ——— disease, such as beriberi.
19. If the brightness of a star increases greatly in a short time, it is called a ———.
20. A ——— is a rock from space that streaks through the sky and strikes the earth.
21. If heat changes the state of a material without making it warmer, it is called ——— heat.
22. The opposite of evaporation is ———.
23. The long-term weather pattern of a certain area is called its ———.
24. Winds moving toward the equator turn to the west because of the ——— effect.

Multiple Choice

Write the letter of the correct choice.

25. The product of the mass and speed of a moving object is called
    a. inertia.
    b. momentum.
    c. centrifugal force.
    d. centripetal force.

26. According to the second law of ———, energy becomes less concentrated.
    a. acceleration
    b. motion
    c. physics
    d. thermodynamics

27. The part of the blood that is important in clotting is the
    a. plasma.
    b. platelets.
    c. red corpuscles.
    d. white corpuscles.

28. The largest organ of the lymphatic system is the
    a. liver.
    b. pancreas.
    c. spleen.
    d. appendix.

29. For the growth and maintenance of cells, the body needs
    a. protein.
    b. fats.
    c. cellulose.
    d. carbohydrates.

30. A lack of iron causes
    a. scurvy.
    b. anemia.
    c. goiter.
    d. pellagra.

31. The magnitude of a star refers to its
    a. size.
    b. distance.
    c. brightness.
    d. temperature.

32. The path of the sun through the stars is called the
    a. declination.
    b. ecliptic.
    c. equinox.
    d. zodiac.

33. A thermometer can be made by using any of the following *except* a
    a. bimetallic strip.
    b. thermocouple.
    c. thermistor.
    d. thermal unit.

34. Heat is transferred through empty space by
    a. conduction.
    b. convection.
    c. radiation.
    d. sublimation.

35. A ——— is a huge, circular storm over the Atlantic Ocean.
    a. hurricane
    b. tornado
    c. monsoon
    d. typhoon

36. Both the hygrometer and the ——— can be used to measure humidity.
    a. altimeter
    b. anemometer
    c. barometer
    d. psychrometer

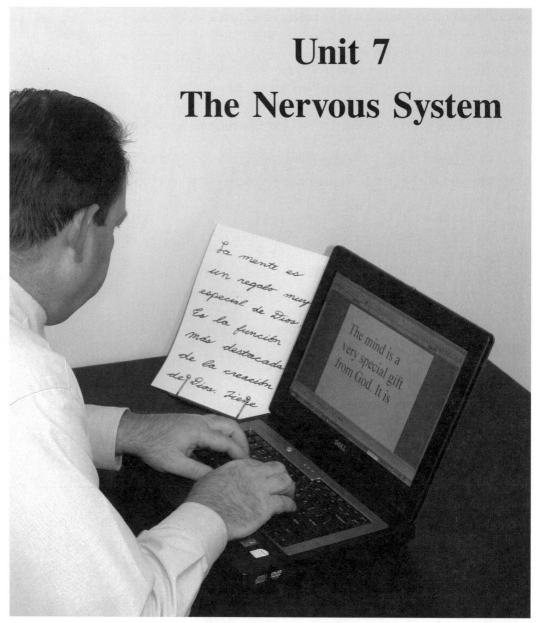

Unit 7

The Nervous System

This person reads Spanish handwriting and types it into English text. As the handwriting scene is projected into the eye, its image travels as nerve signals to the brain. The sight part of the brain reconstructs this information into one three-dimensional image. Due to past learning, other parts of the brain recognize the meaning of this image. Additional parts of the brain translate the meaning into English expression. The brain sends an impulse for each letter through the spinal cord to the fingers. Each finger can type several different letters because a downward motion at different positions activates different keys.

The mind is a very special gift from God. It is the crowning feature of God's creation. Its tremendous capabilities far surpass those of an any man-made computer. Scientists gradually learn more and more about the mind, but they probably will never understand it fully.

Searching for Truth

"Even a child is known by his doings, whether his work be pure, and whether it be right" (Proverbs 20:11). Watch the "doings" (behavior) of any person—even a child—and you will soon know what kind of person he is.

Behavior is controlled by the nervous system. It reveals character because most behavior that others can see is voluntary. God created man with the power to choose between right and wrong. Very soon Eve chose wrong and influenced Adam to choose wrong also. Through their wrong choices, we inherited minds that want to think wrong thoughts and make wrong choices. But by receiving the mind of Christ, we can think right thoughts, make right choices, and produce right actions.

Searching for Understanding

1. How do nerves carry messages through your body?
2. How does your brain know where a pain is?
3. Why do your eyes blink when a quick motion surprises you?
4. Why do you rarely feel tired when you are excited?
5. Why do you shiver when you are cold?

Searching for Meaning

adrenal gland	endocrine system	pituitary gland
autonomic nervous system	equilibrium	reflex
axon	insulin	sensory neuron
brainstem	interneuron	spinal cord
central nervous system	motor neuron	stimulus
cerebellum	neuron	synapse
cerebrum	pancreas	thyroid gland
cortex	parathyroid gland	vestibular apparatus
dendrite	peripheral nervous system	

An Overview of the Nervous System

The nervous system controls all the activities of the body. Some activities, such as breathing, are purely physical; they have no moral or spiritual implications. Other activities do have spiritual implications because they are based on our ability to choose. For example, you can choose to "obey your parents" or to "be kindly affectioned one to another with brotherly love" (Romans 12:10).

As the control system of the body, the nervous system is the physical means by which your choices are translated into behavior. Without the nervous system, no other body part could function, for there would be nothing to tell it what to do.

The nervous system receives and responds to messages from the environment. Some nerves carry incoming messages, and others carry outgoing messages that tell certain parts of the body how to respond. For instance,

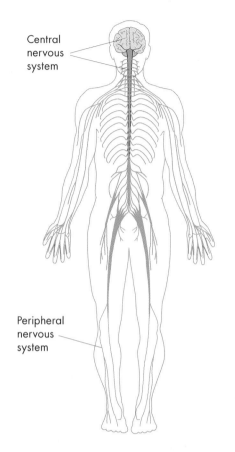

Central
nervous
system

Peripheral
nervous
system

perhaps the telephone rings. Nerves carry this message from the ears to your brain, and then other nerves carry messages telling your legs to take you to the telephone.

The nervous system receives and responds to messages from inside the body. When you drink milk, your stomach tells your nervous system that it has received some milk. The nervous system responds by telling your stomach to produce an acid that breaks down the milk. All these messages are carried by nerves.

The nervous system has two main parts. The *central nervous system* includes the brain and the spinal cord. The *peripheral nervous system* (pə·rif'·ər·əl) is the network of smaller nerves extending from the central nervous system through the muscles and skin of the entire body. The peripheral nervous system provides contact with the outside world by carrying messages from sensory organs to the central nervous system. It also carries messages from the central nervous system to the muscles.

The *autonomic nervous system* (ô'·tə·nom'·ik) is part of the peripheral nervous system, but it controls the automatic functions of involuntary muscles. The autonomic nervous system controls the body parts that function without conscious thought, such as the heart, lungs, and glands.

A nerve cell has three main parts. Nerves are bundles of fibers made of cells called *neurons.* The three parts of each neuron are the cell body, the dendrites, and the axon. The cell body contains the nucleus, which is the control unit of the cell.

The thin, branching *dendrites* gather information and bring it to the cell body. They form more than 80 percent of the neuron. One neuron in the brain may have as many as 50,000 dendrites. Since masses of dendrites and their cell bodies appear gray, scientists call this "gray matter."

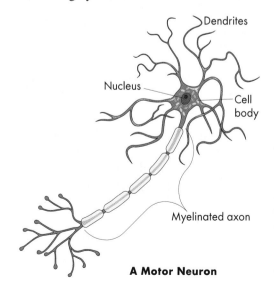

Dendrites

Nucleus

Cell
body

Myelinated axon

A Motor Neuron

A single *axon* extends from each cell body. The axon is thicker than a dendrite, and a single axon may be as long as three feet (0.9 m). Each axon has many endings that pass signals to the dendrites of many other neurons at once.

Nerve signals travel from one cell to another in the form of electrochemical impulses. This means they move like electricity between chemicals with positive and negative charges. Nerve impulses never actually enter the cell body, but travel through the membrane that surrounds it.

A fatty sheath called myelin (mī′·ə·lin) insulates most axons. Myelin keeps nerve impulses from escaping in the same way that insulation on electrical wires prevents a short circuit. The myelin gives axons a white appearance, and for this reason scientists refer to masses of myelinated (myelin-covered) axons as "white matter."

In the disease called poliomyelitis, commonly known as polio, a virus attacks the neurons that control muscles. Myelitis is an inflammation of the myelin of the spinal cord. When it is caused by the polio virus, it is called poliomyelitis.

The disease called multiple sclerosis also inflames and destroys myelin. This blocks nerve impulses or allows them to short-circuit instead of going where they should. Multiple sclerosis interferes with various physical activities, such as walking.

Nerve impulses can travel as fast as 395 feet per second (120 m/s) through myelinated axons. Without this insulation, impulses may travel as slowly as 2½ feet per second (0.8 m/s).

Each neuron acts like a relay runner as it receives signals and sends them on to another neuron. An impulse passes from the axon of one neuron to a dendrite of the next neuron, which transmits the impulse through its own axon to a dendrite of a third neuron. The impulses may be signals from sensory organs or messages to muscles or glands.

Dendrites and axons do not touch one another. At each junction between them is a chemical-filled gap called a *synapse* (sin′·aps′). Like the sparks of a spark plug, nerve impulses jump across these gaps in as little as one-thousandth of a second. Fifty to three hundred impulses may race through one synapse in a second's time.

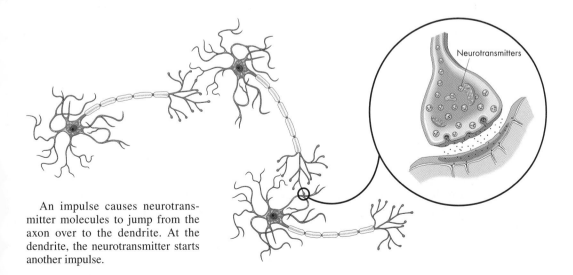

An impulse causes neurotransmitter molecules to jump from the axon over to the dendrite. At the dendrite, the neurotransmitter starts another impulse.

Some neurons carry signals from a sensory organ to the central nervous system. These are called *sensory neurons,* and each cell has a single dendrite acting as a receptor and an axon for transmitting signals. Other nerve cells, called *motor neurons,* carry signals to muscles or glands. These cells have many dendrites branching out from the cell body like rays from a star. The dendrites can receive impulses from other dendrites or axons, and they can transmit impulses to other dendrites or axons.

Major nerves are commonly named for nearby bones or for the regions through which they run. The tibial nerve lies near the tibia in the lower leg. Perhaps you have heard of the sciatic nerve, which is associated with severe pain in the lower back and upper leg. This nerve is the longest one in the body, and its name comes from the Greek word for hip. The sciatic nerve begins at the lower spine, runs through the hip, and extends to the foot.

As you read about the nervous system, your nerves are very busy. You turn a page. Your eyes turn from side to side. Your head drops as your eyes move down the page. At the same time, you are taking signals from the page, processing them in your brain, and storing them for later use. Each of these activities requires a marvelous complex of cooperating nerves. Only God could create such flawless order.

———— Study Exercises: Group A ————

1. The part of the nervous system that interacts with the environment is called the
   a. peripheral nervous system.
   b. central nervous system.
   c. autonomic nervous system.

2. Write *neurons* or *nerves* for each blank.

   The ——— are bundles of fibers made of ——— that carry impulses from one to another.

3. Gray matter is to ——— as ——— matter is to axons.

4. Each neuron has only one axon. Explain how one neuron can send messages to many other neurons.

5. Signals flowing through nerves are most like electricity generated by
   a. the wire coils spinning in the magnetic fields of a generator.
   b. a battery in which chemical reactions cause positive and negative charges.
   c. a solar cell in which light-sensitive materials generate electricity.

6. Myelin is found on
   a. all dendrites.   c. most axons.
   b. all axons.   d. all parts of all neurons.

7. Give two benefits of myelin in relation to nerve impulses.

8. Write the missing words to trace the path of a nerve impulse.

   A message is received by a ———, which transfers the impulse through the ——— that surrounds the cell body. The ——— sends the impulse across a ——— to another dendrite.

Group A Activity

Demonstrating how nerve impulses travel. You can create some weak electrochemical impulses to get an idea of how nerve impulses travel. Unfold and straighten a large paper clip, and cut off 2 inches (5 cm) of wire from one end. Cut off 2 inches (5 cm) of 18-gauge copper wire, and strip the insulation from it. Use coarse sandpaper to smooth the ends of your two pieces of wire.

Get a fresh lemon, and squeeze it (without breaking the rind) so that some of the juice inside is free. Insert the two pieces of wire halfway into the side of the lemon, very close together but not touching. Touch the free ends of the wires with your tongue. You may feel a slight tingle, and you should detect a metallic taste.

The acid of the lemon causes an excess of electrons to leave the one wire and to build up on the other wire. Placing your wet tongue between the two wires closes the circuit and allows the electrons to flow. This is similar to the flow of impulses in the nerves, excited by positive and negative chemical charges. You can think of the acid in the lemon as the synapse between two neurons.

  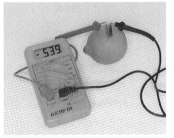

Use a voltmeter to show that the lemon is generating electricity. Touch the red probe to the copper wire and the black probe to the paper clip. To generate a stronger current, use electrodes made with one strip of copper and one of zinc, each measuring about ½ inch by 2 inches (1 cm by 5 cm). The copper electrode can be made by splitting and uncurling a short piece of copper pipe, and the zinc electrode can be made from an old zinc jar lid. This battery can generate up to one volt of electricity.

The Central Nervous System

You are probably familiar with the networks of roads shown on road maps. Roads in rural areas are represented by thin lines, widely spaced. In urban regions the lines are thick and numerous. And within large cities, the roads may run so close together that there is very little space between one line and the next.

The nervous system is much like these maps, with networks of nerves rather than webs of roads. Rural areas are like the peripheral nervous system, and urban areas are like the central nervous system. Not one part of the body is beyond the reach of nerve control. The nerve centers are tightly packed masses of neurons that make your body operate as a unit. When you run, walk, or bat a ball, these nerve centers coordinate many signals from many sense organs and other body parts.

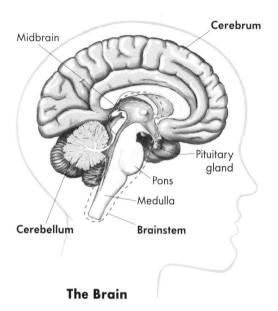

Midbrain

Cerebrum

Pituitary gland

Pons

Medulla

Cerebellum

Brainstem

The Brain

The main nerve centers of the central nervous system are the brain and the spinal cord. Both of these are suspended in a shock-absorbing fluid inside protective bone. The skull protects the brain, and the thirty-three vertebrae of the spinal column protect the spinal cord.

The Brain

The brain is the chief controlling unit of the nervous system. It contains billions of neurons, each one capable of communicating with several thousand other neurons. The brain reaches its full weight of about 3 pounds (1.4 kg) by the time a person is six years old.

The brain has three main parts: the cerebrum (ser′·ə·brəm), the cerebellum (ser′·ə·bel′·əm), and the brainstem. Although each part has its own special purpose, most body functions require the working of neurons in many parts of the brain.

Cerebrum. The *cerebrum* is the main and largest part of the brain, and it processes motor and sensory functions. It has two hemispheres with largely symmetrical functions. For the

most part, neurons in the left part of the cerebrum control the right side of the body, and neurons in the right side of the cerebrum control the left side of the body. This is why a person who has a stroke on one side of his brain suffers impairment on the opposite side of his body.

Each hemisphere is divided further into four main lobes, marked by deep fissures in the outer covering, or *cortex,* of the cerebrum. The cortex consists of ¼ to ⅜ inch (0.6 to 1 cm) of gray matter and forms about 40 percent of the brain. Smaller ridges and folds divide these lobes like the halves of a walnut.

Frontal lobe. This part, lying in the front of the brain, is the site of problem solving, thought processing, reasoning, creativity, and emotion. Try out your frontal lobe activity. Suppose your school has a problem with icy walks. How could you solve this problem? Do not stop with the obvious solutions. Use your imagination to devise any method possible, even if it is expensive. When you cannot think of any more ways, compare your list with the list on page 258.

Of course, the best solution in this case will depend on many factors: the amount of ice, how much the walks are used, the supplies on hand, and similar variables. But your mind can consider many factors. If a certain solution excited you, this emotion too was happening in your frontal lobe.

Parietal lobe. The upper middle part of the cerebrum is the parietal lobe (pə·rī′·i·təl), which lies beneath the parietal bone of the skull. (*Parietal* means "of a wall"; the parietal bones form the "walls" of the skull.) The parietal lobe registers pain, temperature, and touch. When you step out on a balmy April morning and say, "It feels like spring is coming," you are using your parietal lobe. If you jump down the steps and turn your ankle in your excitement, your parietal lobe will register the pain of your sprained ankle.

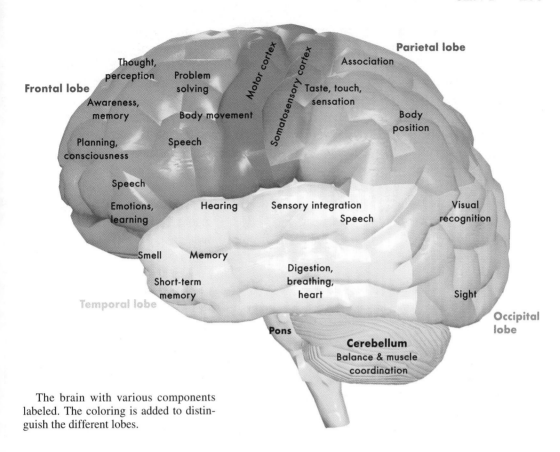

The brain with various components labeled. The coloring is added to distinguish the different lobes.

Temporal lobe. This lobe is named for its position inside the temples, and it processes sound and memory. When did you last hear a robin? Did you sing any songs at the beginning of the day? What was the first song you sang? To answer these questions, you will use your temporal lobe to remember an earlier activity.

Occipital lobe. The word *occipital* (ok·sip′·i·təl) means "at the back of the head," and this describes the location of the occipital lobe. This lobe has the primary function of processing sight, which is why a blow to the back of the head can cause temporary blindness. You are using your occipital lobe at this very moment to read these words. This is a very important lobe of the cerebrum because so much of the information processed in the brain comes through sight.

The functions of the folds and ridges that divide the cerebrum further do not necessarily correspond to the divisions of the lobes. The ridge in front of the middle fold of the cerebrum controls voluntary motions. The ridge behind this fold receives the impulses that come from the skin, skeletal muscles, and abdominal organs. Hearing and smell register in folds of the temporal lobe.

Inside the gray matter, or dendrites, of the cortex lies the white matter, or axons, of the cerebrum. Regions of gray matter intersperse the white matter. A thick plate of white matter connects the hemispheres of the brain, providing communication between the two.

Inside the two hemispheres lie two of the four ventricles (cavities) that produce cerebrospinal fluid. This clear, watery liquid flows into two other ventricles located in the brainstem and connected to the hollow region around the spinal cord, which also contains the fluid. Cerebrospinal fluid surrounds the whole brain, supporting its tissues as water supports the stems of water plants.

Cerebrospinal fluid serves as a shock absorber to protect the brain from injury. However, a sharp blow to the head may cause an injury called a concussion, in which the brain bounces against the inside of the skull. The effects of a mild concussion include headache, dizziness, confusion, and temporary loss of memory. A severe concussion produces unconsciousness; and for some time afterward, the person may have trouble concentrating and may even show changes in behavior.

The central nervous system is surrounded by three coverings called the meninges (mə·nin′·jēz), with cushions of cerebrospinal fluid between them. An inflammation of the meninges is called meningitis, and it causes severe headaches.

Cerebellum. This part of the brain lies behind and beneath the cerebrum; *cerebellum* means "little brain." Like the cerebrum, the *cerebellum* has an outer cortex of gray matter and inner regions of white interspersed with gray. The cerebellum constantly monitors signals that the cerebrum sends to the skeletal muscles, adjusting and smoothing the signals for maximum control. This is what we call coordination.

When you pick up a piece of lead for a mechanical pencil, your cerebellum smoothes out the commands that your cerebrum sends to your muscles. Without the cerebellum, your motions would be very jerky. You would miss the lead several times before you closed your fingers around it. Then your fingers would probably close unevenly and break it. Your cerebellum also helps you to maintain your balance when you walk. With its help you can smoothly and efficiently do many activities, delicate or rough, simple or complicated.

Brainstem. The *brainstem* connects to both the cerebrum and the cerebellum, and it has several parts. Most notable among these parts are the midbrain, the pons, and the medulla. The brainstem is a relay station between various body organs and the cerebrum. It registers appetite, telling you when you are hungry and when you are full. It helps you to go to sleep and wakes you again when you have slept long enough. It senses levels of pressure, oxygen, and other chemicals in the blood, and it adjusts the rate of breathing and heartbeat accordingly. Since it regulates these vital functions, serious injury to the brainstem (especially the medulla) usually causes instant death.

The brainstem also regulates the body temperature and signals the skin capillaries to dilate or constrict to release or conserve heat. It controls the movements of the eyes, face, tongue, and throat. It receives signals from sense organs and uses them to help you maintain your balance. It notifies you of pain in various parts of the body, and it transmits signals from the cerebrum to the parts in distress.

You may notice that many activities of the brainstem also take place in the cerebrum. Usually these activities are initiated in the brainstem, and from there the signals pass to the cerebrum for further processing. This suggests that the brainstem is the seat of the mind, which includes the will, the emotions, and the intellect.

Most major nerves extend from the spinal cord, but twelve pairs called the cranial nerves come directly from the brain. Eleven of these relay motor or sensory information between the brain and the head, neck, and upper chest organs. One pair, called the vagus nerves (vā'·gəs), affects functions such as heart rate, constriction of blood vessels, and contraction of muscles in the digestive system. Thus the cranial nerves control vital life functions—which explains why a person with a spinal cord injury can continue to live.

Neurons die quickly if they do not receive a constant supply of blood. When the body is at rest, 15 percent of the blood flows through the carotid artery to the brain. (It returns through the jugular vein.) If a blood clot or other obstruction blocks the flow of blood to the brain, the result is a stroke. Then brain cells will die, and the functions controlled by those cells will cease.

Man has developed amazing machines; but none of them match the abilities of the brain, which can control the body, respond to the body's needs, and coordinate all these functions with matchless accuracy. God formed man from the dust of the earth and called him good. God never needed to improve anything He created. As you read about your brain, it is doing something that no man-made or mechanical device can do. It is thinking about itself.

—————— Study Exercises: Group B ——————

1. God provided several means of protection for the central nervous system.
   a. What protection serves the brain only?
   b. What protection serves the spinal cord only?
   c. What protection serves both the brain and the spinal cord?

2. The central nervous system is called *central* because
   a. it lies in the center of the body.
   b. it is the center that receives and sends out nerve signals.
   c. it connects the peripheral nervous system with the autonomic nervous system.

3. Name the two nerve centers of the central nervous system.

4. Name the part of the brain that is most closely associated with each function.
   a. coordination
   b. emotion and regulation of body functions
   c. most sensory and motor functions

5. Name the lobe of the cerebrum that a person would use primarily for each activity.
   a. Watching a bird preen its feathers.   c. Reading Braille with the fingertips.
   b. Quoting Psalm 23.   d. Planning a good way to get rid of mice.

6. True or false? A single activity may involve many parts of the brain at the same time.

7. A stroke results when
   a. the brain bounces against the inside of the skull.
   b. a blockage interrupts the flow of blood to the brain.
   c. blood seeps into the brain.
   d. cerebrospinal fluid fails to drain from the ventricles.

Group B Activities

1. *Experimenting with problem solving.* A marvelous aspect of the mind is its ability to reason and solve problems, which is called intelligence. Each of the following problems requires a different kind of intelligence. Try them all. Which one do you think is easiest? Which one do your classmates think is easiest? Most persons excel in one kind of thinking. What brain lobes do you use to solve these problems?

   a. **Symbol sequence.** Tell which symbol comes next in the series.

   b. **Word problem.** Change *tide* to *salt* by changing one letter at a time. Example for changing *time* to *cash: time, tame, came, case, cash.*

   c. **Number problem.** What two-digit number has digits with a sum of 7 and a difference of 5?

2. *Testing your coordination.* Get a flat piece of wood about one foot square, and a piece of heavy, uninsulated wire about two feet long. (A metal coat hanger will do.) Bend the wire into a wavy shape, and also make a closed loop and "handle" with a short piece of the wire. Run the wire through the loop; then fasten the ends of the wire to opposite corners of the wood square, as shown in the picture.

   Get a 6-volt light bulb, a 6-volt lantern battery or a hobby transformer, and three pieces of flexible insulated wire. Attach one piece of insulated wire between one end of the wavy wire and one battery terminal. Attach another piece between the other battery terminal and one terminal of the light bulb. Attach the third piece between the other terminal of the light bulb and the heavy wire loop.

   Test your coordination by trying to move the loop along the wavy wire without making contact between the two. The light bulb will light when the loop touches the wavy wire. Your aim is to keep the bulb from lighting.

3. *Dealing with icy walks.* Compare your list with the one below.

Spread salt on the walks.	Put heat pipes in the concrete.
Build a roof over the walks.	Install moving sidewalks.
Scrape off the ice each morning.	Move the walks to the equator.
Pour hot water on the walks each morning.	Wear cleated shoes.
Shine heat lamps on the walks on cold days.	Stay at home on icy mornings.

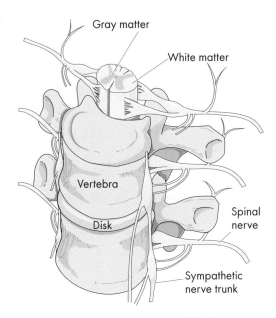

Gray matter

White matter

Vertebra

Disk

Spinal nerve

Sympathetic nerve trunk

The Spinal Cord

The *spinal cord* is the secondary part of the central nervous system. It continues from the base of the brainstem, narrowing to the approximate thickness of the little finger and extending 15 to 18 inches (38 to 46 cm) through the protective tunnel of the vertebrae. Like the brain, the spinal cord includes both white matter and gray matter. A cross section shows an area of gray matter in the center of the spinal cord, shaped somewhat like a swallowtail butterfly. The "tails" are the dendrites that receive sensory impulses from the peripheral nervous system. Unlike the brain, the white matter of the spinal cord is on the outside. This white matter contains the axons that send messages up the spinal cord to the brain or out of the spinal cord to muscles and glands.

Through the center of the gray matter is a vertical canal that contains cerebrospinal fluid. This canal joins at the top to the ventricles of the brainstem and cerebrum.

Thirty-one pairs of nerves extend from the spinal cord, one pair for each vertebral joint. These nerves travel out to the 93,000 miles (150,000 km) of peripheral nerves in the same way that the twelve pairs of cranial nerves carry messages to other parts of the body. Each spinal nerve controls one of thirty-one bands of skin area that run crosswise on the trunk, lengthwise along the arms and legs, and around the scalp.

If a bee stings you on the leg, the nerve responsible for that region carries a message of pain to your spinal cord, where the message travels along a nerve leading to your brain. Judging from the nerve that brought the message, your brain determines that the pain is coming from your leg. From prior experience, you can even decide without any information from your eyes that a bee caused this pain.

Voluntary responses. Some responses of the nervous system are voluntary, and some are involuntary. A voluntary response is an action that is done with conscious thought. It begins in the brain, representing a choice of the will. The most common voluntary responses are learned skills, such as writing and drawing.

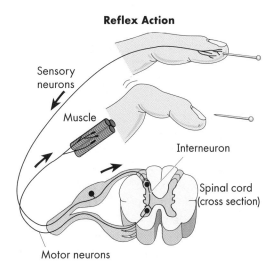

Reflex Action

Sensory neurons

Muscle

Interneuron

Spinal cord (cross section)

Motor neurons

The more you practice a certain skill, the more automatic it becomes. For example, walking is a voluntary response that becomes less voluntary with age. A one-year-old must work carefully to balance himself. If a noise distracts him, he will stop and likely fall. But most people are so skilled at walking that some even walk in their sleep. Although you choose to walk from one place to another, the pattern is learned and therefore becomes automatic.

Involuntary responses. An involuntary response is an automatic action that is done without conscious thought. Many involuntary responses are functions that happen almost without our awareness. Breathing, heartbeat, and other vital organ functions are involuntary responses. However, you have some conscious control of certain involuntary responses. For instance, you can hold your breath and suppress a cough.

The most automatic response is a *reflex.* If an insect suddenly darts straight toward your face, you will blink and perhaps swat at it. God built this response into the body as a protection against danger. By blinking your eyes shut, you shield a very sensitive organ from possible harm.

Some reflexes are more voluntary than others. The most involuntary reflexes are the inborn ones, such as blinking and swallowing, which are present even in an infant. However, the swallowing reflex becomes less automatic with age as higher levels of the central nervous system take control of the action. Other reflexes are conditioned or learned. You learn to jerk your hand away from a hot stove so that you are not burned. Experience has conditioned you to do that.

Reflexes do not even travel to the highest parts of the nervous system. Spinal reflexes travel to the spinal cord, brainstem reflexes to the brainstem, and cortical reflexes to the

cerebral cortex. In this way the body reacts very swiftly without waiting for a decision from the cerebrum. As a result, you receive the quickest possible protection.

A reflex action begins with a *stimulus,* something that initiates action. In the diagram on page 259, the stimulus is a sharp object. A receptor, in this case a nerve ending in the fingertip, receives the stimulus and sends a signal by sensory neurons to a reflex center, in this case the spinal cord. In the reflex center, an *interneuron* receives the signal and sends a response impulse by motor neurons to an organ that responds to the stimulus. In the example, the response organ is a skeletal muscle that jerks the finger away from the sharp object.

A response organ could also be a muscle of the heart, which would beat faster in response to a frightening noise. It could be the muscle of an abdominal organ, such as your intestines, which increase peristalsis in response to excitement; or it could be a gland such as your stomach, which produces extra gastric juices in response to stressful situations.

A complex reflex often occurs in a limb on the opposite side of the body. If you jerk your foot away from a sharp thorn, for example, a muscle in your other leg will contract to help you keep your balance.

Doctors often test one of the stretch reflexes by tapping with a reflex hammer just below the knee, which briefly stretches the tendon there and its attached muscle. If the nervous system is healthy, the upper-leg muscle contracts and the leg jerks.

Each of these reflexes happens at a tremendous speed, too fast for you to even think about it. Imagine what would happen if you did not jerk your hand away from a hot stove. Reflexes are another of the marvelous ways that God provided for the body to care for itself.

———————— Study Exercises: Group C ————————

1. The spinal cord is different from the brain in that most of its white matter is on the (inside, outside) and most of its gray matter is on the (inside, outside).

2. Write whether each response is *voluntary* or *involuntary*.
   a. talking
   b. singing
   c. peristalsis
   d. producing saliva

3. Explain how good reflexes are essential to your safety.

4. Label each of the parts of a reflex response.
   a. The neuron in the reflex center that transfers a signal between two other neurons.
   b. A cause that initiates an action.
   c. A neuron that receives the message and sends it to the reflex center.
   d. A neuron that carries the response away from the reflex center.
   e. The nerve ending that receives the message which initiates the reflex.

5. Choose two: Response organs involved in reflexes include (nerves, muscles, glands, interneurons, receptors).

6. With age, many inborn reflexes become more (automatic, voluntary), but many voluntary responses become (automatic, voluntary).

7. A doctor often tests the ——— reflex by tapping just below the knee.

Group C Activity

Testing the stretch reflex. Sit on a table, and let your legs hang over the edge. Have someone tap just below one knee with a small rubber mallet or similar tool. Be sure to use something soft that will not cause injury. Watch your leg jump.

The Peripheral Nervous System

The Senses

Many of the stimuli carried by peripheral nerves come from our environment. They enter our consciousness through five remarkable senses that God gave us—sight, hearing, touch, smell, and taste. These senses bring to us the beauty of a lily, the music of a wren, the softness of a baby chick, the fragrance of a lily, and the sweetness of a ripe strawberry.

But more than these delights, our senses protect us. They show us the rock in our path. They warn us of the rattlesnake in the underbrush. They restrain us from the hot burner that would sear a hand. They alert us to the electrical wire that is overheating. They tell us about the food that is spoiled.

Even if there were no dangers to avoid, we would still need our five senses for our daily work: hammering nails, cleaning out calf pens, threading needles, hanging up the wash, and flavoring spaghetti. Often we use as many as four of our five senses for a single task, without even thinking about it. Consider how you might use each of your senses to cook a meal. How many senses could you use to run a lawn mower or a tractor?

"The hearing ear, and the seeing eye" are

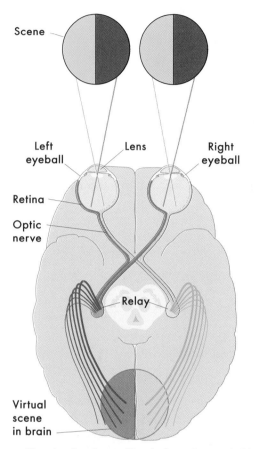

The visual pathway. Signals from the nasal side of the retina cross over and join with the uncrossed signals from the outward side of the retina. The signals are subdivided by a special relay in the midbrain.

our two most important sense organs. "The LORD hath made even both of them" (Proverbs 20:12).

The sense of sight. The eyes are two spheres each almost 1 inch (0.4 cm) in diameter. They receive about one-third of all the sensory perception and more than one-half of all the information that the brain processes each day. The eyes contain 70 percent of the sense receptors in the body, which means they are the most important sense organs.

The optic nerve, one of the twelve cranial nerves, carries sensory signals from the retina at the back of each eyeball to the visual cortex,

a large area at the back of the brain. The more stimuli a sense gathers from the environment, the more of the cerebral cortex it occupies. Around one million axons form the optic nerve, which is about 40 percent of the axons in all twelve of the cranial nerves.

There are about 160 million photoreceptors in each retina, of which 10 million are the rods that perceive black and white. The other 150 million are the cones that perceive color, detail, and bright light.

As you hold out your two hands and look at them, the lenses in your eyes invert the images they receive. This means the images projected onto your retinas are upside down and the right side to the left. The signals from these images are carried by your optic nerves, which join partially behind your eyes and send signals from the left half of each retina to the right part of your brain, and from the right half of each retina to the left part of your brain. Also, different areas of the visual cortex may receive different qualities of the image in view. Your brain puts all this information together as one three-dimensional image.

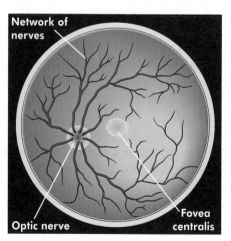

The retina. The fovea centralis is the area of sharpest vision. It enlarges an image 10 times more than other parts of the eye, but the brain automatically compensates for that distortion.

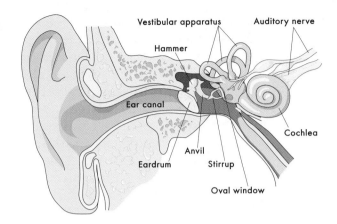

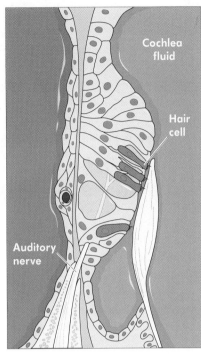

A cross section of cochlea lining

Double vision may indicate some type of disturbance in the brain, sometimes caused by improper drug usage. Optical illusions are said to fool the eyes, but it is actually the brain that makes a false interpretation based on previous experience.

The sense of hearing. Paired like the eyes, the two ears give perception of the direction and distance of the object making a sound. Hearing takes place in three stages. First, sound waves enter the outer ear and strike a membrane called the eardrum. From there the vibrations pass through the hammer, the anvil, and the stirrup, three tiny bones that strengthen the vibrations. (The names describe their shapes.) The stirrup passes the vibrations to the oval window, a membrane at the entrance of a coiled organ called the cochlea (kok´·lē·ə).

In the cochlea, the second stage of hearing takes place. Sound vibrations pass through the fluid in the cochlea, causing specific hair cells to vibrate according to the volume, pitch, and frequency of the sound. The cochlea is lined with about 3,500 of these hair cells, each bearing 40 to 60 hairs.

The third stage of hearing occurs as the hair cells convert the vibrations into electric signals and send them to the auditory nerve. This nerve carries the signals to the brain, where they are interpreted by the auditory cortex in the temporal lobes.

The sense of touch. Touch is probably more important than we normally think. Blind people can live without seeing. Deaf people can live without hearing. In fact, people such as Helen Keller have lived without either seeing or hearing. But it would be very dangerous to live without sensory perceptions from the skin. You could not detect the pain that warns you of danger. You could not sense when your body is becoming too warm or too cold. You could not hold objects or manipulate them without pressure sensors to tell you how hard to grip. You could hardly even walk.

Nerve endings fill every square inch of skin, but they are most concentrated on the fingertips and similar areas critical to fine motor skills. There are at least five different kinds of nerve endings in the skin. Some detect light touch, helping you to do fine and delicate work like typing or picking up fragile glassware. Other nerve endings detect heavy pressure, helping you to grip heavy objects tightly and sense when they are slipping from your grasp. There are separate sensors for detecting heat and cold.

Perhaps most outstanding of our sense of touch are the pain sensors. Acute or ordinary pain registers directly at the site of an injury, calling you to immediately change your behavior. Perhaps you pull your hand away from a sharp knife. Maybe you stay off a sprained ankle until it heals. The pain diminishes or may even stop when you heed its warning.

Chronic pain is usually not as sharp and clearly defined as acute pain. It may register as a widespread or burning sensation, too general to diagnose. A dull ache in your head does not mean that you have just hit your head sharply; it may be a symptom of some problem elsewhere in the body. For this reason, chronic pain

Kinds of Nerve Endings in Skin

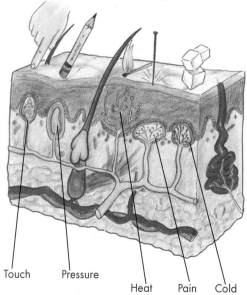

Touch   Pressure   Heat   Pain   Cold

is often difficult to treat. Sometimes a victim may feel extreme pain even when no nerve endings are being stimulated. Often this happens when earlier pain has left memory traces in his brain. There may also be an abnormal activity in his brain in which neurons incorrectly tell the victim that a part of his body is in pain. In such cases, no one can deny the pain, but treatment of the pain site will not help. Somehow medication or therapy must convince his brain otherwise.

Sensations often include signals from several different sensors in many combinations. For example, itching and tickling are often combined signals from lightly stimulated skin receptors.

The sense of smell. Though less important than seeing and hearing, smell is valuable for recognizing dangers, such as a gas leak or a hidden fire. Smell occurs when a cranial nerve called the olfactory nerve (ol·fak′·tə·rē) carries signals from the nose to the olfactory bulb in the brain. This bulb is in the part of the brain responsible for memory and emotion, which is why certain odors can trigger memories of long-ago events. For example, the smell of a new book may suddenly take you back to first grade. Some homeowners have taken advantage of the link between smell and emotion by baking bread just before a visit from a prospective buyer.

Of the five senses, we grow accustomed to smell most quickly. For the smell sensors lose perception after being exposed to a certain odor for some time. This makes it possible for people to work with materials that have a strong, disagreeable odor.

The colorless molecules of scent will not dissolve in water; but when they enter the nose, protein binds them to the mucous membranes that line the nasal cavity. About 10 million smell receptors lie in a small portion of the

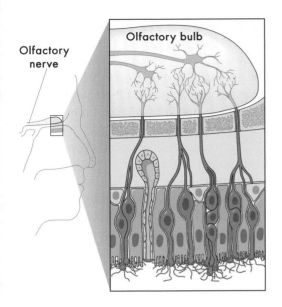

Olfactory nerve

Olfactory bulb

mucous lining of the upper nasal area.

The shapes of these receptors and the shapes of the scent molecules match each other like the pieces of a jigsaw puzzle. When a lilac scent molecule finds a matching receptor, it latches on; a nerve impulse travels along the olfactory nerve to the parietal lobe of the brain; and the brain interprets it as lilac. Though the olfactory bulb has only about 1,000 different receptors, the brain can discern as many as 10,000 different smells by putting combinations together. That is why you can walk into the house and say, "We're having scalloped potatoes and meat loaf for supper"—even before you see the food on the table.

The sense of taste. Like smell, taste is a sense you could live without; but consider how little you would look forward to mealtime if you could not taste or smell. Taste depends greatly on smell; if you cannot smell, you will also have little taste. This is why food seems tasteless when your nasal passages are congested from a cold.

Taste is sensed through chemical receptors on certain areas of the tongue, palate, pharynx, and epiglottis. There are about 10,000 taste buds in the mouth—9,000 in the little bumps on the tongue, and the rest in other parts of the mouth. Each taste bud contains about 100 cells. Chemical flavors of food dissolve in saliva and stimulate these cells, sending nerve impulses to the brain through cranial nerves. At the same time, the smell of the food greatly enhances the perception of taste.

Taste buds at the front of the tongue detect sweet and salty flavors. Those at the sides of the middle of the tongue taste sour flavors. Bitter flavors register at the back of the tongue. This is why a biscuit containing too much baking soda may seem sweet at first. Not until it reaches the back of your tongue do you detect the bitter aftertaste of excess soda. However, the tongue detects bitter flavors more quickly than sour, salty, and sweet flavors.

Though the tongue can sense only four flavors, delicate combinations of sweet, salty, sour, and bitter yield a wide range of flavors, broadened further by the texture of food. Touch receptors on the tongue are closer together than on any other part of the body.

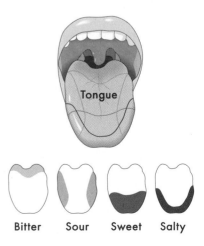

Tongue

Bitter   Sour   Sweet   Salty

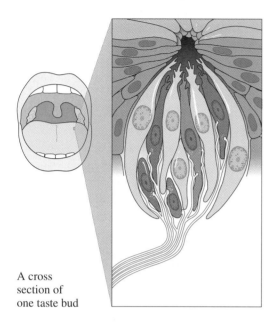

A cross
section of
one taste bud

The sense of balance. This sense, called *equilibrium* (ē′·kwə·lib′·rē·əm), is a combination of various senses. It is perceived mainly by the balance organs in the inner ears, which include fluid-filled loops called the semicircular canals. These loops, along with other small parts, form the *vestibular apparatus* (ve·stib′·yə·lər).

As you rotate your head and move your body, the fluid in the vestibular apparatus stimulates hairs that send signals to the cerebrum to record the position of your head. These signals combine with signals from your eyes and from touch receptors to tell your brain the position of your body. Your brain then sends signals to skeletal muscles to move limbs and other body parts to maintain your balance in relation to gravity. Because there are three arches, each set at a different angle, the brain is certain to receive signals from one of them whenever you turn your head. Equilibrium enables you to sense the position of your limbs without looking at them.

Dizziness results when the equilibrium is disturbed. This is easily caused by spinning rapidly in circles, which causes so much stimulation in the vestibular apparatus that the equilibrium is confused. Dizziness also occurs when internal sensation differs from external reality. Perhaps you have lain on the ground and watched the clouds floating by overhead. Because you were not at the same time seeing objects around you that helped to maintain your equilibrium, it may have seemed that you were moving while the clouds were still. However, you had an internal knowledge that you were not moving. This difference of appearance and actuality can make you feel dizzy.

A similar thing happens with motion sickness. Normally when you are in motion, your body is moving in relation to immediate surroundings. But in a car, you and your immediate surroundings (the car) are moving together. Other cars and trees seem to pass you rather than appearing that you are passing them. You know that you are sitting still inside the car, but you can see that you are also moving. It is this conflicting information that makes you feel dizzy.

─────────── **Study Exercises: Group D** ───────────

1. Which group of words best summarizes the roles of the senses?
   a. warning, judging, general body function
   b. pleasure, warning, general body function
   c. pleasure, sight, warning
   d. sight, hearing, pain

2. a. What percent of the sense receptors in the body are in the eyes?
    b. What percent of the axons in the twelve cranial nerves are in the optic nerve?
    c. These percentages show that the sense of sight is of (high, moderate, low) importance to the body.

3. Name the sense that is served by each kind of nerve.
    a. auditory   b. optic   c. olfactory

4. One person may not see an optical illusion that another person sees because
    a. he has poorer eyesight.   c. his retina registers it differently.
    b. the cones in his eyes are different.   d. his brain interprets it differently.

5. Write the missing words or phrases to describe the process of hearing.

    Sound first causes the *(a)* ——— to vibrate. These vibrations are strengthened by three small bones called the *(b)* ———, *(c)* ———, and *(d)* ———. The last of these causes the *(e)* ——— to vibrate. Then the vibrations pass through a fluid in the *(f)* ———, which has coils lined with tiny *(g)* ———. These send electrical signals through the *(h)* ——— to the brain.

6. Which group of words best summarizes the sense of touch?
    a. pain, pressure, temperature   c. pressure, heat, vibration
    b. heat, cold, pain   d. pressure, cold, pain

7. Provide the missing words by naming two senses.

    The sense of ——— is not very strong without the sense of ———.

8. For each sense organ, choose the stimulus to which it responds. You will use one choice twice.
    a. eyes   *chemical molecules*
    b. ears   *light*
    c. nose   *sound*
    d. tongue

9. a. Why does spinning in circles cause dizziness?
    b. In what situation might you feel dizzy even if you are not moving?

Group D Activities

1. *Experimenting with binocular vision.* When you look at an object, each eye sees a slightly different image; but the images overlap, and the brain merges the two into one. This is called binocular vision. You can merge some strange images by having your eyes transmit different pictures.

    Get a cardboard tube about 1 inch (2 or 3 cm) in diameter, or roll a piece of paper into a tube. Look through the tube with one eye. Hold your hand about 8 inches (20 cm) in front of the other eye, placing the edge against the tube. Look straight ahead with both eyes open. Your binocular vision will merge the view at the end of the tube with the view of your hand, and your hand will appear to have a hole in it.

Hold two pencils horizontally about 12 inches (30 cm) in front of your eyes, and move them toward each other until the points are about ½ inch (1 cm) apart. Focus on something beyond the pencils. You will see a short, double-pointed pencil "floating" between the two pencils you are holding. Again your binocular vision is fusing the two images; but since your focus is beyond the pencils, the images overlap instead of merging into one. You see the right end of the left object and the left end of the right object merged together, as well as a proper image on either side. You can do the same thing by bringing your two index fingers close together. This does not work, however, if you close one eye.

2. *Discovering your dominant ear.* Have one classmate say a word into your right ear at the same time another classmate says a different word into your left ear. Which word did you hear, the one in your right ear or the one in your left ear? You probably heard the word spoken into your dominant ear. This is the ear that leads in detecting sound.

3. *Experimenting with the spacing between nerve endings.* Holding two toothpicks side by side, touch their points to your tongue. Can you feel two points? If you feel only one point, move the toothpicks farther apart. Now try the same thing on your fingertips, on the palm of your hand, and on other skin surfaces. Which surfaces are the most sensitive?

4. *Finding the locations of various taste buds.* (Do this activity at home.) Stick your tongue out. Touch a piece of sweet candy to the side of your tongue. Slowly drag it up toward the tip of your tongue. Where does the candy begin to taste sweet? Try the same thing with a salty cracker. Where are your salt taste buds? Drag a sour pickle along the side of your tongue, beginning at the tip. Where does the pickle begin to taste sour? Now dip a finger in unsweetened cocoa powder. Touch it to the tip of your tongue, the sides of your tongue, and the back of your tongue. Where does it taste bitter?

5. *Noticing the connection between taste and smell.* Pinch your nose shut, and eat a peanut. The peanut probably will not have much flavor. Now release your nose, and eat another peanut. It should have more flavor.

The Endocrine System

Some activities in the body happen swiftly or fairly swiftly. The heart beats about 75 times every minute. The lungs inhale about 25 breaths per minute. Electrochemical impulses zip through the nervous system at speeds faster than 100 miles per hour (160 km/h). Other activities are spread out over hours, days, and even years. The stomach digests the food. The kidneys purify the blood. The body grows in size. These things happen as the body responds to chemicals in the bloodstream much as plants respond to chemicals in fertilizer.

The *endocrine system* is a system of glands that are constantly secreting hormones, which control the body's long-range functions. A hormone is one kind of protein, as you may remember from Unit 3.

Other glands in the body also secrete substances. Salivary glands secrete saliva, and sweat glands secrete sweat. These are not endocrine glands but exocrine glands. *Ex-* means "out"; the exocrine glands have ducts for secreting their fluids, which eventually go *out* of the body. Endocrine glands have no ducts. They secrete their hormones directly into the bloodstream as blood flows through them.

Scientists have identified at least 200 different hormones, which come from about ten different glands. These glands work closely with the nerves to control various functions in the body. A region in the brain called the hypothalamus links the work of the nerves and the glands. This section discusses five main endocrine glands and makes brief mention of a sixth one.

The pituitary gland. The master gland of the endocrine system is the *pituitary gland* (pi·tōo′·i·ter′·ē), which coordinates the work of the other glands. This little two-lobed gland, only ⅓ inch (85 mm) in length, lies in the center of the head just above the pons. The pituitary gland has its own work to do besides its management responsibilities.

The front lobe of the pituitary gland produces a hormone that stimulates the growth cells in the long bones. Too much of this hormone will cause a person to become a giant. The tallest man of modern times, whose height was 8 feet 11 inches (2.72 m), was diagnosed with tumors of the pituitary gland. However, too little of this hormone causes people to be dwarfs. One man was a dwarf of less than 4 feet (1.22 m) at age twenty-one, but he died a giant of 7 feet 8 inches (2.34 m).

The front lobe of the pituitary gland also triggers hormone production in other glands. The back lobe controls mineral levels in the blood by regulating the amount of water that the kidneys reabsorb.

The pineal gland (pin′·ē·əl) is a gland in the brain that is shaped like a tiny pine cone. It produces melatonin (mel′·ə·tō′·nin), a hormone that is involved in regulating sleep, wakefulness, and other functions of the body's biological clock. You have probably heard older people saying they cannot sleep well. One reason is that this gland secretes less melatonin as we grow older.

The thyroid gland. Shaped like a butterfly, the *thyroid gland* wraps around the trachea at the base of the larynx. This gland keeps the calcium level in the blood from rising too high. It also produces hormones that control the rate at which the body uses food, called its metabolism. The thyroid needs iodine to work properly. A lack of iodine will cause goiter, a condition in which the thyroid becomes enlarged. In a severe case, the thyroid swells so much that it restricts vital functions in the throat.

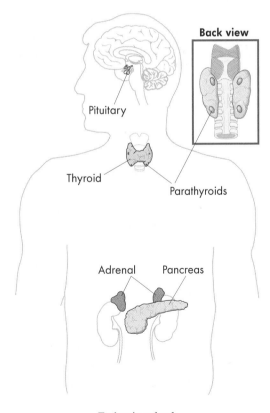

Endocrine glands

If a person's thyroid produces too much hormone, his body will use up food too fast. The person may lose weight and have a higher temperature because his body is working harder. His heart will need to pump faster so that the blood brings more oxygen to "burn" the food faster. The victim becomes tense and irritable.

A lack of thyroid hormone causes a person to become sluggish and listless. If there is a severe deficiency in childhood, the person may become physically stunted and mentally handicapped. Such a dwarf is called a cretin (krēt'·ən).

The parathyroid glands. The four little *parathyroid glands* lie embedded in the back of the thyroid. (*Para-* means "beside.") While the thyroid suppresses the calcium level in the blood, the parathyroids raise it by taking calcium from the bones. If these glands produce too much hormone, bone tissue breaks down from the loss of calcium. Too little hormone will cause bones and teeth to grow abnormally. Remember from your study of food and digestion that a lack of calcium can result in muscle spasms. Too little parathyroid hormone can cause muscle spasms for the same reason.

The pancreas. Behind the stomach lies the *pancreas,* shaped like a triangular flag that points toward the spleen. You may remember from Unit 3 that the pancreas produces pancreatic fluid to help digest fats. In the digestive process, the pancreas is an exocrine gland with ducts.

The pancreas is an endocrine gland in that it secretes the hormone *insulin* directly into the bloodstream. Insulin causes body cells to absorb sugar from the blood, which lowers the glucose level in the blood. A lack of insulin causes diabetes, a condition of high glucose levels, which upsets the body's fluid balance. Diabetic persons restrict the sugar they eat, and many take daily injections of insulin to control their sugar levels.

The pancreas also produces a hormone called glucagon (gloō'·kə·gon), which raises the glucose level by causing the liver to release sugar into the bloodstream. Low sugar levels can starve the brain.

The adrenal glands. Perched on top of the kidneys are the two *adrenal glands* (ə·drē'·nəl), which are shaped like triangular caps and have two layers. The outer layer, called the cortex, produces cortisone and several other hormones. One of these hormones helps the body to break down proteins and release fat and sugar into the blood. Another hormone of the cortex prompts the kidneys to absorb sodium from the blood. This controls the blood pressure and determines the volume of blood that the heart pumps.

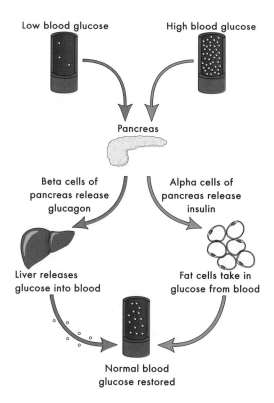

Low blood glucose   High blood glucose

Pancreas

Beta cells of pancreas release glucagon

Alpha cells of pancreas release insulin

Liver releases glucose into blood

Fat cells take in glucose from blood

Normal blood glucose restored

The inner layer of the adrenal glands is the medulla, which produces adrenalin in stressful situations. Adrenalin boosts the sugar level in the blood, raises the blood pressure, and speeds up the heartbeat. These changes provide the means to respond to an emergency with sudden energy and strength, sometimes far beyond one's normal ability. If the grass fire you are tending gets out of control, you will suddenly be able to run faster and beat the flames harder than ever before. After the fire is out, you will suddenly feel very exhausted.

The endocrine system shows how the small marvels of God's creation are extremely important.

Study Exercises: Group E

1. What is the difference between endocrine glands and exocrine glands?

2. *Impulse* is to ——— *system* as *hormone* is to ——— *system.*

3. Why is the pituitary gland called the master gland of the endocrine system?

4. Name the gland or glands responsible for each function.
   a. Helping body cells to absorb sugar from the blood.
   b. Regulating the body's use of food.
   c. Regulating the body's biological clock.
   d. Controlling calcium levels in the blood. (two answers)
   e. Preparing the body to deal with an emergency.
   f. Controlling sodium in the blood, which affects blood pressure.

5. Name the endocrine gland or glands whose malfunction could cause each condition.
   a. giantism
   b. dwarfism (two answers)
   c. goiter
   d. diabetes
   e. mental deficiency

6. Briefly describe the position of each gland or set of glands.
   a. pituitary gland
   b. thyroid gland
   c. parathyroids
   d. pancreas
   e. adrenal glands

7. The endocrine system is closely tied to the function of
   a. the digestive system.
   b. the circulatory system.
   c. the nervous system.
   d. all the systems of the body.

Group E Activity

Controlling sugar in the diet. Most diabetics restrict their intake of sugar. Perhaps you think this would be difficult to do, but anyone can learn to eat less sugar. Try a bowl of cereal without adding any sugar. Make a list of any other foods you ate for breakfast that had sugar in them. Was the sugar added in preparation, or was it a natural part of the food? Sugar added in preparation is the main sugar that diabetics restrict, but they must also limit their intake of foods high in natural sugar.

The Autonomic Nervous System

Many functions of the body work automatically. You do not need to look after your breathing, the beating of your heart, or the working of your digestive system, for God provided the autonomic nervous system to keep these vital functions working. (*Autonomic* means "self-governing.") Consider how dangerous it would be if your heart depended on your conscious control. Suppose you were engrossed in an interesting book and forgot to keep your heart beating. Forgetting would be fatal.

There are two divisions of the autonomic nervous system: the sympathetic division and the parasympathetic division. Remember that the thyroid and parathyroid glands reverse each other's actions. The same thing is true of the sympathetic and the parasympathetic divisions of the autonomic nervous system. For each action of one, there is an opposite action of the other.

The sympathetic division. The nerves in the sympathetic division extend from the middle and lower spinal column, and they prepare the body for action. This division is sometimes called the fight-or-flight reaction because of how it affects the body in emergencies. The sympathetic division also activates in apprehensive situations, such as visiting the dentist, going to a new school, and speaking before an audience (stage fright).

The respiration speeds up to provide more oxygen for the blood. More sugar is released into the blood, the blood pressure rises, and the heart beats faster to carry more fuel and oxygen to the skeletal muscles. The pupils enlarge to let in more light. The mouth grows drier as the salivary glands produce less saliva. The adrenal glands pour out adrenalin. To prevent overheating, the thermostat in the brainstem stimulates more perspiration and suppresses

involuntary muscles whose action would produce unnecessary heat. If the environment is too cold for proper muscle function, the thermostat stimulates these involuntary muscles to generate more heat for the body.

These physical preparations are useful in times of danger, such as occasional emergencies. Normally the danger is over in a short time, and the body returns to a more relaxed state. Through fear and anxiety, however, many people keep the sympathetic division working actively for long periods of time. This is exhausting to the organs and contributes to stomach ulcers, high blood pressure, strokes, and other physical problems. God's solution for stressful situations is, "Be careful [anxious] for nothing; but in every thing by prayer and supplication with thanksgiving let your requests be made known unto God" (Philippians 4:6).

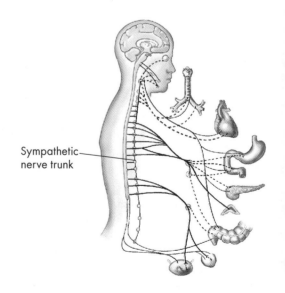

Sympathetic nerve trunk

The sympathetic division serves various parts of the body. The spine diagram on page 259 shows the sympathetic nerve trunk realistically.

The parasympathetic division. Like operations in the work world, the body needs time to stop, refuel, prepare the fuel for use, and get rid of wastes. The parasympathetic division is the part of the autonomic nervous system that takes care of these functions. After meals, extra blood leaves the skeletal muscles and moves to the digestive system. Respiration and heartbeat operate at normal rates. This is like the energy-saving mode of some machines, conserving energy when there is no immediate threat or need for extra fuel and oxygen in the skeletal muscles.

With an understanding of the autonomic nervous system, you can help your body to work better by allowing it time to relax and digest its food. But if you are agitated at mealtime or bedtime, you make demands on both divisions of your autonomic nervous system at the same time. Then neither one can do its job well. You will be much more efficient if you allow time for your parasympathetic division to restore your body.

Besides normal patterns of exertion and relaxation, the autonomic nervous system is at work all the time to maintain equilibrium inside the body. It regulates chemical levels that would spell disaster if they rose too high or sank too low. Consider the carbon dioxide that cells release in the process of combining oxygen with fuel foods. A weak acid forms as carbon dioxide enters the bloodstream and dissolves in the plasma. This acid could cause serious problems by lowering the pH level of the blood. But if the acid level rises, the brainstem calls for faster breathing to accomplish faster removal of carbon dioxide from the blood. As a result, the blood remains at its nearly neutral pH between 7.45 and 7.35.

Although you cannot directly control the functions of the autonomic nervous system, your thoughts and thought patterns influence these processes. Your parasympathetic division cannot do its job well if you have a habit of being tense and worried. But if you are calm and cheerful, your entire muscular system will relax and every part of your body will function better. "A merry heart doeth good like a medicine" (Proverbs 17:22).

Doctors observe that people who are anxious and fearful often have poorer physical health, whereas those who are calm and optimistic tend to have better health. God does not intend that we live in tension and fear. He wants us to enjoy a spirit "of power, and of love, and of a sound mind" (2 Timothy 1:7).

Study Exercises: Group F

1. *Autonomic* means
   a. "self-operating."   b. "self-governing."   c. "self-repairing."
2. Why is it important that the autonomic nervous system function automatically?
3. Tell how the sympathetic division of the autonomic nervous system affects each organ during an emergency.
   a. lungs
   b. heart
   c. eyes
   d. skin
   e. involuntary muscles
   f. salivary glands
   g. adrenal glands

4. It is best for the body if the sympathetic division operates
   a. continually.   b. frequently.   c. occasionally.

5. Tell how the parasympathetic division of the autonomic nervous system affects each organ after a meal.
   a. lungs   b. heart   c. skeletal muscles   d. digestive organs

6. Why should you not do heavy work immediately after eating a large meal?

7. You cannot directly ——— the functions of the autonomic nervous system, but you can ——— them by your thoughts and thought patterns.

Group F Activity

Observing the function of the sympathetic and parasympathetic divisions. Fill in the data sheet below. You will need to borrow a blood pressure meter for blood pressure. Count your own pulse by placing two fingers on the inner side of your wrist just below your thumb. For your breathing rate, you will need to lie down and have someone count the rise and fall of your chest. Have someone else observe your pupil size. Look on your forehead or around your mouth to detect any sweat.

	Sympathetic (after vigorous exercise)	Parasympathetic (while sitting down after a meal)
Blood pressure		
Pulse		
Breathing rate		
Pupil size (large or small)		
Sweat (present or absent)		

Consciousness, Memory, and Emotion

Some very special things happen in the brain, things that cannot all be defined scientifically. Yet they are the defining characteristics of humans. Perhaps we could say this is part of what God breathed into man when he became a living soul. These special features distinguish humans from the animal creation.

Consciousness. You know what it is to be conscious, but how would you define that state? Consciousness includes an awareness of one's surroundings and an ability to think about oneself and one's environment. Many animals are also conscious (especially the mammals), but none have the degree of consciousness that is present in humans.

Although the brainstem and the cerebral cortex work together to maintain consciousness, no single part of the brain is specifically responsible for this state. Consciousness involves thoughts, feelings, desires, moods, and more. If you could take every part of the brain and add up each of its functions

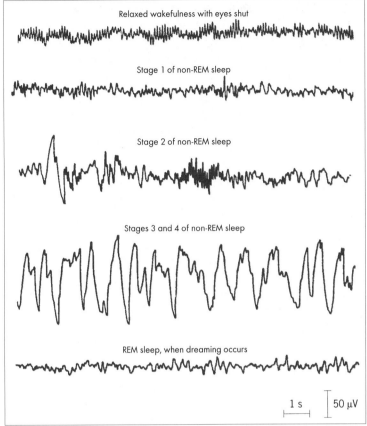

Relaxed wakefulness with eyes shut

Stage 1 of non-REM sleep

Stage 2 of non-REM sleep

Stages 3 and 4 of non-REM sleep

REM sleep, when dreaming occurs

1 s | 50 μV

These electroencephalogram (EEG) tracings show electrical activity in the brain during different stages of sleep. Stage 2 has mixed frequencies. Stages 3 and 4 have higher and longer waves because the neurons in the brain are less active. The brain waves of an alert person are similar to those of a person in REM sleep.

also a handicap to be so distracted by our environment that we cannot focus on a necessary task. Healthy consciousness includes deciding what is most important to think about now and shutting out distracting stimuli.

Sleep is an important and interesting state of suspended consciousness. During sleep the heartbeat, breathing, metabolic rate, and temperature decrease. The body benefits as the voluntary muscles rest and the various hormones and other proteins do their building and repairing.

When we first fall asleep, we lapse into a light slumber that grows progressively deeper. After about 1½ hours, we enter a lighter sleep called REM sleep. (REM stands for "rapid eye movement" and is pronounced "rem.") During this period the eyes move under closed eyelids much as they would when we are awake and thinking.

Periods of REM sleep return every hour or two, each one becoming longer until the length is about half an hour. During these periods the cerebral cortex is as active as when we are awake; but because we are not fully conscious, we are not aware of sensory signals. Dreaming occurs during REM sleep; however,

individually, you would come up with less than what the brain really does. The total function of the brain is far greater than the combined working of all its parts.

These cooperating parts help us to focus our attention by disregarding sensory signals that do not relate to the task of the moment. Sometimes people are said to be absent-minded when they are so absorbed with something, that they have little awareness of their surroundings. This can be a handicap, but it is the brainstem blocks signals to the motor neurons that would act out our dreams.

Infants may have eight hours of REM sleep per day. Scientists think that during this time, new neuron connections develop in their brains. For older children and adults, REM sleep is a benefit to the mind because this is when new material from the previous day is processed and recorded. This is apparently how things are permanently stored in the memory.

Memory. Scientists do not understand how the mind stores information, but they do know that learning causes new connections to form between specific brain cells. You recall a memory by reactivating the appropriate circuit; however, the information comes into your consciousness only when it is recalled. You do not have to think about something all the time to remember it. For example, you have hundreds of mathematical facts in your memory. You cannot think of all the addition, subtraction, multiplication, and division facts at once, but you can recall them one at a time.

Two categories of memory are short-term memory and long-term memory. Both are based in the cerebral cortex; but without the other parts of the brain, you cannot store memories effectively. Visual spatial memory remembers things from seeing where they are. This type of memory helps you to remember where your car is in a large parking lot. Loop memory remembers by repeating things over and over. This type of memory helps you to learn number facts and new words. The working memory makes use of stored information. It remembers where your car is and decides how far you will need to walk, or it recalls number facts and applies them.

Memory can also be categorized as declarative versus nondeclarative memory,

or specific versus implicit memory. Declarative or specific memory deals with facts and events. For example, you may remember that you started going to Sunday school at age four, and that in this first class the children sang "Jesus Loves Me." You remember that Columbus discovered America in 1492. Declarative memory occurs in the temporal lobe.

Nondeclarative or implicit memory deals with skills, habits, and attitudes. A habit is an action repeated so often that it is automatic. You learn to tie your shoes by habit, and this skill stays with you for life. After you learn to ride a bicycle, you will be able to ride one even after not seeing a bicycle for years. Nondeclarative memory involves the cerebral cortex and the cerebellum.

Some people classify individuals as being right-brained or left-brained. They assume that if the right hemisphere predominates, a person is better at dealing with shapes and symbols. Such a person is supposed to do well in art and music, to think creatively, to perceive emotions distinctly, and to have more negative feelings. If the left hemisphere predominates, a person is supposed to do well with language, to enjoy analytical thinking, and to have more positive feelings.

While it is true that certain abilities seem to predominate in one side or the other, that does not mean we are unable to develop a particular ability, with persistent effort. Each person has special abilities that can make a worthwhile contribution to the home, the school, and the church. In particular, any person can learn to overcome negative feelings and focus on the positive aspects of life.

The layers of the brain have varying thicknesses corresponding to the type of activity that occurs in each. Much of the cerebral cortex has six layers. Here is where conscious thoughts and actions are processed. A lower section of the brain, called the limbic system, has only three layers of cortex. Emotions, desires, and feelings are processed in the limbic system. However, there is no distinct division between these two regions.

It is important that we use our memories for good purposes. "A good man out of the good treasure of the heart bringeth forth good things: and an evil man out of the evil treasure bringeth forth evil things" (Matthew 12:35). Satan will have much less opportunity to tempt you with evil if your heart (your store of memories) is filled with good things.

Emotion. Five basic emotions are love, joy, fear, anger, and grief. These usually occur when the brain receives signals by sight, sound, taste, or other sensation. It decides whether the signal indicates something good or bad, and this determines the resulting emotion. Scientists believe that emotions occur largely in the temporal lobe of the cerebrum. However, this section sometimes blocks feelings of pain to keep us from feeling the emotion of the moment.

Mood is a general, long-term emotion, often affected by physical needs or desires. However, no one needs to be controlled by moods and emotions. There is a right and proper way to respond to the varying experiences of life. The Spirit of God can help us to find this way and follow it.

The mind is a very special gift from God. It is the crowning feature of God's creation. Scientists gradually learn more and more about the mind, but they probably will never understand it fully. Let us not pollute our minds with wrong thoughts, but rather think on things that are true, honest, just, pure, and lovely, as described in Philippians 4:8. Then God will set our minds at peace, and we can practice truly God-honoring behavior.

Study Exercises: Group G

1. Consciousness involves
   a. primarily the cerebrum.
   b. mostly the frontal cortex.
   c. mainly the cerebrum and the brainstem.
   d. most or all parts of the brain.

2. How can we focus our attention on a certain task when we are surrounded by distracting stimuli?

3. Compare REM sleep with consciousness.
   a. How are they alike?
   b. How are they different?

4. Sleep is important for us both physically and mentally.
   a. What is one physical benefit?
   b. What is one mental benefit?

5. Memory is most like
   a. an electrical circuit whose current grows weaker or stronger according to use.
   b. a nail pounded in place by repeated blows of a hammer.
   c. a book to which you add more material each time you learn something new.

6. For each question, write *long-term* or *short-term* to tell which type of memory an eighth grader would use to give the answer.
   a. Where did I put my books when I came home from school yesterday?
   b. What is six times eight?
   c. What color are my father's eyes?
   d. Did I have orange juice for breakfast?
   e. What season comes after winter?

7. Write *declarative* or *nondeclarative* to tell which type of memory you would use to answer each question.
   a. What is the capital of Maryland?
   b. Who was the first king of Israel?
   c. How do you tie a square knot?
   d. When did the Pilgrims land at Plymouth?
   e. How do you peel a potato?

8. Although emotions, desires, and feelings are processed in a part of the brain called the ———— system, they depend on input from (the temporal lobe, the frontal lobe, various regions of the brain) and can be influenced by conscious thought.

Group G Activities

1. *Experimenting with REM sleep.* Ask permission before doing this experiment. Observe a sleeping person. When you see his eyes rolling under his eyelids, waken him and ask if he remembers a dream.

2. *Comparing visual spatial memory and loop memory.* Have someone put 10 small items in a box and show them to you for several seconds. After he removes the box, draw simple sketches of the items in the box, and label them. To discover your visual spatial memory score, count the number of items you remembered correctly and divide by 10.

   Next, have someone write a line of 10 digits in random order on the chalkboard. Read them aloud. Erase one digit, and read the digits aloud again, inserting the missing digit from memory. Repeat this process, erasing one digit at a time until no digits remain. Now try to write all the numbers from memory. Count the number of digits you remembered correctly and divide by 10. This is your loop memory score.

   Which is better, your visual spatial memory or your loop memory? How do your scores compare with those of your classmates?

3. *Experimenting with a memory game.* In this game, one classmate names a state of the United States. (Any other familiar group of words can be used.) A second classmate repeats the state name and adds a second name. A third classmate repeats both names and adds a third name. Continue this process as long as you can, seeing how many state names you can remember in the order they are given. For best results, look at each classmate as he says the name of a state. Associate that name with him, his position in the room, or some other helpful detail.

Unit 7 Review

Review of Vocabulary

All the activities of the body are controlled by the nervous system, which has two main parts. The —1— is the part that receives messages from the environment. These messages travel through —2—, or nerve cells, each having several —3— to receive signals and a single —4— to pass signals on to another nerve cell. To get from one nerve cell to another, a signal crosses a chemical-filled gap called a —5—. The —6— is part of the peripheral system, but it controls automatic functions of the body.

The —7— is the system that receives and sends out nerve signals. This system includes the brain, which has three main parts. The —8— is identified by its deeply fissured outer covering, called the —9—; it controls voluntary responses. The —10— coordinates skeletal muscle activity, and the —11— controls the function of involuntary organs.

The —12— is the relay center for signals traveling to and from the brain, and it also controls some involuntary responses. The simplest involuntary action is called a —13—, which begins with a —14— that initiates action. The signal travels by —15— to a reflex center, where an —16— sends a response impulse by —17— to a muscle or some other response organ.

Besides the five senses that are commonly recognized, the —18— can be considered another sense. This is the balance that the body maintains through the —19— of the inner ear.

The —20— interacts closely with the nervous system by secreting hormones into the bloodstream. The —21— in the middle of the head coordinates the other glands of this system as well as the body height. The —22— in the throat region controls the body's metabolism. Embedded in this gland are the four —23—, which raise calcium levels in the blood. The —24—, a gland lying behind the stomach, secretes —25— to control the sugar level in the blood. Another pair of glands, the —26—, perch on top of the kidneys. They help to prepare food for the body's use, and they provide extra energy for emergencies and strenuous work.

Multiple Choice

1. Which statement in *not* true?
   a. Most of the mass of a neuron is composed of dendrites.
   b. Axons and dendrites are separated by synapses.
   c. There are more dendrites than axons in the body.
   d. Only axons send impulses, and only dendrites receive impulses.

2. Nerve signals travel from one cell to another
   a. through the myelin of the nerves.
   b. as electrical impulses generated by chemicals with opposing charges.
   c. as electrical impulses generated by magnetism in the body.
   d. when dendrites and axons make contact.

3. *Stimulus* is to *sense organ* as *response* is to (choose two)
   a. cerebrum.
   b. gland.
   c. spinal cord.
   d. reflex.
   e. muscle.

4. The chief control of the nervous system takes place in
   a. the central nervous system.
   b. the peripheral nervous system.
   c. the autonomic nervous system.
   d. the endocrine system.

5. Voluntary actions are generally initiated in the
   a. cerebrum.
   b. cerebellum.
   c. brainstem.
   d. spinal cord.

6. The cerebellum is most important as a center of
   a. gray matter.
   b. skeletal reflexes.
   c. internal organ control.
   d. coordination of muscular movements.

7. Removing the brainstem would put an end to
   a. thinking.
   b. living.
   c. voluntary muscle functions.
   d. reflex actions.

8. Reflex actions are valuable because
   a. they keep you from touching hot or dangerous objects.
   b. they save time by making quick responses possible.
   c. they contribute to long-term memory.
   d. they indicate a healthy nervous system.

9. If the spinal cord were severed, a person would lose his
   a. voluntary responses.
   b. endocrine gland function.
   c. reflex actions.
   d. memory.

10. Which group gives the correct sequence for a reflex action?
    a. receptor, sensory neurons, spinal cord, brain, response organ
    b. stimulus, sensory neurons, interneuron, motor neurons, response organ
    c. stimulus, sensory neurons, brainstem, motor neurons, response organ
    d. receptor, stimulus, interneuron, motor neurons, response organ

11. Reflex actions can best be classified as
    a. voluntary or involuntary.
    b. learned or inborn.
    c. blinking or swallowing.
    d. protective or nonprotective.

12. *Optic nerve* is to *sight* as
    a. *olfactory nerve* is to *tasting.*
    b. *cochlea* is to *hearing.*
    c. *auditory nerve* is to *hearing.*
    d. *retina* is to *eye.*

13. The balance achieved by the vestibular apparatus depends most on the senses of
    a. sight, hearing, and touch.
    b. sight, hearing, and smell.
    c. hearing, touch, and smell.
    d. touch, smell, and taste.

14. The gland that is both an endocrine gland and an exocrine gland is the
    a. pituitary gland.
    b. thyroid gland.
    c. pancreas.
    d. adrenal gland.

15. Insulin, adrenalin, and cortisone are all
    a. diseases caused by malfunctioning endocrine glands.
    b. materials used by the endocrine glands to produce hormones.
    c. hormones produced by the adrenal glands.
    d. hormones produced by the endocrine glands.

16. The general action of hormones is to ——— functions in the body.
    a. suppress
    b. increase
    c. regulate
    d. excite

17. The autonomic nervous system is most closely associated with
    a. conscious thought.
    b. involuntary responses.
    c. the cerebellum.
    d. work activities.

18. On the basis of your study of the autonomic nervous system, tell which pair of actions would be *least* likely to occur at the same time.
    a. active stomach muscles, slow heartbeat
    b. extra adrenalin in the blood, low saliva secretion
    c. inactive intestinal muscles, low saliva secretion
    d. enlarged pupils, slow heartbeat

19. Which statement about the cerebrum and the autonomic nervous system is true?
    a. Both control voluntary responses.
    b. Both are protected by cerebrospinal fluid.
    c. The cerebrum has an indirect influence on the autonomic nervous system.
    d. The autonomic nervous system is controlled by conscious thought from the cerebrum.

20. Consciousness, memory, and emotion involve
    a. the limbic system only.
    b. the cerebrum only.
    c. all parts of the central nervous system.
    d. many parts of the brain.

Unit 8

Health and First Aid

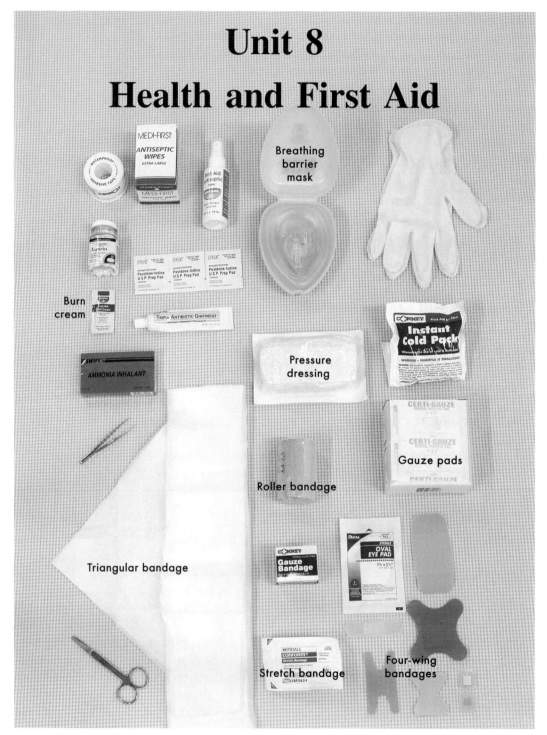

Burn cream

Breathing barrier mask

Pressure dressing

Gauze pads

Roller bandage

Triangular bandage

Stretch bandage

Four-wing bandages

A sampling of supplies in some first-aid kits

Searching for Truth

God created the body with natural drives to care for itself. When you are hungry, you eat. When you are tired, you sleep. When danger approaches, you step aside. Paul recognized this in Ephesians 5:29 when he said, "For no man ever yet hated his own flesh; but nourisheth and cherisheth it, even as the Lord the church." It is right that you should care for your own body. This is God's personal gift to you, a marvel of His handiwork. You should care for your body conscientiously and not abuse it, because God wants our bodies to be the temple of the Holy Spirit. "Therefore glorify God in your body, and in your spirit, which are God's" (1 Corinthians 6:20).

Searching for Understanding

1. Is there any connection between living the Christian life and having good health?
2. Are you obligated to help if you see a person who is injured?
3. What should you do if a person is choking or stops breathing?
4. How can bleeding be stopped?
5. When should you call 911?

Searching for Meaning

abrasion	dressing	pressure point
amputation	heat exhaustion	shock
avulsion	heat stroke	simple fracture
bandage	hygiene	splint
cardiopulmonary resuscitation	hypothermia	sprain
compound fracture	immobilize	strain
compressions	incision	tourniquet
dislocation	laceration	

Guiding Principles for a Healthy Body

It takes some scrubbing to dislodge dirt between the tiny rows of papillae in the skin.

Our bodies are "fearfully and wonderfully made" (Psalm 139:14). What man-made machine can repair itself? What modern engine can function day and night on so little fuel? But this marvelous organism needs our care to work at its best. Many people dishonor their Creator by misusing their bodies. They rob themselves of sleep and proper diet. They indulge in smoking and other practices that God never intended for them.

Some people become so concerned about their bodies that they damage their health by fear, worry, and anxiety. This lack of faith is also dishonoring to the Creator. You can best care for your body by following some basic principles of health. Then relax, knowing that you are acting wisely. If God chooses to allow illness in your body, you can then accept it as His will, knowing that you were doing what you could.

Other units in this science course deal with disease, food, body systems, and behavior. This unit gives an overview of practices that are helpful for maintaining good health, along with first-aid procedures to use in the event of sickness or injury.

Elements of Good Health

Genuine health includes three main elements as described below.

Spiritual well-being. "Honour thy father and mother; . . . that it may be well with thee, and thou mayest live long on the earth" (Ephesians 6:2, 3). Obedient, respectful children usually are healthy, happy children.

Sickness, disease, and death are part of the curse that God placed upon sin. Righteous living does not exempt Christian people from all the effects of this curse. Christians understand that their lives and health are in God's control and that their bodies are "the temple of the Holy Ghost." Therefore they observe healthful practices that are consistent with God's Word and that respect the gift of life and health that God has given.

Mental and emotional well-being. An emotionally healthy person can accept challenges and failure with a positive outlook. He learns to handle the stresses of life with cheerfulness. "A merry heart doeth good like a medicine: but a broken spirit drieth the bones" (Proverbs 17:22).

On two occasions, Christ told His disciples,

"Be of good cheer." But we cannot be cheerful in our own strength. We need the Spirit of God to bring us "love, joy, peace, longsuffering, gentleness, goodness, faith, meekness, temperance" (Galatians 5:22, 23).

Doctors recognize that hatred, anger, strife, impatience, worry, and self-pity work against good health. These feelings overstimulate the sympathetic division of the autonomic nervous system. Food digests poorly, causing stomach ulcers and abdominal pains. Muscles grow tense, leading to headaches and weariness. Increased blood pressure can lead to heart attacks and strokes. The immune system weakens, predisposing the body to cancer and arthritis. Cheerfulness cannot guarantee good health, but it greatly increases the likelihood.

Physical well-being. The remainder of this section focuses on this aspect. You can contribute to your physical health by observing good ***hygiene*** (practices that affect health).

Essentials for Good Health

Cleanliness and sanitation. Disease germs and other harmful substances surround you in the soil, in the air, and on the surfaces you touch. You cannot live in a dirt-free environment, but you can reduce infection and disease by bathing and washing regularly. Especially remember to wash your hands with soap and water before eating or handling food. Cooks should take special care to keep kitchens and cooking utensils clean.

Proper washing also removes the dirt of perspiration and dead skin cells. Brush your teeth twice daily to remove the food particles that feed the bacteria which cause tooth decay.

God is interested in cleanliness. For example, whenever the Old Testament priests offered a sacrifice or went into the tabernacle, they were to wash their hands and their feet (Exodus 30:20). This physical cleansing symbolized being spiritually clean in their service as priests.

God still wants us to be clean today. Hebrews 10:22 says, "Let us draw near with a true heart in full assurance of faith, having our hearts sprinkled from an evil conscience, and our bodies washed with pure water." This verse suggests both physical and spiritual cleansing.

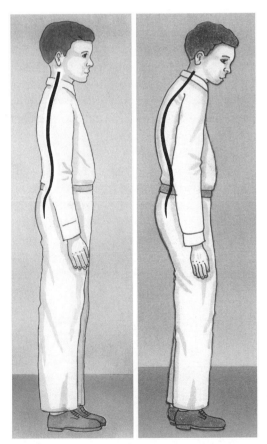

Compare the unnatural curves of the spine in poor posture with the natural curves in good posture.

Good posture. Posture refers to the way a person holds his body. The heart, lungs, and digestive organs can function better if they are not crowded by slouching or hunching. Poor posture also puts extra strain on the muscles, forcing them to do what the skeleton should be doing. Good posture helps you to stay relaxed but alert. Practice holding your head erect, your back straight, and your shoulders back. Good posture coupled with the ornament of a meek and quiet spirit makes you look neat, respectful, and secure.

Regular exercise. Exercise strengthens all the muscles, including the skeletal, heart, and respiratory muscles. It causes the heart to pump more oxygen to the brain, which improves mental function, speeds reaction time, and helps one to deal with stress. Exercise also builds strong bones and burns calories, thus helping to prevent disorders linked to excessive weight. It helps you to breathe better, tire less quickly, and sleep better at night.

Some exercise is aerobic (â·rō′·bik) and some is anaerobic (an′·ə·rō′·bik). Aerobic exercise involves rhythmic, sustained activities, such as walking, bicycling, and working in the garden. This kind of exercise causes the muscles to use oxygen at a steady rate. Anaerobic exercise involves a short burst of activity, such as a quick dash at top speed. The muscles use energy so rapidly that the body cannot supply them with oxygen fast enough, which leaves the person gasping for breath. In general, a person should have thirty to sixty minutes of moderate exercise every day. The best kind of exercise comes from honest, useful work.

However, exercising too strenuously in very warm weather can result in two serious conditions. One is *heat exhaustion,* which results when prolonged sweating removes body fluids faster than the body can replace them. This condition is most likely to occur if a person is not acclimated to very warm weather.

In heat exhaustion, the victim's skin becomes cold and clammy. His pulse races, his temperature rises, and he may feel dizzy, nauseated, and short of breath. Muscle cramps may develop in his abdomen or limbs.

To deal with heat exhaustion, first move the victim to a cooler place. Then if he is conscious and breathing naturally, give him a weak electrolyte solution and bathe his face with cold water. Heat exhaustion can usually be prevented by drinking plenty of water if it is necessary to exercise in very warm conditions.

A more severe condition is a *heat stroke,* in which the body becomes so overheated that it loses the ability to regulate its temperature. The person no longer sweats, and his skin becomes red and hot. He has a rapid heartbeat and a high temperature, and he may become delirious and even unconscious. A victim of a heat stroke needs emergency medical attention. Until such help arrives, move him to a cooler place and sponge or spray him with cold water. A fan may also be used to help lower his body temperature.

Exercising in extremely cold weather brings the danger of frostbite, a condition in which the fluids in body tissues freeze. Frostbite usually affects the nose, ears, fingers, or toes, which are especially subject to losing body heat. The affected tissue first tingles and later becomes white and numb. A person who notices these symptoms should find shelter immediately and warm the frostbitten area gradually without rubbing it.

Rest and sleep. Most adults need seven or eight hours of sleep every night. You probably need eight to ten hours because you are young and growing. To rob yourself of sleep hinders the body and mind from doing their best work during the day. Sleep renews all the body systems and the nervous system in particular. "The sleep of a labouring man is sweet" (Ecclesiastes 5:12).

You need times of rest even when you are awake. You may have noticed that when you use a muscle very hard for several minutes, it becomes weaker and sometimes hurts. This tiring of a muscle from overexertion is called muscle fatigue. A short rest will allow the blood to remove the waste materials that accumulate during exercise. Then you will be ready to work again.

Rest is as important to your health as exercise. It allows the parasympathetic division of the autonomic nervous system to work. Relaxation during and after mealtime allows the body to digest its food.

A balanced diet. Unit 3 of this book discusses food in detail. Every day you should eat

Sleeping Patterns at Various Ages

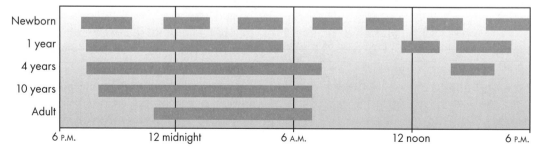

food from each of the basic five food groups: bread and cereal, vegetable, fruit, meat, and milk. This rule will assure that the body receives a good balance of proteins, carbohydrates, minerals, and vitamins, without excessive fats. Sufficient water is also important. God provided hunger and thirst as signals to tell when your body needs food and water and when it has had enough. If you heed these signals, you can avoid many problems and be a more productive person.

Proper clothing. Young people often neglect adequate clothing in cold weather. It is no sign of hardiness to go outside in winter without a coat and a cap or hood. Large amounts of body heat are lost through the head, so you need to wear proper headgear for outdoor winter activities. Being cold is not a direct cause of illness, but it lowers your resistance to disease.

Properly covering the head and body is also important in hot weather because it protects against the harmful ultraviolet rays of the sun.

Many people deliberately expose large areas of skin to sunshine in order to get a "healthy tan," but this is not wise. Prolonged exposure to sunshine causes premature aging of the skin, destroys its elasticity, and may cause skin cancer.

However, if you dress modestly and wear proper head protection, you need not worry about getting too much sunshine through honest labor in gardening, farming, or construction. Only small areas of skin will be exposed to the sun; and if sunburn is a problem, you can protect the exposed areas by applying a sunscreen lotion.

Adequate clothing also protects you from invaders such as disease germs, chemicals, and the irritations of insects and plants. Proper footwear protects your feet from these dangers. Items such as raincoats and boots are important in wet weather to keep your body dry. Moisture evaporating from wet skin causes chilling that can lower resistance to disease in the same way that cold weather does.

Children must be dressed well for snowy activities.

Study Exercises: Group A

1. Explain how extreme concern about the care of the body can dishonor God.

2. Give three main elements of genuine health.

3. Explain how proper cleanliness promotes good health.

4. Good posture aids good health because
   a. it makes you think more positively about yourself.
   b. it gives your organs sufficient space to function properly.
   c. it makes others respect you more.
   d. it keeps your muscles tense and ready for action.

5. Tell how regular exercise aids each of these body systems.
   a. skeletal system
   b. muscular system
   c. respiratory system
   d. circulatory system
   e. nervous system

6. Exercising in very warm weather may cause heat exhaustion or a heat stroke.
   a. How does heat exhaustion develop?
   b. How does a heat stroke develop?

7. What is frostbite?

8. Tell how proper rest benefits each of the following systems.
   a. nervous system
   b. muscular system
   c. digestive system

9. Why is it important that you eat foods from each of the five food groups daily?

10. Tell why proper clothing is important in each of these kinds of weather.
    a. cold weather
    b. sunny weather
    c. rainy weather

Group A Activities

1. *Showing the importance of washing hands.* Sprinkle 1 tablespoon of unflavored gelatin over ½ cup of water. Let it stand for 5 minutes to soften the gelatin. Meanwhile heat ½ cup of water to a boil. Reduce the heat to low, and stir the softened gelatin into the hot water until the gelatin dissolves.

   Remove the gelatin mixture from the heat, and divide it equally among 3 petri dishes or small glass jars. Cover and refrigerate the containers several hours until the gelatin sets. Remove the containers from the refrigerator.

   Cough or breathe on your hands; then put your fingerprints on the surface of the gelatin in one dish. Cover the dish, and label it "Dirty hands." Wash your hands with water but no soap, and put your fingerprints on the surface of the gelatin in the second dish. Cover it, and label it "Hands washed with water only." Wash your hands with soap and water, and put your fingerprints on the surface of the gelatin in the third dish. Cover it, and label it "Hands washed with soap and water."

Label each dish with today's date, and place the dishes in a warm, dark place. Check them each day. Note on each the date when you first see mold growing in your fingerprints. Which dish develops mold first?

2. *Keeping a record of exercise.* For one week, keep a record of your daily activities. About how many hours per day do you spend in each activity, including sleeping, eating, sitting, working, and playing? How much of your exercise is aerobic? How much of it is anaerobic? Do you exercise 30 to 60 minutes per day? Do you sleep 8 to 10 hours per night?

Safety Rules

Safety rules are not a needless nuisance that someone devised to spoil our fun and make life inconvenient. Instead, they are designed for our own well-being. Even the Law of Moses has some safety rules. In Deuteronomy 22:8, God told the Israelites to build banisters around their house roofs to prevent falls. If we ignore safety rules, we will be the ones who suffer for it.

Protecting the eyes. God gave you a pair of eyes; but if you ruin that pair, you cannot get another. For this reason you should wear goggles when doing something that could harm your eyes. A flying chip of wood or metal from a saw, a wood splitter, or a grinder can quickly embed itself in an unprotected eye. Also wear goggles when using corrosive sprays and chemicals, such as battery acid and strong cleaners.

If a piece of dirt gets into your eye, do not rub it. First blink your eye. If that does not remove the irritation, wash your hands and pull your lower eyelid down. Look upward while you or another person checks for a speck of dirt under your lower lid. Lift out any loose, foreign object with the corner of a clean cloth or the tip of a moist cotton swab. If you cannot find the object, try the same thing with your upper lid, looking down.

You may be able to remove dirt from your eye by holding it under a water faucet and allowing water to wash across your eyeball from the inner corner out. However, you will need medical aid if a foreign object is embedded in your eye.

If you are welding, wear a shield that protects your whole face against flying sparks and other flying particles. A welding helmet should also have a tinted lens rated for the type of welding that is being done. The

Welding with adequate face protection

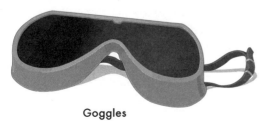

Goggles

intense rays and ultraviolet rays of welding can cause temporary or partial blindness. Some helmets have a lens that darkens automatically.

Protecting the ears. Extremely loud noise can deafen you partially or completely. Wear earplugs or sound-deadening earmuffs around power machinery, such as chainsaws, wood- and metal-working equipment, lawn mowers, and weed trimmers. In addition, continuous loud noise from heavy traffic, large equipment, or loud music can permanently damage the nerves in the ears. If you cannot reduce the noise or get away from it, wear ear protection.

Protecting the respiratory system. Pollen, sawdust, and airborne mold can severely irritate the nasal passages and lungs. God equipped you with hairs and mucous membranes to filter out much of this dust, but they do not remove it all. A simple facemask will protect you from these fine particles. The allergies and respiratory infections it prevents is worth your trouble in learning to use one.

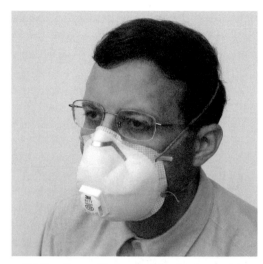

While a disposable facemask has its place, a deluxe mask (shown above) is more comfortable and more effective in filtering out harmful substances.

Other kinds of respiratory hazards require a special respirator with perhaps a supply of fresh air or oxygen. Avoid spraying materials with strong vapors, such as paint, without wearing a respirator designed to filter out irritating chemicals. You should also work in a well-ventilated area with a fan or an exhaust system to reduce both the risk of lung damage and the possibility of fire from combustible vapors.

Protecting the skin. Wear protective clothing such as a heavy, long-sleeved shirt when you work around flying sparks or other flying particles. If you use an irritating chemical such as paint thinner or dry-cleaning fluid, wear gloves to protect your skin. Thinners evaporate quickly, drying and chapping your skin. Also, harmful chemicals can enter your body through your skin.

Never work around hazardous equipment in your bare feet or with light footwear. Steel-toed shoes will help to protect your feet from heavy falling objects.

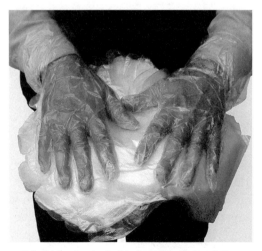

It is fairly economical to keep hundreds of disposable gloves handy in the shop or house.

FAILURE TO READ AND FOLLOW INSTRUCTIONS ON THIS LADDER MAY RESULT IN INJURIES OR DEATH.

Protecting against falls. Prevent falls by keeping stairways clear and lighted. Pick up toys from the floor, and wipe up spills immediately. Do not leave low objects, such as footstools, where someone might trip over them. Extension cords should not be laid in places where people normally walk. Buildings should have safety rails along stairways and ramps. Any deep holes on a property should be well guarded.

When using ladders and stepladders, avoid falls by obeying signs such as those that caution you not to stand too close to the top. Also resist the temptation to use an unstable stack of stools or boxes as a makeshift ladder. And never stand on a rolling piece of furniture.

Protecting against fires. Always keep a fire extinguisher near your workspace for any activity that could present a fire hazard. This includes cooking spaces and workshops. Place a bucket of water near sites of welding or similar activities. A quick splash of water will sometimes extinguish a small flame with less damage than using a fire extinguisher. However, be sure not to pour water on grease or electrical fires. Suffocate those fires with a heavy blanket or an appropriate fire extinguisher. Baking soda will put out small fires around a stove.

It is good to keep a fire blanket handy. If your clothes catch fire, wrap in that blanket and roll to put out the flames. Another good precaution is to have smoke detectors that will alert you to fire before it traps you in a building.

Protecting against electrical dangers. Standard household current has only 120 volts, but this can be enough to cause electrocution. When you work with electrical appliances, keep your hands and all electrical connections dry. Wet clothes, hands, and appliances increase the danger of electrical shock.

Unplug an appliance by grasping the plug, not the cord. Pulling on the cord may break the wires, resulting in the potential for electric shock or a fire caused by a short circuit. Always keep electrical cords in good repair, and never place them under rugs. Repeated walking across a rug-covered cord will eventually break the insulation and expose the wires.

Fuses and circuit breakers are designed to keep electrical circuits from being overloaded. If a fuse blows, never replace it with a fuse rated at a higher amperage. This could allow more current to flow through the circuit than the wire can safely handle, thereby overheating the wire and causing a fire.

Using automobiles and power equipment safely. Develop the habit of always fastening your seat belt when you ride in an automobile. It is true that a seat belt may cause injury in a serious crash, but this danger is minor compared with the danger of being hurled against the dashboard or windshield or being thrown out of the vehicle.

Never operate power equipment with shields removed from gears, belts, blades, or other moving parts. Many people have lost fingers, hands, and even arms for this reason. Do not disconnect safety devices that are intended to disable a machine when the operator is not in control. Finally, do not operate power equipment if you are too young, too inexperienced, or too tired to use it safely. The danger of being maimed or even killed is too great to be worth the risk.

Shields for PTO drivelines and for other moving parts should be attached securely before each use.

Study Exercises: Group B

1. Which statement does *not* give a good rule for eye protection?
   a. Wear goggles for protection against flying particles or strong chemicals.
   b. If a piece of dirt gets into your eye, avoid rubbing the eye.
   c. Carefully remove any foreign object embedded in your eye.
   d. When welding, wear a helmet with a properly tinted lens.

2. a. Both ——— loud noise and ——— loud noise can cause hearing loss.
   b. What are two ways to protect your ears from loud noise?

3. Write the missing words for these sentences about protecting the respiratory system.
   a. A ——— will protect against pollen, sawdust, and airborne mold.
   b. A special ——— is needed for protection against irritating chemicals.

4. One kind of danger that requires skin protection is ——— ——— or ——— particles, and another kind is harmful ———.

5. If you cannot remove a hazard that might cause a fall (such as a deep hole), how can you reduce the risk?

6. You should use different methods to fight different kinds of fire.
   a. What can you use to extinguish most fires?
   b. What kinds of fire should be extinguished with a blanket or a suitable fire extinguisher?

7. Give three rules for avoiding fire or electrical shock when using electrical cords.

8. Never replace a blown fuse with a fuse rated at a ——— amperage.

9. What is a good answer to the argument that a seat belt may cause injury in a crash?

10. Tell when you should avoid operating power machinery because of the condition of
    a. the machine.   b. the operator.

Group B Activities

1. *Measuring noise levels.* Find a sound-level meter that measures decibels. These are available in most science catalogs, but you may be able to borrow one. Record the decibels of each of the following sounds: whispering, normal conversation, a slamming door, a vacuum cleaner, a circular saw, and a lawn mower. Even brief exposure to noises over 120 decibels can damage your hearing, but you should wear ear protection for extended exposure to noises measuring over 85 decibels. How many of the noises you measured require ear protection?

2. *Observing wear of an electrical cord.* Find a length of heavy electrical cord that is no longer in use. Place it under a rug in a place of constant travel, being sure that it will not be a hazard. After a week, remove the cord and peel away the outer covering. What do the wires look like? Compare this cord with a cord in good condition.

3. *Collecting safety warnings.* Look for safety warnings on containers, machines, tools, and equipment in your environment. Copy them into a notebook. Classify them according to the type of harm they are intended to prevent.

First-aid Procedures

A very familiar Bible story involves first aid. The Good Samaritan had compassion on a half-dead, wounded traveler, "and went to him, and bound up his wounds, pouring in oil and wine" (Luke 10:34). His love moved him to administer first aid.

First aid is immediate care given to an ill or injured person until professional medical care can be provided. No one is legally bound to provide first aid unless he is a parent or legal guardian of the victim. But Christian love moves one to help. Laws called Good Samaritan laws protect responsible, unpaid persons

from liability if they attempt first aid in good faith but cause further injury.

General Principles

In any emergency, the first step is to assess the situation for danger to yourself. The danger may be fire, hazardous chemicals, electricity, or the potential of falling or drowning. Remember that Christian love does not mean you should be reckless about your own safety. You cannot help another person if you become a victim yourself. If you cannot approach the victim safely, call for professional help and stay at a safe distance.

The main goals of first aid are to preserve life, prevent worse injury, and promote recovery. Summon emergency medical care in the event of choking or severe breathing difficulty, severe bleeding, drowning, electrocution, poisoning, severe burns, paralysis, spinal injury, or possible heart attack. In some cases you may need to call for medical help before administering any first aid. But if the victim is likely to die or be injured worse before you can do this, try to give aid first. Perhaps someone else can call for medical service while you try to help the victim.

In most areas, 911 reaches emergency medical services; otherwise, simply dial 0 to reach the operator. Before calling, learn what happened, what kind of injury is involved, and whether the victim is breathing. Be ready to give the address and telephone number at the scene, along with appropriate landmarks to describe the location. Do not hang up until you are told that you may do so.

With medical help on the way, stay at the scene and calmly give any aid that is needed.

Calmness will help you to make wiser decisions and to better comfort the victim. If possible, use gloves to protect yourself and the victim from infection. Except when a greater danger threatens, never move a victim. You could injure him further.

The A-B-C-D-E of First Aid

When you reach the victim, first find out if he is conscious. Tap or lightly squeeze his shoulders, and ask, "Are you all right?" If he is conscious, tell him what you plan to do, and obtain his consent before administering first aid. If the victim does not respond, begin what is called the A-B-C-D-E of first aid.

A: Check the victim's <u>airway</u>. Blood, food, or vomit can block the airway of an unconscious person, or his tongue may block his throat. Carefully sweep your finger through his mouth to remove obstructions.

B: Check the victim's <u>breathing</u>. Observe his chest. Is it rising and falling? Listen for the sounds of breathing. Put your cheek close to his mouth to feel his breath on your cheek.

If the victim is not breathing or you cannot tell, then . . .

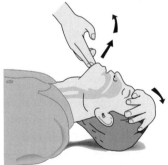

1. Begin by tilting the victim's head back and lifting the chin. Loosen any clothes around the neck.

2. Look, listen, and feel for breathing by putting your ear close to his mouth and watching his chest. If he is not breathing, go to Step 3.

3. Fit your mouth over the victim's mouth. Lift the chin and pinch his nostrils together. Blow gently until his chest rises. Repeat this step.

If the victim still does not breathe, go to Step 4.

(Check pulse in the groove between Adam's apple and side of the neck.)

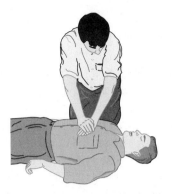

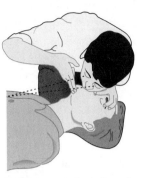

4. Check the pulse. If a pulse is present, give 12 breaths in 1 minute. Recheck for pulse and breathing. Continue these cycles until breathing returns. But go to Step 5 if the pulse is absent.

5. Find your hand position on his lower breastbone with your shoulders directly above your hands. Depress his chest 1½ to 2 inches (4 to 5 cm) for 30 times in 20 seconds.

6. Give two slow breaths. Complete 5 cycles of compressions and breaths in 2 minutes, and check again for a pulse.

Continue until his pulse and breathing return, or until medical help arrives.

If the victim is not breathing, begin *cardiopulmonary resuscitation,* or CPR. The *pulmonary* part of the procedure is also known as rescue breathing or artificial respiration. With one hand on the victim's forehead and the other under his chin, tilt his chin up. Lift his lower jaw to move the tongue away from the back of the throat. Pinch the victim's nose shut to prevent your breath from escaping through his nose. Then seal your mouth over his mouth. (For an infant, seal your mouth over his mouth *and* nose.) Blow slowly and gently until his chest rises. Release his nose, remove your mouth from his, and listen for escaping air. Repeat this process a second time. If the victim does not begin breathing on his own, go to the next step.

C: Check the victim's <u>circulation</u>. The normal adult pulse is around seventy-five beats per minute. If his pulse is good, continue rescue breathing until the victim begins breathing on his own or until help arrives. If you grow tired, try to find another person to continue the rescue breathing.

If the victim's pulse is weak or absent, begin the *cardio* part of CPR by doing *compressions* of his heart. Put one hand on top of the other and place the heel of the lower hand just above the base of the victim's breastbone. With your arms perpendicular to his body, press vertically and depress his chest 1½ to 2 inches (4 to 5 cm). Repeat this thirty times in a third of a minute (30 times in 20 seconds); then give 2 rescue breaths. Complete five cycles of compressions and breaths in two minutes, and check again for circulation. Continue until the pulse returns and the victim begins breathing on his own, or until medical help arrives.

The timing of the cycles is the same for an infant or a young child, but use only two fingers and less force in the compressions.

D: Check the victim's <u>disability</u>. This means evaluating the victim's ability to respond. Trained rescue workers do this by using a scale called AVPU. *A* means that he is *alert;* he knows who he is, where he is, and what time it is. *V* means that he responds to *verbal* stimulus but is not fully alert. *P* means that he responds only to *painful* stimulus, such as a pinch on his collarbone. *U* means that he is *unresponsive* to any stimulus.

E: Try to <u>expose</u> any hidden injury. You may need to cut away clothing and cover the victim with a sheet or blanket. Look and feel carefully to check for DOTS, an acronym that stands for *deformity, open wounds, tenderness,* and *swelling.*

First Aid for Choking and Drowning

Choking can cause death in just a few minutes. It is caused by a blockage, often food, in the victim's airway. A person can usually go four minutes without oxygen and suffer no brain damage. Up to six minutes may mean minor damage, but ten minutes or more will cause almost certain brain damage or death.

A choking person often clutches his throat (this is the universal choking sign). He may make weak, high-pitched sounds and try to cough, but he cannot speak. He will likely turn blue.

First sweep your finger carefully through his mouth, seeking for a lodged object. Be careful not to push an object farther down his throat. If that does not clear his airway, do not pound him on the back. This may drive the obstruction farther down. Instead, proceed with abdominal thrusts (also called the Heimlich maneuver) as described below. These thrusts are designed to dislodge the obstruction by forcing air out of the lungs.

Stand behind the victim, and wrap both arms around him. Clench one fist with your thumb turned inward and upward under his breastbone. Grasp this fist with your other hand. With your arms beneath his rib cage, pull sharply inward and upward. Repeat this four times in rapid succession; then check for results. If the victim is still choking, repeat the procedure three more times before calling for emergency medical help.

First Aid for Choking

(Conscious)

(Unconscious)

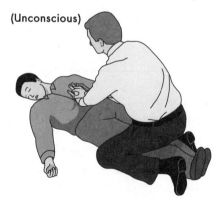

(Infant)

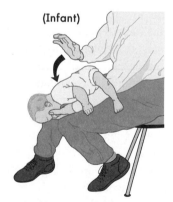

Place clasped fist with thumb against the abdomen just above the navel. Give quick, upward thrusts. Repeat until object is coughed out. If victim loses consciousness, call 911 and do the next procedure.

Loosen any clothes around the neck. With victim lying flat, tilt his head and give 2 breaths, as shown on page 294.

If air does not go in, place heel of one hand just above the navel. Give up to 5 abdominal thrusts. Lift his jaw, and sweep out the mouth. Retilt the head and give breaths. Repeat whole sequence as needed.

Place infant face-down on your forearm with the head lower than the chest. Give 5 back blows between the shoulder blades. Turn infant over, and give 5 chest thrusts at the center of the breastbone. Repeat blows and chest thrusts as needed. Intersperse with breaths and mouth sweeps if infant loses consciousness.

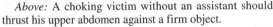

Above: A choking victim without an assistant should thrust his upper abdomen against a firm object.

Left: Recovery position for an unconscious person. The hand should not block the mouth but elevate the mouth away from a dusty or dirty surface.

If you are the choking victim, thrust your chest with your own fist, or do abdominal thrusts by pushing your abdomen against a chair back.

If an infant is choking, place him face down on your forearm, supporting his head and neck on your hand. Hold his head below his chest. Give five blows with the heel of your hand between his shoulders. Then turn him over and give five chest thrusts with two fingers. Continue alternating back blows and chest thrusts for one minute or until he stops choking. If the choking continues, call for emergency medical help.

An unconscious victim is in danger of choking even if he has normal respiration and circulation. Unless you suspect a head or spinal injury, turn him to the recovery position. This is a side position that keeps his airway open by allowing any secretions to drain from his mouth. Turn the person on his side with his lower arm extending in front of him. Bend his legs to keep him from rolling farther, and bend his upper arm around so that the hand is palm down under his face. You should end with his thighs at a right angle to his body.

Drowning is another leading cause of accidental death. Most drownings happen because persons, especially children, venture too near to deep water without proper precaution and without the ability to swim well.

When you try to help a drowning victim, reach from solid ground with a pole or rope if possible. Call for professional help rather than risking your own life, for a drowning person can pull even a strong swimmer under the water. If you must swim to the victim, keep a buoyant device between yourself and him. A spare tire makes a good life preserver if nothing else is available.

Water does not enter a drowning person's lungs until he becomes unconscious. First the air passages close, and he suffocates from lack of oxygen. Water enters his stomach, sometimes forcing the stomach contents into his throat and choking him. Thus the first step is to open the victim's airway. If he is then breathing, place him in the recovery position.

If he is not breathing, begin rescue breathing. Rarely should you perform chest compressions or abdominal thrusts on a drowning victim.

A drowning person frequently suffers **hypothermia** (abnormally low body temperature) from prolonged exposure to cold water. Cover him with a warm, dry blanket, and move him to a warm place as soon as possible.

The best way to prevent drowning is to be cautious along banks, observe posted signs, and avoid spending time in the water by yourself. Sometimes a child falls headfirst into a large bucket of water and drowns. Prevent this hazard by emptying water buckets that are within reach of small children.

Try to stay calm in any situation that requires first aid. Pray for God's help as you approach an emergency scene, for He has more wisdom and power than any trained emergency personnel.

Study Exercises: Group C

1. In an emergency, why should you consider the danger to yourself before trying to help someone else?

2. Give the three main goals of first aid.

3. Which group best summarizes the facts that you should be ready to give when you call 911?
   a. cause, condition, location
   b. address, telephone number, landmarks
   c. name, age, type of injury
   d. how, when, where

4. Give the five words represented by the A-B-C-D-E of first aid.

5. Fill in the missing words in these sentences about cardiopulmonary resuscitation.
   a. The *cardio* part refers to the heart. In this part of the procedure, the rescuer tries to restore circulation by the use of ———.
   b. The *pulmonary* part refers to the lungs. In this part of the procedure, the rescuer tries to restore breathing by the use of ——— ———.

6. Give the words represented by the letters in each acronym.
   a. AVPU   b. DOTS

7. Why is choking a serious emergency?

8. What is the purpose of abdominal thrusts?

9. Which one is *not* a good way to help a drowning person?
    a. Reach out with a pole, or throw a rope to him.
    b. Call for professional help.
    c. Swim to him, and keep his head above water.
    d. Push a buoyant device to him.

10. Answer the following questions with these words: *abdominal thrusts, chest compressions, rescue breathing.*
    a. Which may you need to perform on a person rescued from drowning?
    b. Which should you *not* usually perform?

Group C Activities

1. *Learning to administer CPR.* Invite trained CPR personnel to demonstrate this procedure. You may also practice with a CPR demonstration unit rented from a local health department. It is not recommended that you practice the CPR procedure on fellow classmates.

2. *Practicing first-aid procedures.* Find a life-size rag doll, and use it to practice abdominal thrusts. Also use the doll to practice the recovery position. It is not recommended that you practice these procedures on fellow classmates.

    Use an infant-size doll to practice the first-aid procedure for choking infants. It is not recommended that you practice this on real infants.

First Aid for Wounds and Bleeding

In the story of the Good Samaritan (Luke 10), the robbers left the man by the side of the road, wounded and bleeding. The Good Samaritan knew how to bind up those wounds. Basic knowledge of wound care may enable you also to spare someone from death.

If you need to deal with bleeding, first wash your hands with soap and water. This will avoid further contaminating the wound with dirt or germs from your hands. If you do not know the victim, it is wise to wear gloves to protect yourself from any disease that may be in his blood. Summon emergency medical help in cases of severe bleeding.

Control of bleeding. There are three types of bleeding. Arterial (är·tir′·ē·əl) bleeding, involving an artery, is associated with very severe accidents. The blood spurts out rapidly under high pressure. Venous (vē′·nəs)

bleeding, involving a vein, is also serious but is less critical. The blood flows out in a steady stream. Capillary bleeding is the most common and least serious type. The blood flows more slowly in capillary bleeding.

To control any type of bleeding, first apply pressure directly over the wound with a clean or sterile cloth. This direct pressure will usually stop the blood. If it does not, elevate the affected part, if possible, to reduce the pressure of the blood at the site of the wound.

If heavy bleeding still continues, apply pressure to a ***pressure point*** located between the wound and the victim's heart. By this method, you will cut off the blood supply to the affected part of the body, similar to cutting off the water flow by kinking a hose. The pressure point diagram on page 300 shows those places where an artery can be pressed against

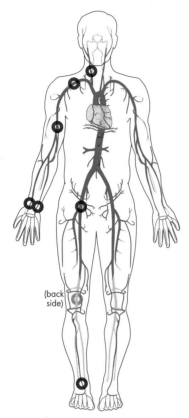

Pressure points on one side of the body

a nearby bone to stop bleeding.

These methods will stop most bleeding, but in rare cases you may need to apply a **tourniquet** (tûr′·ni·kit). This is a band of cloth twisted tightly around an arm or a leg to cut off the blood flow completely. Since lack of blood quickly causes death in a body part, you should use a tourniquet only as a last resort. This method is so risky that it should be applied only if a limb has been amputated or will probably need to be amputated.

If a person's nose begins bleeding, have him sit down quietly and lean forward. While he breathes through his mouth, he should pinch the sides of the nose together firmly without letting go. If any blood drains into his mouth,

have him spit it out. After five to ten minutes, check to see if the bleeding has stopped. If the bleeding persists for more than twenty minutes, get medical help.

Categories of wounds. Wounds are placed in several classes according to their nature.

Abrasions. An **abrasion,** commonly known as a brush burn, is generally a minor wound. The skin is scraped by a hard, rough surface, such as concrete, which breaks the small blood vessels. Dirt and bacteria easily enter an abrasion.

Incisions. An **incision** is a deep, clean cut. Though it may bleed little, sometimes it injures tendons and underlying tissue.

Technique to control nosebleeds.

Lacerations. A **laceration** is like an incision except that it has jagged edges due to tearing of the flesh. There is greater risk of infection with a laceration than with an incision.

Punctures. A puncture is a deep injury caused by a pointed object, such as a nail. Though the hole is often small and bleeds little externally, there is a likelihood of internal bleeding. Punctures also carry a high risk of infection. If a large pointed object is still embedded in the wound, do not try to remove it, for that may cause uncontrollable bleeding. Get the person to a doctor as soon as possible.

Avulsions. An **avulsion** involves a body part that is partly or completely torn away. When a finger, toe, or limb is cut off, the injury is called an **amputation.** Bleeding is severe. If the part is still attached, do not sever any remaining tissues. Instead, realign the parts as best you can, and call for help immediately. Do not discard any tissue from the wound site.

If the part is completely severed, do not wash it but wrap it in a clean plastic bag. Wrap the bag in a clean cloth, and submerge this in ice in a plastic container. The cloth and plastic keep the part from freezing, but it must remain cold. Clearly label the container with the victim's name and the time of the injury. A doctor can often attach a severed part if he receives it within several hours, but the part must be cooled right away.

Treating minor wounds. If a wound is shallow, begin by washing it with soap and running water. Do not use high pressure, and do not scrub the wound. Gently remove small pieces of dirt by using a tweezers.

After the wound is clean, apply an antibiotic ointment and then cover it with a ***dressing.*** This is a sterile piece of gauze or cloth applied directly over the wound to control the bleeding and to protect the wound from contamination. A dressing should be larger than the wound, and it should be thick, soft, and lint-free.

Next apply the ***bandage.*** A bandage is a piece of cloth or tape used to hold the dressing in place. It should be clean, but does not need to be sterile. A bandage also puts pressure on the wound, which helps to control swelling. Some larger bandages, such as triangular bandages, are used to support and ***immobilize*** the wounded part.

A Band-Aid (or similar product) is a commercially prepared dressing and bandage that is applied as one piece. These items come in a number of shapes and sizes for treating a variety of wounds.

When you bandage a hand or a foot, leave the fingers or toes exposed unless they are wounded. This will allow you to gauge whether you have made the bandage too tight. Starting below the wound with a roller bandage, wind around the affected limb and work upward. Make the bandage snug, but do not cut off circulation. Never apply a circular bandage to a victim's neck lest you cut off his airway.

A cold compress helps to control swelling. This may be an ice pack, a cold pack, or a chemical pack that becomes cold when an inner capsule is broken and two chemicals react. Do not apply a cold compress directly to the wound, but wrap it and place it over the bandage or on adjacent skin for about twenty minutes.

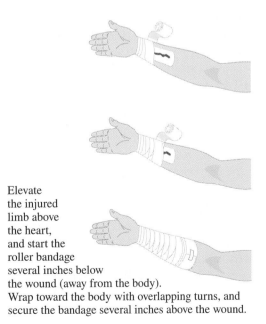

Elevate the injured limb above the heart, and start the roller bandage several inches below the wound (away from the body). Wrap toward the body with overlapping turns, and secure the bandage several inches above the wound.

Treating major wounds. You will need medical help if a wound reaches the bone, if it bleeds profusely, if it is on the face and may leave a large scar, or if it is full of dirt that could cause an infection such as tetanus. Tetanus is a life-threatening infection that can cause a condition called lockjaw.

A head wound may cause a concussion, which means the brain has been bruised by bumping into the skull. Symptoms include pale skin, dizziness, blurred vision, nausea, headache, and sometimes partial loss of consciousness.

If a victim is bleeding at the mouth, have him lean forward and spit out the blood to avoid swallowing it or choking on it. Frothy blood may indicate internal bleeding involving the lungs. If a tooth has broken off, find it, wash it off, and save it for possible implanting. Have the victim bite down on a roll of gauze to control bleeding from the tooth socket.

Ears may bleed for several reasons. Bright red blood flowing from the ear may indicate a ruptured eardrum. If the blood is thin and watery, it possibly contains cerebrospinal fluid, which suggests a skull fracture. Contact emergency medical services immediately.

When an eye is injured, bandage both eyes to keep the affected eyeball from moving. Do not apply any pressure, and do not remove any embedded objects. In this event, cover the affected eye with an inverted paper cup before applying a bandage.

Chest and abdominal wounds may be accompanied by internal injury and bleeding. Do not try to replace any protruding organs or tissue. Cover chest and abdominal wounds with a large dressing that is secured by a bandage taped on three sides.

Moderate bleeding tends to reduce the risk of infection by carrying some dirt away from the wound. But the risk still remains with any open wound. Signs of later infection include increased pain, swelling, redness, discharge or unpleasant odor from the site, red streaks leading toward the victim's heart, swollen glands, and poor healing. If any of these symptoms appear, contact a doctor immediately.

Internal bleeding may occur apart from external wounds. Symptoms include bruising, rigid or swollen tissue, and bleeding at the mouth, nose, or ears. When you suspect internal bleeding, call emergency medical services immediately and monitor the victim's pulse and respiration. Be alert for shock.

First Aid for Fainting

1. Check for possible injury.

2. Lay victim in a comfortable position with legs raised.

3. Loosen any tight clothes.

4. Blow fresh air over him.

5. Put a cold, wet cloth on his face.

6. Upon recovery, give him a cool, sweet drink.

Dealing with shock and fainting. A victim often goes into *shock* because of the pain and stress of an accident, especially after a massive loss of body fluids. The symptoms are pale, clammy skin, a weak and rapid pulse, rapid or irregular breathing, and a drop in body temperature. Nausea and mental confusion develop.

Shock is a serious condition because the circulatory system is not supplying sufficient blood to the vital organs. When you recognize its symptoms, call emergency medical help immediately. Have the victim lie down, keep him warm, and elevate his legs. This will cause more blood to flow toward his vital organs. Do not give him anything to drink unless he is fully conscious. Try to reassure him, and keep his airway open.

Fainting is a less serious condition in which a person becomes unconscious due to decreased blood flow to the brain. This may be caused by a serious wound or by pain, hunger, or standing for a long time in a warm place. Before fainting, the person may experience dizziness, weakness, nausea, sweating, or blurred vision. Then suddenly he loses consciousness.

Keep the fainting victim from falling, if possible. If he has fallen, check for possible injury and then lay him in a comfortable position. Loosen any tight clothes, raise his legs, and blow fresh air over him. A cold, wet cloth on his face may help him to recover, but do not pour water directly on his face. After he recovers, give him a cool, sweet liquid to drink.

———————— Study Exercises: Group D ————————

1. Why is arterial bleeding more serious than venous or capillary bleeding?

2. Complete the following about first aid for bleeding.
    a. Most bleeding can be stopped by ——— ———.
    b. If bleeding persists, ——— the affected part.
    c. If bleeding still continues, what should you do next?

3. Which of the following is *not* recommended first aid for a nosebleed?
    a. Tilt the victim's head back.
    b. Pinch the sides of the nose together.
    c. Spit out any blood that drains into the mouth.

4. Name the wound that matches each description.
    a. A deep, clean cut, as by a knife.
    b. A hole caused by a pointed object.
    c. A wound caused by tearing the flesh.
    d. A minor wound caused by scraping skin on a rough surface.
    e. An injury involving a severed body part. (two answers)

5. Explain the difference between a dressing and a bandage.

6. What is beneficial about moderate bleeding?

7. Which *two* sentences describe shock?
    a. It often results from the pain and stress of a serious accident.
    b. It is a temporary loss of consciousness due to decreased blood flow to the brain.
    c. It is a condition that chiefly affects the nervous system.
    d. It is dangerous because vital organs are deprived of sufficient blood.

8. What is the basic cause of fainting?

9. Which of the following is *not* recommended first aid for fainting?
    a. Lay the victim down, and elevate his legs.
    b. Blow fresh air over the victim.
    c. Pour cold water on the victim's face.
    d. Loosen any tight clothes.

Group D Activities

1. *Locating the nearest urgent-care center.* Although calling 911 will contact emergency medical care, many injuries do not require emergency transport. For such injuries, you should know where the nearest hospital or urgent-care center is located. Use the telephone directory to find the appropriate number, and post the number near the telephone.

2. *Learning about first-aid kits.* Find a first-aid kit that is completely equipped. If you do not have one available, you may want to ask someone from the local health department to bring a kit to your school and give a demonstration. Examine the different types of dressings and bandages in it.

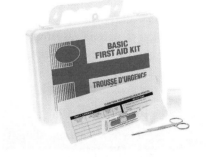

3. *Applying a dressing and a bandage.* Use a red pen to mark a "wound" in the palm of a class-mate's hand. Wash the area with soap and water, and then cover it with a piece of sterile gauze. Secure it with a roller bandage by placing the end of the bandage over the gauze and wrapping the bandage three or four times around the hand between the fingers and the thumb. Then bring the bandage across the hand, wrap it several more times below the thumb, and secure it at the wrist. Be sure not to cut off the blood circulation.

    You may want to borrow a first-aid book and study different ways to wrap bandages.

First Aid for Fractures, Sprains, and Dislocations

God provided the skeleton as a strong but lightweight structure to support and protect the body. But if that structure falls, collides, or twists, bones can break. The remarkable thing is that with proper care, bones will heal again. A person needs sufficient calcium, phosphorus, and vitamin D for strong, healthy bones and for broken bones to heal quickly. Even a minor fall can break malnourished bones.

Have you ever heard of a person who had a fractured ulna or a fractured femur? Perhaps you have wondered whether that was as serious as a break. A fracture *is* a broken bone.

Types of fractures. Broken bones are placed in two main categories—simple fractures and compound fractures.

Simple fractures. In a **simple fracture,** the broken bone does not protrude through the skin; so there is little risk of infection. Simple fractures are also called closed fractures.

Compound fractures. In a **compound fracture** (also called an open fracture), the broken bone protrudes through the skin. A compound fracture involves a serious risk of infection.

Fractures are further classified according to the extent of damage to the bone.

A *greenstick fracture* is much like the partial breaking of a green stick. The fracture extends only partway through the bone, without breaking the bone into separate pieces.

A *transverse fracture* is a clean square break across the bone. Such a fracture has less bonding area and may require surgery with pins to hold the bone in place.

A *comminuted fracture* (kom'·ə·nōō'·tid) breaks the bone into a number of small pieces. These fractures usually require surgery that involves steel plates and screws. Sometimes the steel parts are removed after the bone heals, and sometimes they stay permanently in the victim's body.

Dealing with fractures. Simple fractures are often hard to diagnose. Symptoms include pain, deformity, swelling, and tenderness. Sometimes the victim cannot use the injured part. He may say he heard or felt a snap or a grating sensation. Since you cannot tell, it is safest to assume that a bone is broken until an x-ray proves otherwise. If you carelessly move a person with a broken bone, you may cause a worse injury by turning a simple fracture into a compound one. A broken vertebra can sever a victim's spinal cord and paralyze him for life.

Fractures rarely require emergency transport unless they involve serious wounds. However, never attempt to set a bone or to straighten a crooked part. If you suspect a fracture, take the victim promptly to a medical clinic or hospital for evaluation and professional treatment.

The first thing to do is to immobilize the affected part with a *splint.* A splint reduces pain by restricting motion. It prevents further damage and also reduces bleeding and swelling.

There are several kinds of splints. Factory-made splints have the shapes of the injured parts, but these are rarely on hand in an emergency. You can improvise a splint using rolls of newspaper, heavy cardboard, or a padded board. The splint should be long enough to extend well above and below the break.

Before applying a splint, cover any open wound with a sterile dressing and a bandage. Place splints on both sides of the affected part, and bind them in place with narrow bandages. Be careful not to cut off circulation. If the injury affects the upper or lower thirds of a bone, extend the splint across the joint.

Left: A splint keeps a broken limb in a fixed position.

Right: A triangular bandage can be used as a sling to support a splint.

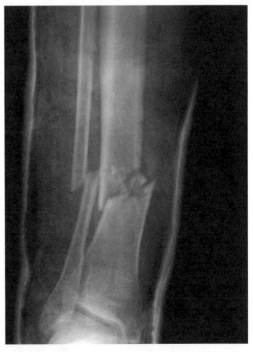

An x-ray gives a clear picture of a broken bone. Here, the lower leg is in a plaster cast, but the remedy is not working. Casts are not as effective with transverse-like fractures.

Types of Fractures

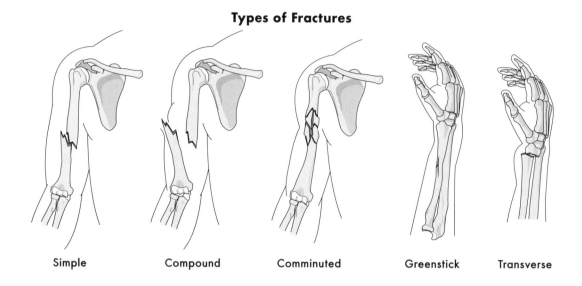

Simple	Compound	Comminuted	Greenstick	Transverse

When you have immobilized the fracture, you can transport the victim. The location of his injury and the number of helpers available will determine your transport method. A conscious victim with an arm fracture will probably be able to walk on his own. If a leg is injured, support his injured side. You may allow him to lean on you, supporting him like a human crutch. For other methods of transport, see the activities on pages 308–309.

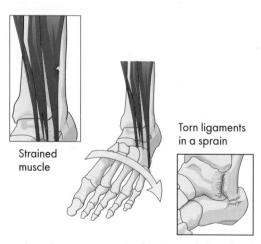

Strained muscle

Torn ligaments in a sprain

A strain versus a sprain. Sprains are joint-related injuries, but strains are muscle-related injuries.

Dealing with other skeletal injuries. Skeletal injuries other than fractures include sprains, strains, and dislocations.

Sprains. A **sprain** is an injury to the ligaments and tissues surrounding a joint. If you have ever twisted your ankle and felt a sharp stab of pain, you have felt a sprain. Tissues swell, and the joint becomes difficult to bend.

Strains. A **strain** is an injury to a muscle or tendon, caused by overstretching it. You have felt the sharp pain of this injury if you have ever stretched too far to reach something or strained too hard to lift a heavy object. A victim of muscle strain sometimes talks of something "flying" into his back.

First aid is the same for both sprains and strains. The acronym RICE should help you to remember the four parts in the treatment of these injuries.

Part 1 is *Rest.* Never think that a good run will heal a sprained ankle. Nor can you help a strained muscle by giving it a workout. Exercise will only damage the tissues more.

Part 2 is *Ice.* Wrap ice or a cold pack in a clean cloth, and place it on the injury for twenty minutes. Then remove the ice for twenty minutes. Continue this on-and-off procedure for several hours. As the pain and swelling become less, gradually lengthen the time between the cold applications. Do not apply any heat until the pain and swelling have subsided. Heat increases swelling.

Part 3 is *Compression.* If possible, wrap the affected area in a light elastic bandage for at least two days. Elastic bandages obviously are not practical for strained muscles in the back or neck.

Part 4 is *Elevation.* Elevate the affected area so that it is higher than the heart, especially at night. Cold, compression, and elevation all help to reduce swelling.

Dislocations. A ***dislocation*** occurs when the end of a bone slips out of its normal position at a joint, such as a shoulder or hip. This often causes tearing of the ligaments surrounding the joint.

Symptoms of dislocation include swelling, tenderness, pain upon motion, and deformity, which are also symptoms of a fracture. Do not try to put a dislocated bone back in place, for you may cause even greater injury. Immobilize the affected part, and get medical help as soon as possible.

Rest

Ice

Compression

Elevation

Study Exercises: Group E

1. Give another name for a broken bone.
2. What is the difference between a simple fracture and a compound fracture?
3. Give a name for each of the following types of fractures.
   a. The bone is broken into a number of small pieces.
   b. The bone is broken only partway through.
   c. The bone is broken straight across.
4. Give four symptoms of a fracture.
5. Which of the following is *not* proper first aid for a suspected fracture or dislocation?
   a. Straighten any crooked limb, and put any dislocated bone in place.
   b. Treat the victim as though he has a fracture.
   c. Immobilize the affected part before transporting the victim.
   d. Cover any open wound with a sterile dressing and a bandage.

6. What are the benefits of applying a splint to a fracture and any associated wound?

7. When an injured person needs to be carried, what are two things that determine how you will transport him?

8. How does a sprain differ from a strain?

9. Name the four parts in the treatment of sprains and strains.

Group E Activities

1. *Practicing with splints.* Make a newspaper splint for an arm by rolling a newspaper into a long, firm roll and flattening it. Use cloth bandages to tie the splint to a classmate's lower arm with the end extending beyond the fingers and the elbow.

   Make a padded wooden splint for a leg by using a board about three inches wide and slightly longer than a classmate's leg. Wrap the board in an old sheet. With the classmate lying down, lay his "injured" leg on the board. Tie the splint in place with four strips of cloth.

2. *Practicing one-person transports.* If you are alone and the victim is light, you can transport him by four different methods. One method is the cradle carry, in which you place one arm under his knees and the other under his back. The victim wraps one of his arms around your neck.

   A good method for a longer distance is the pack-strap carry. With the victim behind you and facing forward over your one shoulder, hoist him onto your back by pulling his arms around either side of your neck. Cross the arms, and grasp them in front of you.

   Another good method for longer distances is the fireman's carry. It can be used for unconscious victims if no back or neck injury exists. With the victim facing you, pull him upright onto his knees while you squat down with your face close to the victim's right shoulder. Place his right arm over your right shoulder and then drape his upper body across your right shoulder and around your neck to the other side. Wrap your right arm around his right leg and grasp the victim's arm that hangs from your left shoulder.

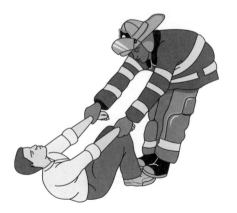

The fireman's carry. If possible back or neck injury exists, only professional help should move a victim. In extreme circumstances, you may move him if that saves his life. In such a case, drag the victim by the feet or pants. Or with the head and neck supported in a very straight line, roll him onto a blanket and drag the blanket headfirst.

Use the piggyback carry if the victim can hang onto you with his arms. Position him as for the pack-strap carry, but pull his legs around your waist and lock your hands under his knees. Let him wrap his arms around your shoulders.

3. *Practicing two-person transports.* When two persons are present, you can use five other methods. In the first method, one person supports each side of the victim while he places one arm around each neck.

In the two-handed seat carry, two assistants face each other and grasp each other's opposite wrists. The victim sits on their hands and places one arm around each neck.

The four-handed seat carry is similar. Each assistant grasps his own left wrist with his right hand and grasps the other assistant's right wrist with his left hand. The victim sits on their hands and places one arm around each neck.

In the extremity carry, the two assistants stand with one following the other. The assistant in the back holds the victim in front of him and wraps his arms around the victim's chest beneath his armpits. The front assistant reaches backward and grasps the victim's legs under the knees, bringing each leg forward on either side of his body.

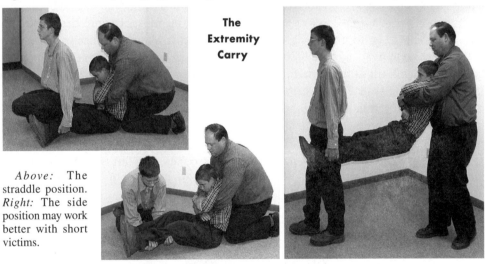

The Extremity Carry

Above: The straddle position. *Right:* The side position may work better with short victims.

Two persons can also transport a victim on a chair. This method is particularly useful for narrow halls or stairways. The victim sits on the chair. Then one assistant lifts the chair by its back, and the other assistant lifts the chair by its front legs.

4. *Practicing three- to six-person transports.* Using the hammock carry, three to six people can transport a victim who must lie flat. The assistants kneel on one knee at opposite sides of the victim, facing each other. Reaching under the victim, each assistant grasps the left wrist of an opposite assistant with his left hand and the right wrist of another opposite assistant with his right hand. Assistants at either end support the victim's head and feet with a free hand. The person at the head gives the cue to rise, and all rise at once.

First Aid for Burns

God covered the body with a marvelous layer of durable skin, which gives protection against dehydration and disease germs. The skin can be harmed in various ways, but few injuries do as much damage as a serious burn.

Thermal burns. Burns caused by fires or hot materials are called thermal burns and are divided into three classes according to their depth.

First-degree burns. A first-degree burn (superficial burn) affects only the epidermis. The skin turns red and tender, and mild swelling and pain develop.

Cool a first-degree burn with cold running water for about ten minutes or until the heat and pain subside; then pat it dry. This is a mild burn that usually heals within a week.

You may apply a light layer of antibiotic cream or moisturizer to a first-degree burn, but it is unwise to treat any kind of burn with home remedies, such as butter, lard, or mustard. These remedies may increase the danger of infection. And you should never apply heat to a burn, for that may turn a mild burn into a more severe burn.

Second-degree burns. A second-degree burn (partial-thickness burn) affects the dermis. It causes blistering, patchy red skin, and severe pain. The area swells, and a watery fluid oozes from cells and exposed capillaries.

Cool a second-degree burn with cold running water; then cover the affected area with a dry, sterile dressing. To reduce the risk of infection, do not rub the area, break a blister, or remove any loose skin. Get medical attention unless the burn affects only a small area.

Third-degree burns. A third-degree burn (full-thickness burn) damages the epidermis, the dermis, and the underlying fat and muscle. The remaining skin turns leathery, waxy, and gray. Often it is charred and dry because blood vessels are destroyed. There is little pain in the immediate area because the nerve endings have been destroyed. Any pain comes from surrounding second-degree burns.

You can cool a third-degree burn with cold running water; but since less pain is involved, it is more important to cover the burned area with a dry, sterile dressing and get immediate medical care. Never try to remove stuck clothes or charred skin. Third-degree burns usually require plastic surgery.

The danger of shock often accompanies second- and third-degree burns. Treat the victim for shock if the symptoms are present, but do not give him any fluids by mouth. When

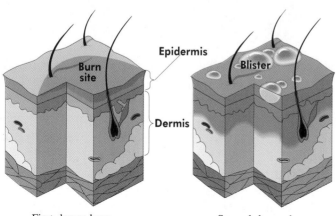

First-degree burn   Second-degree burn

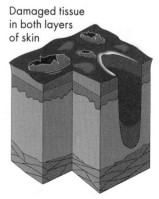

Third-degree burn

a large part of the body is affected, be sure not to chill the victim in cooling the burn. Aspirin or ibuprofen may be given to reduce the pain and swelling of first- and second-degree burns, but it is best to leave the care of third-degree burns to medical professionals.

Severity as determined by area. Burns are also classified according to the percentage of body area that they affect. For an adult, the percentage can be estimated with the following figures.

> head = 9%
> front trunk = 18%
> back trunk = 18%
> each upper leg = 9%
> each lower leg = 9%
> each arm = 9%
> palms = 1%

A second- or third-degree burn that covers more than 15 percent of the body is very serious. If a burn circles the body, it may cut off circulation. With any severe burn, the pulse and respiration of the victim should be carefully monitored.

Other types of burns. Some burns are caused by radiation, chemicals, or electricity.

Radiation burns. The most common burn of this kind is sunburn. Long exposure to the sun can cause a second-degree burn that swells and blisters painfully. The victim may experience shock and breathing difficulty. Sunburn is treated in the same way as a thermal burn, but prevention is the wisest route. Always wear sufficient clothes when working in the sun. Apply a sunscreen lotion, and avoid unnecessary exposure.

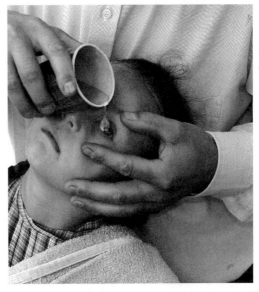

To wash an eye, pour water on the inner corner of the eye and let it run across the eyeball and onto a towel on the shoulder.

Chemical burns. Most of these are caused by acids and alkalis, of which the latter is more serious because alkalis burn longer and deeper. Petroleum products can also cause chemical burns.

Before trying to help the victim of a chemical burn, put on protective clothes and see that the area is well ventilated. Remove any affected clothing, and flush the area with running water for at least twenty minutes. Be sure the water runs away from the affected area and not toward other skin surfaces. When the pain subsides, cover the burn with a sterile dressing and get medical help.

For a chemical burn in the eye, hold the victim's eye open and pour water on the inner corner of his eye so that it runs across the eyeball to the outer corner. Continue this for at least thirty minutes. Then get medical treatment immediately.

Electrical burns. An electrical arc can cause a thermal burn by its heat or a flash burn by its intense light. More serious is the contact burn, which often reaches the second or third degree. A portion of this burn is visible on the outside of a victim's body, but much more harmful are internal burns. Electricity enters the body at a point of contact and follows nerves and blood vessels to a point of exit, burning all the way. It can put a temporary or permanent stop to activity in the heart, brain, and involuntary muscles. Death by electrocution often comes from a current of 1,000 volts or stronger, but even the 120 volts of household current can be fatal.

Never touch a victim of electrical burn unless he is no longer in contact with the electricity. First turn off the main breaker if you can; otherwise, use a long piece of wood or plastic to move the victim or the source of electricity. However, do not try this around high voltage.

Treat an electrical burn survivor with the A-B-C-D-E of first aid. When his vital functions are working, dress his injuries and get medical help.

———— Study Exercises: Group F ————

1. Copy the following table, and describe each burn in the ways listed. A sample is done for you.

	First-degree	Second-degree	Third-degree
Other name	superficial burn		
Appearance of skin	red, tender, mild swelling		
Degree of pain	mild		
Deepest tissue affected	epidermis		

2. What is the first thing to do in treating first- and second-degree burns?

3. Write *yes* or *no* to tell whether each of the following is recommended first aid for a burn.
   a. Run cold water over the area.
   b. Massage the area.
   c. Break the blisters.
   d. Apply a dry, sterile dressing.
   e. Remove loose skin, charred skin, or stuck clothes.
   f. Watch for symptoms of shock.
   g. Apply a home remedy.

4. a. What are the two ways that thermal burns are classified?
   b. A second- or third-degree burn is very serious if it covers more than —— percent of the body.

5. The most common type of radiation burn is ——.

6. What are three things that cause chemical burns?

7. List three steps in giving first aid for chemical burns.

8. Of the different kinds of electrical burns, the greatest injury is caused by
   a. a thermal burn from the heat of an arc.
   b. a flash burn from the intense light of an arc.
   c. a contact burn along an electrical path.

9. What should you do before touching the victim of an electrical burn?

Group F Activities

1. *Interviewing a burn victim.* Find a relative or neighbor who has experienced a severe burn. If he is willing to talk about the experience, ask him the following questions: What caused the burn? How could such burns be avoided? What did the burn look like? What first aid was applied? What medical procedures were used to treat the burn? Give a report to the class, or have the person come to school and give a talk.

2. *Listing acids and alkalis around your home.* Find cleaners around your home and property. Copy their names and ingredients, and classify them as acids or alkalis. Copy the first-aid instructions printed on the bottle in case of contact or ingestion.

3. *Observing the effect of acid on cloth.* Rub an old nylon stocking against the post of an automobile battery. Immediately you should notice a white spot. Several days later, you will see a hole developing in that area. Acid will do something similar to your skin.

First Aid for Poisoning

A poison is any substance that causes injury, illness, or death when it touches or enters the body. Poisons can harm the body in four different ways.

Poisoning through the digestive system. Children are curious and may swallow such liquid poisons as cleaners, medicines, or pesticides. Adults sometimes poison themselves by an intentional or unintentional overdose of drugs or alcohol. Symptoms of poisoning by mouth include abdominal pain, nausea, diarrhea, abnormal breathing or heartbeat, and drowsiness or unconsciousness. Other clues are a strange odor on the breath and stains or burns around the mouth.

If you suspect poisoning, look for a container that indicates what kind of poison the victim swallowed. Take any such container with you to the telephone, and contact the nearest poison control center. These centers provide 24-hour help and will tell you what to do for specific poisons. If it is a corrosive poison, such as an acid or alkali, the center may tell you to give a conscious victim one or two glasses of water or milk to dilute the poison.

CAUTION: KEROSENE HEATERS USE OXYGEN AND THEREFORE MAY CAUSE A HAZARD OF ASPHYXIATION IF USED IN SMALL, CLOSED AREAS SUCH AS BATHROOMS AND TIGHTLY CLOSED CAMPERS. ALWAYS PROVIDE A SOURCE OF FRESH AIR.

Another treatment for poisoning is to have the victim drink something that induces vomiting, such as warm salt water or syrup of ipecac (ip′·i·kak′) mixed with water. However, do not induce vomiting unless the poison control center tells you to do so. The poison may get into the victim's lungs; or if it is a corrosive substance, it will burn his esophagus again on its return trip.

Until help arrives, position the victim on his left side to delay the poison's entrance into his small intestine. Monitor his breathing and heartbeat.

Poisoning through the respiratory system. This is caused by breathing in a harmful substance. Sometimes the poison works very slowly, and the symptoms may mimic those of other disorders.

One of the most common poisonous gases is carbon monoxide, which is invisible, tasteless, odorless, and nonirritating. This gas binds to red blood cells more readily than oxygen does, thus causing asphyxiation by robbing the victim's body of oxygen. Carbon monoxide is formed by incomplete burning of carbon in motor vehicles, gas stoves, and heaters. To avoid this kind of poisoning, make sure all gas or liquid heaters are well vented according to their instructions, and never operate a gasoline engine in an enclosed space. Running a car or a generator in a closed garage has taken many lives.

Symptoms of carbon monoxide poisoning include headache, ringing ears, chest pain, muscle weakness, nausea, dizziness, and double or blurred vision. These are followed by unconsciousness, respiratory failure, and heart failure. First aid starts with moving the victim to a place of fresh air. If his vital signs are normal, turn him on his side and summon medical care.

Other gases that cause poisoning and respiratory illness are chemical solvents and the nitrous oxides in the exhausts of vehicles and factories. Respiratory function is also disturbed by the tiny particles in dust, soot, and smoke, as well as by certain molds and pollens.

Poisoning by contact with the skin. A mild kind of poisoning is caused by an oily resin in plants known as poison ivy, poison

Many poisonous snakes, including this copperhead, inject their venom through the two upper fangs. A victim may receive a single bite that leaves two small puncture wounds. But during that quick bite, venom is pumped through each tubed fang into the victim.

Poison ivy leaves are always grouped in threes. The leaves can be hairy, or they can be smooth and glossy. The oil-like poison can be passed on by a tool or a glove up to a year afterward if it is not washed.

oak, and poison sumac. Contact with these plants causes many people to develop a skin rash with water-filled blisters, which is really an allergic reaction. To prevent this, wear protective clothing in the woods, and use soap and water to thoroughly wash any skin that has contacted one of these plants. Avoid burning the plants, for the smoke can cause a rash on any skin that it touches.

A more serious kind of poisoning occurs when a hazardous substance, such as a pesticide, is absorbed through the skin. Prevent this by storing such chemicals properly, handling them carefully, and wearing protective clothing when using them. If a poisonous chemical gets on your skin, promptly wash the affected area with soap and water.

Poisoning by injection through the skin.

This poison comes from the bite or sting of a creature such as an insect or a poisonous snake or spider. If the stinger of a bee is stuck in the skin, scrape it off without squeezing it, to avoid forcing more venom into the wound. A paste of baking soda and water will help to reduce the pain and swelling. If the victim develops an allergic reaction, such as respiratory trouble, get medical help immediately.

Keep a snakebite victim calm to slow the circulation of the poison through his body. If he must move, have him walk very slowly. Try to keep the affected part below his heart. Do not administer any drug, for it may interact dangerously with the snake venom. Also do not apply ice or heat. Ice may damage tissues, and heat will spread the poison.

The old remedy of cutting the wound

and sucking out the venom by mouth is not advisable. However, some first-aid kits have an extractor that can be safely used to suction out some of the poison. Wash the bite with soap and water, and obtain medical help as soon as possible.

Other threats come from ticks that carry diseases, such as Rocky Mountain spotted fever or Lyme disease, and from animals infected with rabies. Promptly remove any tick that has fastened itself to the skin. Use a tweezers to grasp it by the mouthparts as close to the skin as possible; do not grasp it by the body, or its head may break off. Pull firmly and steadily until the tick comes loose; then wash the tick bite with soap and water. Be alert afterward for rash, fever, aches, leg weakness, and sensitivity to bright lights, which are symptoms of Lyme disease.

Rabies is a viral disease that attacks the nervous system. The disease can be prevented by vaccination, but there is no known cure after the symptoms appear. Thoroughly wash an animal bite with soap and water. Then cover the bite with a sterile dressing and get medical help. If the responsible animal behaves strangely, report it to authorities but do not try to capture it. You can protect yourself from animal bites by not teasing animals and by not trying to pet strange animals.

Emergency Numbers

Ambulance	911
Fire	911
Police	911
Poison Center	1-800-222-1222

Post the Poison Center number near the phone, or program it into the phone. This center gives very helpful advice in cases of uncertainty.

Study Exercises: Group G

1. Tell whether each item gets into the body by means of the digestive system (*D*), the respiratory system (*R*), contact with the skin (*C*), or injection through the skin (*I*).
   a. carbon monoxide
   b. cleaning liquid
   c. poison ivy
   d. pesticide (two answers)
   e. rabies or Lyme disease
   f. mold
   g. snake or insect venom

2. For each set of symptoms, tell what kind of poisoning you would suspect.
   a. headache, ringing ears, chest pain, dizziness
   b. skin rash with water-filled blisters
   c. two small puncture wounds within a red, swollen area

3. If someone has swallowed a poison, what information from the container should you be able to give to the poison control center?

4. Which of the following is *never* recommended as first aid for a person who has swallowed a poison?
   a. Give him several glasses of water or milk to dilute the poison.
   b. Induce him to vomit.
   c. Lay him on his right side.
   d. Monitor his breathing and heartbeat.

5. How can you avoid carbon monoxide poisoning?

6. For each item below, tell how you can protect yourself in case of skin contact.
   a. poison ivy or poison oak
   b. a pesticide or other hazardous substance

7. For each snakebite remedy, write *yes* or *no* to tell whether you should use it.
   a. Keep the victim calm.
   b. Give him a medication.
   c. Apply heat to the bite.
   d. Keep the affected part below the victim's heart.
   e. Suck out the poison by mouth.
   f. Wash the bite with soap and water, and get medical help.

8. Which sentence is true?
   a. All ticks carry Lyme disease.
   b. An attached tick should be removed by using a tweezers to grasp it by the mouthparts.
   c. Jerking sharply is the best way to make a tick come loose.
   d. A tick bite should be covered with a paste of baking soda and water.

9. Rabies can be cured only if
   a. the animal that bit the victim is identified.
   b. the responsible animal exhibits strange behavior.
   c. a vaccination is given before symptoms appear.
   d. the bite bleeds well.

Group G Activities

1. *Posting the poison control telephone number.* Post the number of the National Poison Center Network near your telephone for use in case of accidental poisoning. The number is 800-222-1222, and the center is open 24 hours a day.

2. *Listing poisons around your home.* Make a list of pesticides, cleaners, and other poisonous substances around your home. Are any first-aid procedures given on the containers? Be sure each container is labeled and is out of the reach of children.

3. *Identifying poisonous plants.* Find pictures of poison ivy, poison oak, and poison sumac in a plant identification guide. Take a walk in the woods to see if you can find these plants in your community. Poison ivy has rather attractive whitish berries in the fall after its leaves are gone. These berries are just as harmful as the leaves. Make a list of any other poisonous plants in your vicinity. Some common houseplants are poisonous.

4. *Identifying poisonous animals.* Use an identification guide to learn about the appearance of poisonous spiders and insects. Learn which ticks carry diseases. Do any of these live in your area? Look up poisonous snakes in a snake identification guide. Are any of these native to your area?

Unit 8 Review
Review of Vocabulary

To maintain a healthy body, you should follow the principles of good —1—, such as cleanliness, good posture, rest, and exercise. However, if you exercise in very warm weather, you can develop a condition called —2— from losing too much water and salt through sweating. You might even suffer a —3— because the body loses its ability to regulate its temperature.

It is valuable to understand some basic first-aid procedures. If a person is not breathing and has no pulse, you may be able to revive him with —4—, or CPR. This includes rescue breathing to restore his respiration, and chest —5— to restart his circulation. A drowning person rescued from cold water often needs to be treated for low body temperature, called —6—.

Many injuries are accompanied by severe bleeding. If you cannot stop the bleeding by direct pressure, you may need to stop the flow of blood from the heart by applying pressure at a suitable —7—. As a last resort, you may need to use a tight band called a —8— to totally cut off the flow of blood.

Wounds are classified according to their nature. The —9— occurs when the skin is scraped by a hard, rough surface. The —10— is a deep, clean cut, but the —11— is a jagged wound, caused by tearing of the flesh. The puncture is a deep injury caused by a pointed object. The —12— is the partial or complete tearing away of a body part; this is called an —13— if a limb is severed.

If a wound is small, clean it and then apply a —14— to control the bleeding and protect the wound. Hold this in place by applying a —15—, which will put pressure on the wound and may help to —16— the injured part (keep it from moving). If there is a serious accident, the victim often needs to be treated for —17—, especially if he loses large amounts of body fluids.

Broken bones are placed in two main categories: the —18— if the broken bone does not protrude through the skin, and the —19— if the broken bone is exposed. Immobilize a fracture with a —20— before trying to move the victim. Not all skeletal injuries involve broken bones. A —21— is an injury to the ligaments and tissues surrounding a joint. A —22— is caused by overstretching a muscle or tendon. A —23— occurs when the end of a bone slips out of its normal position at a joint.

Multiple Choice

1. Which statement best describes the kind of care the body should have?
   a. If the basic needs of the body are supplied, the body will be healthy.
   b. The body must be protected against environmental dangers, such as cold, excessive sunshine, germs, and chemicals.
   c. The body should not be misused but cared for as a valuable gift from God.
   d. The body requires careful, daily attention to many detailed rules, since even a slight mistake will cause it to become ill.

2. Which statement expresses a popular idea that may not be true?
   a. Sunshine on the skin produces a healthy suntan.
   b. Good exercise helps to promote good rest.
   c. Poor posture can lead to fatigue.
   d. Inadequate clothing lowers one's resistance to disease.

3. Both a poor diet and a lack of cleanliness can be a direct cause of
   a. weight loss.   b. disease.   c. fatigue.   d. poor posture.

4. Which phrase does *not* describe proper observance of safety rules?
   a. wearing goggles around dangerous equipment
   b. wearing ear protection around noisy machinery
   c. removing shields that are a nuisance when operating machinery
   d. wearing a facemask in dusty situations

5. Which one of the following is *not* a main goal of first aid?
   a. Keep the victim alive.   c. Promote recovery.
   b. Prevent worse injury.   d. Give help regardless of personal risk.

6. Which sentence correctly states a step in rescue breathing?
   a. Repeat the cycle one time per minute.
   b. Raise the victim's arms between each breath.
   c. Hold the victim's nose shut.
   d. Blow air into the victim's lungs, and then draw it out again.

7. If an unconscious victim has neither respiration nor pulse, the rescuer should
   a. alternate chest compressions with rescue breaths.
   b. lay the victim on his side.
   c. perform abdominal thrusts.
   d. do nothing until medical help arrives.

8. A person giving first aid should
   a. calm and reassure the victim.
   b. tell the victim how serious his condition is.
   c. tell the victim what he should have done to avoid the injury.
   d. say as little as possible to the victim.

9. In an effort to stop bleeding, you should first
   a. apply pressure to the artery between the wound and the heart.
   b. elevate the wounded part.
   c. apply a tourniquet to the limb above the wound.
   d. apply direct pressure to the wound.

10. If you find the severed body part of a victim, you should
    a. wash it carefully.
    b. tie it in place until you can get medical help.
    c. wrap it in plastic and cloth, submerge it in ice, and label the container.
    d. transport it in a plastic container of cold water.

11. To treat or prevent shock,
    a. give the victim plenty of cold water.
    b. keep the victim warm and lying down.
    c. keep the victim awake and active.
    d. keep the victim in a seated position.

12. The purpose of a splint is *not* to
    a. immobilize the affected part.
    c. set the broken bone.
    b. reduce swelling.
    d. reduce the risk of further injury.

13. Applying ice is good first aid for
    a. shock.
    b. a sprain.
    c. a snakebite.
    d. a chemical burn.

14. A primary goal of the RICE procedure is to
    a. exercise the affected part to avoid stiffness.
    b. reduce swelling.
    c. enable the victim to use the affected part.
    d. immobilize the affected part.

15. Second-degree burns are generally more painful than third-degree burns because
    a. second-degree burns affect more blood vessels.
    b. second-degree burns generally cover a larger percentage of the body.
    c. third-degree burns destroy the nerve endings that sense pain.
    d. third-degree burns involve less tissue damage.

16. The kind of burn most likely to be accompanied by immediate death is the
    a. thermal burn.
    c. chemical burn.
    b. radiation burn.
    d. electrical burn.

17. In most cases of chemical poisoning by mouth, the victim should *not* be encouraged to
    a. lie on his left side.
    c. vomit.
    b. drink water.
    d. explain where he got the chemical.

18. In general, poisoning by ——— works the most slowly and poses the least threat to life.
    a. swallowing
    b. breathing
    c. skin contact
    d. injection

Final Review: Units 1–8

Matching

Write the letter of the word that matches each description.

1. A quantity of matter with a definite shape and volume.
2. The red substance in blood that helps to carry oxygen.
3. A rock from space that streaks through the sky and burns up.
4. A poisonous gas produced by a gasoline engine.
5. The gap between two nerve cells.
6. The brightness of a star.
7. The outer layer of the skin.
8. A building with equipment for viewing heavenly bodies.
9. Disorder and randomness in energy and materials.
10. The amount of heat needed to raise 1 gram of water 1 degree Celsius.

a. calorie
b. carbon monoxide
c. entropy
d. epidermis
e. hemoglobin
f. interneuron
g. liquid
h. magnitude
i. meteor
j. observatory
k. planetarium
l. solid
m. synapse

11. The long-term weather pattern of an area.

12. Projections that increase the surface area of the small intestine.

13. The changing of a gas to a liquid.

14. The scientific study of the stars and other heavenly bodies.

15. A huge, circular storm that covers a large area.

16. Tiny air sacs in the lungs.

17. A hormone produced by the pancreas, which causes body cells to absorb sugar from the blood.

18. An organ that removes wastes by taking material out of the blood and then returning most of it.

19. The changing of a solid directly to a gas.

20. An involuntary response such as automatically jerking away from a hot object.

a. adrenalin
b. alveoli
c. astrology
d. astronomy
e. climate
f. condensation
g. evaporation
h. hurricane
i. insulin
j. kidney
k. liver
l. reflex
m. sublimation
n. tornado
o. villi

Completion

Write the correct word to complete each sentence.

21. The tendency of an object to remain at rest or in motion is called ———.

22. The ——— gland coordinates the work of the other glands in the endocrine system.

23. If air contains all the moisture that it can hold, it is said to be ———.

24. Snow, rain, hail, and sleet are all forms of ———.

25. The sun passes through the twelve constellations of the ——— each year.

26. The process of changing a liquid to a gas and then back into a liquid is called ———.

27. An enzyme in saliva changes starch into ———.

28. A ——— is a clear mixture formed when one material spreads molecule by molecule through another material.

29. A ——— is an instrument for measuring air pressure.

30. In the condition called ———, the thyroid gland is enlarged due to a lack of iodine.

31. Blood vessels that carry blood away from the heart are called ———.

32. Muscles that cause joints to bend are called ——— muscles.

33. One's ——— can be found by measuring the angle of the North Star above the horizon.

34. If a bone is broken, first aid includes using a ——— to immobilize the bone.

35. The principles of good ——— include cleanliness, good posture, rest, and exercise.

Multiple Choice (1)

Choose the correct words in these sentences.

36. The body system that includes the lungs and the diaphragm is the (circulatory, respiratory, integumentary) system.

37. An agent that allows oil particles to become suspended in water is (an emulsion, a solvent, a surfactant).

38. Centrifugal force is caused by (acceleration, inertia, gravity).

39. Coordination of muscular movements is done by the (brainstem, cerebrum, cerebellum).

40. The relative humidity of air can be raised by (heating it, cooling it, putting it in motion).

41. Heat is transferred by (conduction, convection, radiation) when it results from a liquid or gas circulating because of uneven temperature.

42. The uniting of elements to form compounds is an example of a (physical, chemical, nuclear) change.

43. The boiling point of water at the top of a mountain is (higher than, equal to, lower than) the boiling point at the foot of the mountain.

44. The rate at which the body uses food is called its (organism, metabolism, synchronism).

45. If the end of a bone slips out of place, the injury is called a (sprain, strain, dislocation).

46. A nerve cell is called (an axon, a neuron, a dendrite).

47. A cloud that produces rain is called a (cirrus, cumulus, nimbus) cloud.

48. The vestibular apparatus is associated with the (ears, eyes, lungs).

49. A bimetallic strip can be used to make a (barometer, hygrometer, thermometer).

50. A disease caused by the lack of vitamin D is (beriberi, scurvy, rickets).

Multiple Choice (2)

Write the letter of the correct choice.

51. Which statement about kinetic energy is *not* true?
    a. A falling rock provides an example of kinetic energy.
    b. Kinetic energy must be changed to potential energy before it will produce change.
    c. An object that has the inertia of motion also has kinetic energy.
    d. A swinging boy has the greatest kinetic energy when he is closest to the ground.

52. On December 21 the sun is directly above
    a. the ecliptic.   c. the Arctic Circle.
    b. the equator.   d. the Tropic of Capricorn.

53. Which weather sign is stated correctly?
    a. Cirrus clouds are a sign of fair weather.
    b. A decrease in air pressure is a sign of coming rain.
    c. Sundogs are a sign of coming warm weather.
    d. In the Northern Hemisphere, wind from the west is a sign of coming rain.

54. The capillaries are most important in the function of which system?
    a. circulatory
    c. respiratory
    b. lymphatic
    d. excretory

55. All of the following are matter *except*
    a. air.
    c. solids.
    b. light.
    d. compounds.

56. A malfunctioning organ affects
    a. only the surrounding tissues.
    b. only the nearby organs.
    c. only the system of which it is a part.
    d. the entire body.

57. Fronts are associated with all the following *except*
    a. stratus clouds.
    c. clear skies.
    b. thunderstorms.
    d. falling air pressure.

58. A large cloud of glowing gases in outer space is called
    a. an aurora.
    c. a nebula.
    b. a galaxy.
    d. a nova.

59. A volatile liquid
    a. evaporates easily.
    c. freezes quickly.
    b. absorbs heat slowly.
    d. puts out fires effectively.

60. Refrigeration produces cold temperatures by allowing a gas to
    a. condense.
    c. expand.
    b. evaporate.
    d. increase pressure.

61. To learn the schedule of the tides, it would be best to know
    a. the phase of the moon.
    c. the positions of the planets.
    b. the day of the year.
    d. the tilt of the earth.

62. Food passes through the alimentary canal by a muscular action called
    a. digestion.
    c. absorption.
    b. osmosis.
    d. peristalsis.

63. A deep, clean cut is called
    a. an abrasion.
    c. a laceration.
    b. an incision.
    d. a puncture.

64. Orange and yellow vegetables, such as carrots, are rich in
    a. vitamin A.
    c. vitamin C.
    b. vitamin B.
    d. vitamin D.

65. Which of the following is *not* a goal of first aid?
    a. Preserve life.
    c. Prevent worse injury.
    b. Give help at any risk.
    d. Promote recovery.

Glossary

This glossary lists the vocabulary words with their definitions, as used in this book. The numbers in brackets give the pages in the text.

abrasion, a wound caused by scraping the skin on a hard, rough surface. [300]

acceleration, any change of speed, including speed reduction (negative acceleration). [23]

adrenal glands (ə·drē′·nəl), two glands on top of the kidneys, which produce several hormones, including adrenalin. [270]

alimentary canal (al′·ə·men′·tə·rē), the passage by which food goes through the body. [109]

alloy, a mixture of two or more metals. [40]

alveoli (al·vē′·ə·lī′), *sing.* **alveolus,** the tiny sacs in the lungs where oxygen and carbon dioxide are exchanged. [67]

amino acid (ə·mē′·nō), one of the units that make up a protein molecule. [94]

amputation, the cutting off of a finger, toe, or limb of the body. [301]

anemometer (an′·ə·mom′·i·tər), an instrument for measuring wind speed. [239]

antibody, a substance in the blood that fights a certain germ or group of related germs. [72]

antigen (an′·ti·jən), any item, such as a disease germ, that stimulates the production of antibodies. [73]

appendix, a narrow, dead-end tube connected to the large intestine, which apparently destroys antigens. [77]

artery, a blood vessel that carries blood away from the heart. [70]

astrology, the practice of fortunetelling by studying the stars. [148]

astronomy, the scientific study of the stars and other heavenly bodies. [121]

atom, the smallest particle of an element. [33]

aurora australis (ə·rôr′·ə ô·strā′·lis), the lights in the southern sky produced when the charged particles of solar wind enter the earth's magnetic field and cause gases in the upper atmosphere to glow; also called southern lights. [165]

aurora borealis (ə·rôr′·ə bôr′·ē·al′·is), the lights in the northern sky produced when the charged particles of solar wind enter the earth's magnetic field and cause gases in the upper atmosphere to glow; also called northern lights. [165]

autonomic nervous system (ô′·tə·nom′·ik), the part of the nervous system that controls the automatic functions of involuntary muscles. [250]

autumnal equinox (ē′·kwə·noks′, ek′·wə·noks′), the time around September 22 when the sun is directly above the equator and autumn begins in the Northern Hemisphere. [151]

avulsion, the partial or complete tearing away of any body part. [301]

axon, the single strand of a nerve cell that carries signals away from the cell body. [251]

bandage, a piece of cloth or tape that holds a dressing in place. [301]

barometer, an instrument for measuring atmospheric pressure. [238]

bimetallic strip (bī′·mə·tal′·ik), a strip made with two metals fastened together, which bends with temperature changes. [190]

brainstem, the part of the brain that regulates vital functions and is thought to be the seat of the mind. [256]

British thermal unit, a unit for measuring heat energy, equal to the amount of heat needed to raise 1 pound of water 1 degree Fahrenheit. [184]

bronchus (brong′·kəs), *pl.* **bronchi,** either of two tubes that carry air from the windpipe to a lung. [67]

calorie, a unit for measuring the energy in foods, equal to 1,000 thermal calories. [104]

calorie, a unit for measuring thermal energy, equal to the amount of heat needed to raise 1 gram of water 1 degree Celsius. [183]

capillary (kap'·ə·ler'·ē), a tiny blood vessel that carries blood to individual cells. [70]

carbohydrate, one of the main fuel foods, which includes sugars and starches. [88]

cardiopulmonary resuscitation, a procedure in which rescue breathing and heart compressions are used in an effort to revive an unconscious person. [295]

cellulose (sel'·yə·lōs'), the complex carbohydrate that makes up the strong supporting membrane of plant cells. [90]

Celsius, the kind of temperature scale on which water freezes at 0 degrees and boils at 100 degrees. [174]

central nervous system, the brain and the spinal cord. [250]

centrifugal force, the outward force produced when a moving object is made to travel in a curved path. [21]

centripetal force, for an object moving in a curved path, the inward force that keeps it from moving in a straight line. [21]

cerebellum (ser'·ə·bel'·əm), the part of the brain lying behind and beneath the cerebrum, which regulates the signals that the cerebrum sends to the skeletal muscles. [256]

cerebrum (ser'·ə·brəm, sə·rē'·brəm), the main part of the brain, which processes motor and sensory functions. [254]

chemical change, a change in materials that results in new materials. [12]

chemical formula, two or more chemical symbols joined together to represent a compound. [34]

chemical symbol, one or two letters used to represent an element. [33]

chemistry, the study of the properties and changes of matter. [12]

circulatory system, the body system that includes the blood, the heart, and the blood vessels. [70]

cirrus cloud, a wispy, featherlike cloud composed of ice crystals. [221]

climate, the general weather pattern of a region. [214]

colon, the main part of the large intestine. [115]

comet, a small heavenly body that travels in a very oblong orbit and forms a tail when it passes near the sun. [164]

compound, a material formed by combining two or more elements. [34]

compound fracture, a broken bone that protrudes through the skin, also called an open fracture. [304]

compressions, the procedure of trying to restore the heartbeat by repeatedly putting heavy pressure on the chest. [295]

condensation, the changing of a gas to a liquid. [177]

conduction, the transfer of heat by energy passing from one molecule to another. [193]

constellation, a group of stars that forms a picture in the sky. [128]

convection, the transfer of heat by a liquid or gas that circulates because of uneven temperature. [194]

Coriolis effect (kôr'·ē·ō'·lis), the change of wind direction caused by the earth's rotation. [210]

cortex, the outer covering of the cerebrum. [254]

cumulus cloud (kyōōm'·yə·ləs), a fluffy, billowy, piled-up cloud. [221]

declination (dek'·lə·nā'·shən), the distance north or south of the celestial equator, measured in degrees. [131]

deficiency disease, a disease caused by the lack of a vitamin or mineral in the diet. [99]

dendrite, one of the thin, branching strands of a nerve cell that carry signals to the cell body. [250]

dermis, the inner layer of skin on the human body. [54]

dew point, the temperature at which a body of air becomes saturated as it is cooled. [217]

diaphragm (dī'·ə·fram'), the dome-shaped sheet of muscles that contracts to bring air into the lungs. [67]

digestion, the process of changing food so that the body can use it. [109]

dislocation, an injury that occurs when the end of a bone slips out of its normal position at a joint. [307]

distillation, the process of changing a material into a gas by heating and then condensing it into a liquid by cooling. [179]

dressing, a sterile piece of gauze or cloth applied directly over the wound to control the bleeding and to keep it clean. [301]

ecliptic (i·klip'·tik), the apparent path of the sun through the stars. [132]

element, one of the basic materials that cannot be divided into other materials. [33]

El Niño (el nēn′·yō), a disturbance of normal weather patterns at various places, caused by changes in the currents of the Pacific Ocean. [214]

emulsion, a suspension of one liquid in another liquid. [38]

endocrine system (en′·də·krin, en′·də·krēn′, en′·də·krīn′), a system of glands that secrete hormones, which control various body functions. [268]

energy, the power to produce physical or chemical changes. [12]

entropy (en′·trə·pē), disorder and randomness in energy and materials. [17]

enzyme, a protein that speeds up chemical reactions in the body. [94]

epidermis (ep′·i·dûr′·mis), the outer layer of skin on the human body. [53]

epiglottis (ep′·i·glot′·is), the flap that covers the glottis in swallowing, to keep food and water from entering. [67]

equilibrium (ē′·kwə·lib′·rē·əm), the sense of balance, perceived mainly by the vestibular apparatus. [266]

esophagus, the tube that carries food from the pharynx to the stomach. [111]

evaporation, the changing of a liquid to a gas. [179]

excretory system (ek′·skri·tôr′·ē), the system that includes the kidneys and associated organs which remove wastes from the bloodstream. [78]

extensor, a muscle that causes a joint to unbend. [63]

Fahrenheit, the kind of temperature scale on which water freezes at 32 degrees and boils at 212 degrees. [174]

fat, a fuel food such as animal fat or plant oil, which is a concentrated source of energy. [90]

flexor, a muscle that causes a joint to bend. [63]

fluid, a material that flows; a liquid or gas. [31]

front, the boundary between two air masses of different temperatures. [225]

galaxy, a huge group of stars orbiting around a common center. [124]

gas, matter that takes both the shape and the volume of its container. [31]

glottis, the opening at the top of the voice box. [66]

glucose (gloo′·kōs′), a simple sugar with the chemical formula $C_6H_{12}O_6$. [90]

goiter, the enlargement of the thyroid, caused by a lack of iodine. [102]

heart, the organ that pumps blood throughout the body. [70]

heat, a form of energy caused by molecules in motion. [174]

heat exhaustion, a serious condition in which prolonged sweating removes body fluids faster than they can be replaced, with symptoms that include cold skin, rapid pulse, and nausea. [285]

heat stroke, a severe condition in which the body becomes so overheated that it can no longer regulate its temperature, with symptoms that include fever, rapid pulse, and red, hot skin. [286]

hormone, a protein that regulates cell activities. [94]

humidity, the moisture in the air. [216]

hurricane, a huge, rotating tropical storm with winds of at least 75 miles per hour (119 km/h). [231]

hygiene, the group of practices that promote good health, especially by cleanliness and sanitation. Hygiene also includes proper nutrition, exercise, and rest. [284]

hygrometer (hī·grom′·i·tər), an instrument that measures the relative humidity of the air. [238]

hypothermia (hī′·pə·thûr′·mē·ə), abnormally low body temperature. [298]

immobilize, to keep (something) from moving. [301]

immunity, the resistance to infection by a certain disease germ. [77]

incision, a deep, clean cut. [300]

inertia, the resistance of an object to changing its state of rest or motion, which causes an object at rest to remain at rest and an object in motion to continue in straight-line motion. [19]

insulin (in′·sə·lin), a hormone that causes body cells to absorb sugar from the blood. [270]

integumentary system (in·teg′·yoo·men′·tə·rē), the system that includes the skin of the human body. [53]

interneuron, a nerve cell that receives a signal from a receptor and immediately sends a response impulse. [260]

jet propulsion, the forward thrust caused by a backward-escaping fluid. [28]

joint, a place where two bones meet. [59]

kidney, one of the two main organs that filter urea and other wastes out of the blood. [78]

kinetic energy, energy acting to produce physical or chemical change; energy in motion. [13]

laceration (las′·ə·rā′·shən), a cut with jagged edges. [300]

large intestine, the final section of the alimentary canal, which removes water from undigested food material. [114]

latent heat (lāt′·ənt), heat that changes a material from one state to another but makes no change in its temperature. [186]

ligament, a strong band of cartilage that holds a joint firmly in place. [59]

light-year, the distance that light travels in one year, equal to about 5.88 trillion miles (9.46 trillion km). [123]

liquid, matter that has a definite volume and takes the shape of its container. [31]

liver, the largest organ in the body, which produces bile and performs various other functions. [114]

lunar eclipse, a darkening of the moon that results when it passes through the earth's shadow. [158]

lung, one of the two main organs of the respiratory system, consisting of a spongy mass of tubes and air sacs. [67]

lymph (limf), fluid drained from body tissues, which contains protein, salt, sugar, and nitrogen wastes. [75]

lymphatic system (lim·fat′·ik), a network of vessels and nodes that follow the circulatory system and provide drainage and defense. [75]

lymph node, one of the smallest organs of the lymphatic system, which produce a kind of white corpuscle and remove foreign materials. [76]

macronutrient (mak′·rō·nōō′·trē·ənt), one of the nutrients that the body needs in large amounts every day, which include carbohydrates, fats, and proteins. [88]

magnitude, the brightness of a star, indicated by a number on a scale. [130]

mass, the amount of matter in an object. [23]

matter, any material that has mass and volume. [30]

meteor, a streak of light caused by a swiftly moving rock from space as it enters the earth's atmosphere. [164]

meteorite, a rock from space that strikes the earth's surface. [164]

meteorology (mē′·tē·ə·rol′·ə·jē), the scientific study of weather. [207]

micronutrient, one of the nutrients that the body needs in very small amounts every day, which include vitamins and minerals. [98]

mineral, a micronutrient that occurs naturally in the earth's surface. [101]

molecule, the smallest particle of a compound. [34]

momentum (mō·men′·təm), the strength of motion in a moving object, which is the product of its mass and its speed. [24]

motor neuron, a nerve cell that carries signals from the central nervous system to a muscle or gland. [252]

mucous membrane, a membrane that lines body passages and secretes mucus to remove and destroy disease germs. [77]

muscular system, the system that includes the muscles of the human body. [63]

nebula (neb′·yə·lə), a large body of glowing gases in outer space. [162]

nephron (nef′·ron), one of the tiny purifying structures in the kidneys. [78]

neuron (nŏŏr′·on′), a nerve cell. [250]

nimbus, a kind of cloud that produces rain. [222]

nova, a dim star that explodes and becomes many times brighter. [161]

observatory, a building with a large telescope and other equipment for observing heavenly bodies. [135]

organ, a body unit consisting of various kinds of tissues that work together. [51]

organism, a whole body, consisting of all the systems that work together to perform the processes of life. [52]

pancreas, an organ that produces several digestive enzymes and the hormone insulin. [114, 270]

parathyroid glands, four small glands in the back of the thyroid, which produce a hormone that raises the calcium level in the blood. [270]

peripheral nervous system (pə·rif′·ər·əl), the network of nerves connecting to the spinal cord and extending through the entire body. [250]

peristalsis (per′·i·stôl′·sis), the wavelike motion of muscles that push food through the alimentary canal. [111]

phase, one of the different apparent shapes of the moon or a planet because of its position in relation to the earth and the sun. [154]

physical change, a change that affects the shape, form, or position of a material but produces no new material. [12]

physics, the study of energy and motion. [12]

pituitary gland (pi·tōō′·i·ter′·ē), a small gland in the center of the head which coordinates the work of the other glands in the endocrine system. [269]

planet, one of the large bodies, including the earth, that orbit the sun. [139]

planetarium, a device for demonstrating the movements of the heavenly bodies. [133]

plasma, a pale yellow fluid that is the main liquid in blood. [72]

platelet, a tiny, disk-shaped body that is important in the clotting of blood. [72]

potential energy, energy at rest, producing no physical or chemical change; stored energy. [13]

precipitation, moisture that condenses out of the air in a form such as dew, rain, or snow. [217]

pressure point, a point on the body where pressure can be applied to an artery to control bleeding. [299]

protein, a food from which the body makes protoplasm to build new cells. [93]

psychrometer (sī·krom′·i·tər), an instrument for measuring the relative humidity by the use of a wet-bulb and a dry-bulb thermometer. [238]

radiation, the transfer of heat by light and similar rays that move out from hot objects. [195]

red corpuscle, a red blood cell, which carries oxygen and carbon dioxide. [72]

reflex, a response that occurs without conscious thought, often to give quick protection to the body. [260]

refrigerant, a fluid that is used to produce cold, usually one that vaporizes and condenses at a low temperature. [200]

refrigeration unit, a device that produces cold by circulating a fluid that absorbs heat in one place and releases heat in another place. [200]

relative humidity, the amount of moisture in the air, stated as a percent of the total amount it could hold at that temperature. [216]

respiratory system, the body system that includes the lungs and other organs used in breathing. [66]

retrograde motion (ret′·rə·grād′), the apparent backward motion of a planet in its path among the stars. [140]

rickets, a deficiency disease resulting from a lack of vitamin D, which causes the bones to become soft. [100]

right ascension, the distance east of the vernal equinox, measured in hours. [131]

saliva (sə·lī′·və), the watery secretion in the mouth, which contains an enzyme that changes starch into sugar. [111]

satellite, a body such as the moon, which orbits a planet. [139]

scurvy, a deficiency disease resulting from a lack of vitamin C, which causes bleeding of the gums and other soft tissues. [99]

sensory neuron, a nerve cell that carries signals from a sensory organ to the central nervous system. [252]

shock, a condition resulting from pain and stress, with symptoms that include pale, clammy skin, rapid pulse, abnormal breathing, and reduced body temperature. [302]

simple fracture, a broken bone that does not protrude through the skin, also called a closed fracture. [304]

skeletal system, the system that consists of the 206 bones of the human skeleton. [57]

small intestine, the section of the alimentary canal where final digestion occurs and dissolved food is absorbed into the bloodstream. [112]

solar eclipse, a darkening of the sun that results when the moon passes in front of it. [155]

solid, matter that has a definite shape and a definite volume. [31]

solute (sol′·yōōt), a substance, usually a solid, that dissolves in another substance. [39]

solution, a mixture in which one material spreads molecule by molecule through another material. [38]

solvent, a substance, usually a liquid, in which another substance is dissolved. [39]

specific heat, the number of calories needed to raise 1 gram of a material 1 degree Celsius. [187]

spinal cord, the bundle of nerves inside the spinal column, which carry signals between the brain and the peripheral nerves. [259]

spleen, the largest organ of the lymphatic system, which removes old blood cells from circulation and produces new ones. [76]

splint, a rigid device used to immobilize an injured body part. [305]

sprain, an injury to the ligaments and tissues surrounding a joint. [306]

starch, a complex carbohydrate with very large molecules, which must be changed to glucose molecules before the body can use them. [90]

stimulus, something that initiates a reflex action. [260]

stomach, the hollow, muscular organ in which food is stored and mixed with digestive juices. [112]

strain, an injury to a muscle or tendon, caused by over-stretching it. [306]

stratus cloud (strā′·təs, strat′·əs), a cloud in the form of a layer covering a large area. [221]

sublimation (sub′·lə·mā′·shən), the changing of a solid directly to a gas. [179]

sugar, a carbohydrate with simple molecules of carbon, hydrogen, and oxygen. [89]

summer solstice (sol′·stis, sōl′·stis), the time around June 21 when the sun reaches its northernmost point, directly over the Tropic of Cancer, and summer begins in the Northern Hemisphere. [151]

surfactant (sər·fak′·tənt), a surface-active agent, which allows particles of oil or grease to become suspended in water. [42]

suspension, a mixture in which very fine particles of one material are suspended (held up) in another material. [38]

synapse (sin′·aps′), a chemical-filled gap that transmits impulses between nerve cells. [251]

system, a group of body organs working together to perform a major function. [51]

temperature, a measure of the internal energy level of a material. [174]

tendon, a tough tissue that connects a muscle to a bone. [63]

thermistor (thûr′·mis·′tər), a thermometer with an electrical conductor that changes resistance with changes of temperature. [191]

thermocouple, a thermometer with two kinds of metal wire joined together, which produce a varying electrical current with changes of temperature. [191]

thermometer, a device for measuring temperature. [174]

thermostat, a device that automatically regulates temperature with a bimetallic strip that operates a switch. [190]

thymus, a gland between the heart and the sternum, which makes lymph cells that destroy viruses and cancer cells. [76]

thyroid gland, a gland in the throat that regulates the metabolism of the body and suppresses the calcium level in the blood. [269]

tissue, a group of cells that specialize in the same function. [50]

tonsil, one of six lymphatic organs in the throat, which destroy disease germs that enter through the mouth and nose. [76]

tornado, a whirling funnel of wind moving up to 300 miles per hour (480 km/h), which extends downward from a storm cloud. [233]

tourniquet (tûr′·ni·kit), a band of cloth twisted tightly around an arm or a leg to cut off the blood flow. [300]

trachea (trā′·kē·ə), the windpipe, which carries air to the bronchi. [66]

typhoon, a hurricane that occurs in the western Pacific Ocean. [232]

vein, a blood vessel that carries blood toward the heart. [70]

vernal equinox (ē′·kwə·noks′, ek′·wə·noks′), the time around March 21 when the sun is directly above the equator and spring begins in the Northern Hemisphere. [150]

vestibular apparatus (ve·stib′·yə·lər), a structure in the inner ears that includes the fluid-filled semicircular canals, which help to provide the sense of balance. [266]

villi (vil′·ī), *sing.* **villus,** tiny, fingerlike projections lining the small intestine, which absorb dissolved food into the bloodstream. [114]

vitamin, a naturally occurring organic substance that enables macronutrients and minerals to do their work. [98]

volatile (vol′·ə·təl, vol′·ə·tīl′), evaporating easily at ordinary temperatures. [179]

white corpuscle, a white blood cell, which fights disease germs. [72]

winter solstice (sol′·stis, sōl′·stis), the time around December 21 when the sun reaches its southernmost point, directly over the Tropic of Capricorn, and winter begins in the Northern Hemisphere. [151]

zodiac (zō′·dē·ak′), the group of twelve constellations through which the ecliptic passes. [146]

Index

Diagram of Opposite High Tides

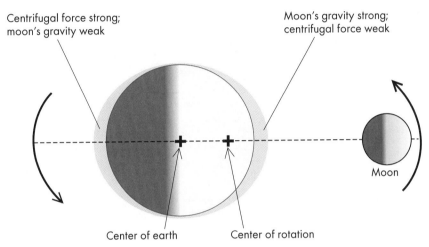

Centrifugal force strong;
moon's gravity weak

Moon's gravity strong;
centrifugal force weak

Moon

Center of earth   Center of rotation

   Why do high tides occur on the side of the earth opposite from the moon? The reason is that the earth and moon rotate around a common center about three-fourths of the way between the earth's center and its surface. This rotation causes centrifugal force, which is weak on the side of the earth nearest to the moon and strong on the side opposite from the moon. The result is tides of about equal height on opposite sides of the earth, caused mainly by the moon's gravity on one side and by centrifugal force on the other side.

Illustration Credits

American Meteorological Society©. Reprinted with permission (*Weather & Forecasting*, Dec. 2007, p. 1221, "Spatial and Temporal Analysis of Tornado Fatalities in the United States: 1880–2005" by Walker Ashley): 233 (left).

Christy Collins: 48.

Comstock.com Images©: 10 (bottom right), 50 (tissue), 120, 194, 232.

Corbis©: 86.

Corel Corporation©: 20 (top), 42 (molecule), 73, 89 (top), 90, 105, 134, 190, 213 (top), 289 (left).

Corel/Image Club: 179.

Corel/Nova: 154 (bottom), 208.

Corel/Samuel Hoover: 24, 31, 33, 34, 94, 150 (top), 151, 154 (top), 159.

Dorcas Mast: 311 (bottom).

Dover Publications: 218.

Dynamic Graphics©: 311 (left).

Dynamic Graphics/PhotoDisc/Digital Vision/Getty Images/Samuel Hoover: 226 (top).

Dynamic Graphics/Stockbyte/Corel/Brand X Pictures/Digital Vision/Getty Images/Samuel Hoover: 221.

EclectiCollections©: 130 (bottom).

FBI: 56.

Hemera©: 96 (bottom), 283, 304.

Hemera/Corel/Samuel Hoover: 202.

Hemera/Samuel Hoover: 177, 200.

Image Club/Samuel Hoover: 21, 305 (center).

iStockphoto.com/Alexey Stiop©: 158.

Jerry Nissley/Samuel Hoover: 223 (bottom).

Jim Reed/Wild Weather by Digital Vision/Getty Images©: 165, 228, 229, 234.

Lester Miller: 224, 285, 305 (left).

Lester Showalter: 19 (top right), 20 (bottom), 29, 88, 141, 146, 185, 192, 193, 197, 212, 242.

LifeART© 2011 Lippincott Williams & Wilkins (from 10 CD collections): 50 (not tissue), 58, 59, 60, 61, 63, 64, 66 (not insets), 69 (bottom two), 70, 71, 72 (not blood cells), 75, 76, 78, 79, 109, 110, 112, 113 (not salivary glands), 250, 251, 255, 259, 262, 263, 265, 266, 269, 270, 294, 295, 296, 300 (left), 301, 306 (top), 308, 310.

LifeART/Corel: 306 (bottom).

LifeART/Samuel Hoover: 23 (top).

Liquidlibrary/Dynamic Graphics©: 223 (top).

Lisa Weaver: 54, 219, 264.

Map Resources/Bennie Hostetler: 214, 215, 231, 233 (right).

Mary Jane Miller: 66 (insets), 113 (salivary glands).

NASA: 163 (top).

NASA and The Hubble Heritage Team (STScI/AURA): 163 (middle).

NASA, ESA and W. Harris (McMaster University, Ontario, Canada): 163 (bottom).

NASA, ESA, S. Beckwith (STScI), and The Hubble Heritage Team (STScI/AURA): cover.

National Cancer Institute: 72 (blood cells).

NOAA: 225.

Nova/Corel: 254.

Nova Development Corporation©: 27 (top), 57, 191, 272.

Nova/Samuel Hoover: 123, 143, 147, 153 (top), 211, 213 (bottom), 227.

Peter Balholm: 180.

Phoebe Hoover: 290 (top), 311 (top).

PhotoDisc by Getty Images©: 153 (bottom), 157, 161, 162, 172, 206, 223 (middle), 230.

PhotoDisc by Getty Images/Samuel Hoover: 226 (bottom).

PHOTOTAKE© Watney Collection/Phototake: 305 (right).

Rod and Staff Publishers, Inc.: 126–127, 128, 130 (top), 132–133, 136, 148–149.

Samuel Hoover: 10 (top and left), 11, 14, 16, 17, 19 (top left and bottom), 22, 23 (bottom), 26, 27 (bottom), 28, 30, 35, 39, 40, 41, 42 (grease diagram), 65, 69 (top), 89 (bottom), 91, 92, 95, 96 (top), 101, 107, 124, 131, 135, 139, 142, 144, 155, 164, 174, 176, 181, 182, 183, 188, 189, 195, 198, 201, 209, 220, 237, 238, 239, 240, 248, 253, 258, 282, 286, 287, 289 (right), 290 (bottom), 291, 292, 297, 298, 300 (right), 307, 309, 313, 314 (top), 315, 316, 335.

Samuel Hoover/Dale Yoder: 150 (bottom), 210.

Thomas Child: 314 (bottom).